Communication, Computation and Perception Technologies for Internet of Vehicles

Yongdong Zhu · Yue Cao · Wei Hua · Lexi Xu
Editors

Communication, Computation and Perception Technologies for Internet of Vehicles

 Springer

Editors
Yongdong Zhu
Zhejiang Lab
Hangzhou, Zhejiang, China

Wei Hua
Zhejiang Lab
Hangzhou, Zhejiang, China

Yue Cao
Wuhan University
Suzhou Research Institute of Wuhan
University
Wuhan, Hubei, China

Lexi Xu
China Unicom & Beijing University
of Posts and Telecommunications
Beijing, China

ISBN 978-981-99-5438-4   ISBN 978-981-99-5439-1   (eBook)
https://doi.org/10.1007/978-981-99-5439-1

This Springer imprint is published by the registered company Springer Nature Singapore Pte Ltd.
The registered company address is: 152 Beach Road, #21-01/04 Gateway East, Singapore 189721,
Singapore

Paper in this product is recyclable.

Acknowledgement

Suzhou Municipal Key Industrial Technology Innovation Program under Grant SYG202123. National Key Research and Development Program of China (No. 2021YFB2900200). Key Research and Development Program of Zhejiang Province (No. 2021C01197)

Contents

Chapter 1
Modeling Microscopic Traffic Behaviors for Connected and Autonomous Vehicles

Tu Xu and Yongdong Zhu

Abbreviations

AVs	Autonomous Vehicles
HVs	Human-driven vehicles
SD	Standard deviation
TIT	Time integrated time-to-collision
PR	Penetration rate

1 Introduction

Autonomous vehicle (AV) technology is growing rapidly all over the world. All major car manufacturers as well as several technology companies have announced that fully AV models will be available between 2019 and 2025. The penetration rate of AVs on road is predicted to increase from 0.1% in 2021, to 0.84% in 2025 and 12% in 2030. The prediction indicates that before AVs fully take place of human drivers, the mixed traffic flow consisting of AVs and human vehicles (HVs) will exist for a long time.

Roadway users have high expectations of the potential benefits of AVs. It is commonly believed that the existence of AVs can improve traffic throughput because they tend to maintain a small headway. However, a recent empirical study tells a different story [1]. The study uses Level 2 semi-autonomous vehicles to perform

T. Xu · Y. Zhu (✉)
Zhejiang Lab, Interdisciplinary Innovation Research Institute, Hangzhou, China
e-mail: zhuyd@zhejianglab.com

T. Xu
e-mail: xutu@zhejianglab.com

driving experiments, and the results indicate that all existing AV systems are string unstable. This means that small perturbations will grow into full stop-and-go motions, causing traffic instability phenomena such as traffic oscillations and capacity drop. Moreover, human drivers are likely to react differently when following an AV, possibly inducing more instability to the mixed traffic flow, which could cause serious safety, congestion, and emissions implications [2]. Hence, there is a critical need for a deeper understanding of currently deployed AV technologies to predict their impacts.

Toward that end, several studies have investigated the potential impacts of AVs on traffic flow. Most of them show that AVs make improvements on roadway capacity and/or traffic flow stability [3–5]. Nevertheless, recent studies have shown that the impact of AVs on traffic flow characteristics may be more complex. The capacity of mixed traffic is formulated in [6], based on vehicle spacing characteristics, platoon size, and AV penetration rate. The results show that the optimal policy and AV distribution depend on the AV penetration rate. Furthermore, this research team reveals the mechanism of traffic void generation, which suggests heterogeneous acceleration and car-following behavior in mixed traffic flow diminish traffic throughput [7]. To understand the relationship between traffic throughput and density for mixed traffic flow, the Fundamental Diagram (FD) for mixed Human Vehicle (HV) and AV traffic considering the stochastic headway is derived in [8]. The results show that an increase in the AV penetration rate can reduce the scattering of FD. To study the impact of different AV control algorithms on traffic, the trade-offs between AV following characteristics on safety, mobility, and stability is predicted in [9], based on a parsimonious linear car-following model. Later on, their study [10] shows that commercial AV following control becomes more unstable as the headway is set to a smaller value, which helps them validate several theoretical predictions in [9].

1.1 Motivation

To better understand the potential impacts of AVs on traffic flow. There are three key research challenges. Firstly, a model to explain different types of traffic instabilities needs to be developed. Secondly, the impact of AVs on the longitudinal behavior of HVs need to be studied. Thirdly, simulation studies need to be performed to quantify the potential benefit of AVs on traffic flow.

1.2 Main Contributions

Main contributions to this chapter are as follows:

(1) This chapter summarizes the recent studies on the impact of AVs on transportation system from traffic flow perspective and pointed out potential future research directions.

(2) This chapter develops a stochastic car-following model to capture the traffic instabilities in traffic flow.
(3) This chapter analyses the difference between the car-following behavior between AVs and HVs and how AVs impact the behavior of mixed traffic flow.

1.3 Structural Organization

The Sect. 2 introduces the history of car-following models and recent research progress and gaps in modeling traffic instabilities. The Sect. 3 presents a novel car-following model for mixed traffic flow. The Sect. 4 introduces the differences between different car-following scenarios in mixed traffic flow. The Sect. 5 shows the results of numerical simulations and the Sect. 6 summarizes the whole chapter.

2 Background

2.1 Car-Following Models

To the best of our knowledge, the concept of car-following is first mentioned in [11], who study the dynamics of traffic based on "law of separation", where the desired "following distance" is a linear function of velocity of the following vehicle. In the late fifties, Chandler et al. [12], Herman et al. 13] study the Gazis-Herman-Rothery (GHR) model. The model is calibrated and validated with empirical data. Also, this microscopic model is able to connect with macroscopic characteristics in traffic by integration [14]. On the other hand, Michaels [15] notes that human factors could have a great impact on the driver behaviour. This study builds a connection between car-following models and psychology. Later, many car-following models are developed to explain the nonlinear characteristics in congested traffic [16, 17]. With the development of artificial intelligent, neural networks are used to model car-following behaviour [18].

Most car-following models can be classified into the following categories: (i) stimulus–response models as in [14], (ii) psycho-physical models, such as Widemann Model [19], (iii) crash-avoidance models, such as IDM Model [16], (iv) optimal-velocity models such as Optimal Velocity Model in [20] and Full Velocity Difference Model in [21]. However, most of these models do not consider the variation among the behaviour of different drivers while studies [22, 23] show that timid and aggressive drivers are responsible for traffic oscillations. The model proposed in this chapter is not classified into the four traditional categories listed above but extends Newell's model with bounded accelerations [17] such that stochastic vehicle acceleration process can be incorporated to model the variation among drivers.

2.2 Modeling Traffic Instabilities

As mentioned earlier, recent study uses Level 2 semi-autonomous vehicles to perform driving experiments, and the results indicate that all existing AV systems are string unstable. This means that small perturbations will grow into full stop-and-go motions, causing traffic instability phenomena such as traffic oscillations and capacity drop. Therefore, the key challenge to understand the mixed traffic flow is to model traffic instabilities.

2.2.1 Traffic Oscillations

Stop-and-go driving conditions in traffic oscillations disturb motorists. Mauch and Cassidy [24], Zheng et al. [25] have demonstrated that lane changing is a major factor causing traffic oscillations. However, a convincing explanation for this phenomenon on a one-lane road is still lacking. Stop-and-go traffic on a one-lane ring road without lane changing movements has been empirically found by [26].

To fill this gap, several analytical studies have proposed that a small driving instability may grow into mature traffic oscillations in the absence of lane changing [27, 28]. However, these models require either further model validation with empirical data or a careful selection of parameter values.

On the other hand, Laval and Leclercq [22], Chen et al. [23] argue that timid or aggressive car-following behaviors caused the formation and propagation of traffic oscillations. These models are validated with empirical data, but they require at least four parameters that are not directly observable. Laval et al. [29] assume a constant standard deviation of driver acceleration, resulting in a Brownian motion (BM) process only one unobservable parameter. The models successfully replicate realistic traffic oscillations at uphill segments but are unable to explain capacity drop phenomenon introduced in the next section.

2.2.2 Capacity Drop

It has been empirically observed for years that the queue discharge rate at bottlenecks is lower after the queue forms. This phenomenon is called capacity drop [30] has long been correlated with lane changing activity [31, 32]. But researchers have observed that capacity drop can occur in the absence of lane changing [33]. To explain this, Wong and Wong [34] takes the distribution of drivers' desired free-flow speeds in a traffic into account. However, empirically validation of this assumption is lacking. Yuan et al. [35] argue that empirical data reveals that the standard deviation of the driver desired acceleration is not constant, as in the BM [29], but decreases linearly with speed, leading to a geometric Brownian acceleration process. This geometric Brownian acceleration process successfully explains capacity drop but does not produce periodic traffic oscillations [36].

2.2.3 Concave Growth of Platoon Oscillation

The concave growth of the vehicle platoon oscillation was first observed by [37], where the standard deviation of vehicle speed grows concavely along the platoon. Tian et al. [38] find that the growth pattern in different oscillations collapses into a single concave curve, which indicates the law of the oscillation growth is universal. Treiber and Kesting [39] shows that the Brownian motion model in [29], and the IDM model with noise and indifference regions are all able to reproduce the concave growth pattern, indicating strongly that a stochastic component is vital for replicating this important phenomenon.

3 The Proposed Model

A general two-regime stochastic car-following model may be formulated as follows. Let the random process $X_j(t)$ represent the position of vehicle j at time t, and let $f(x; \Theta)$ be the probability density function of $X_j(t)$ given the set of model parameters $\Theta = (\theta_1, \theta_2, \dots)$ and the data $\mathbf{x} = \{x_{[j]}(t_i)\}$, which represents the observed trajectories of all vehicles $j = 1, 2, \dots$ at times $t = 1, 2, \dots$ up to time t. The general form of these models will be formulated as a two-regime car-following model, where vehicle positions are the minimum of two random variables: Y, Z, given by a stochastic bounded acceleration model and a congestion component that includes perception/measurement errors, respectively:

$$X_j(t) = \min\{Y(\mathbf{x}; \Theta), Z(\mathbf{x}; \Theta)\}$$

If random processes Y and Z are Bivariate normally distributed (BVM) with a correlation ρ_0:

$$(Y, Z) \sim \mathrm{BVN}(\mu_Y, \sigma_Y, \mu_Z, \sigma_Z, \rho_0),$$

where μ_- and σ_- are the mean and standard deviation of the variable in subscript, one can write the probability density analytically, shown by [40].

This result is notable because the model is analytical and lends itself nicely to be estimated using maximum-likelihood estimation (MLE). In addition, it should have a much better explanatory power compared to existing models that are additive.

3.1 The Congestion Term

Based on our previous discussion, it is advantageous to consider both the free-flow (Y) and the congestion (Z) terms normally distributed, as opposed to the original model where Z is deterministic. As demonstrated empirically in [41] parameters τ and δ can be assumed to follow the bivariate normal (BVN) distribution [41], i.e.

$$(\tau, \delta) \sim \text{BVN}(\mu_\tau, \sigma_\tau, \mu_\delta, \sigma_\delta, \rho),$$

Instead of taking the parameters directly from [41], here we estimate them in conjunction with the free-flow regime parameters, which are formulated next.

3.2   The Free-Flow Term

We start by rewriting the proposed car-following model as:

$$X_j(t) = \min\{x_j(t - \tau') + \xi_j(\tau'), x_{j-1}(t - \tau) - \delta\}$$

to emphasize that the time lag in free-flow, τ', and in congestion, τ, do not have to be the same, as customary in the literature, and should be statistically independent. Furthermore, we argue that τ' should not be interpreted as a parameter because it simply defines the initial conditions for the free-flowing component. For simplicity, we fix the value of τ' in the parameter estimation process to $\tau' = 1.2$ s, which is a typical value for the time step in Newell-type car-following models, but any comparable value could be used instead. The main ingredients of the free-flow regime formulation is the displacement ξ_j, as explained next.

3.2.1   The Extended Desired Acceleration Model

Section 2 showed that there is a trade-off between the specification of the acceleration error (i.e. BM or g-BM) and the traffic feature that can be replicated (i.e. realistic traffic oscillations or speed-capacity relationship). In this section, we introduce a single extra dimensionless parameter $m \geq 1$ that produces models in a scale from the g-BM model ($m = 1$) to the BM model ($m \gg 1$), and therefore generalizes the models in the previous section. The hope is that a suitable value of m will produce a model with the best features of the BM and g-BM models, namely:

- **realistic traffic oscillations**: the BM model produces traffic oscillations even at high speeds because the acceleration error $\text{SD}[a(v(t))]$ is independent of the speed. In contrast, the g-BM model cannot because the $\text{SD}[a(v(t))]$ is assumed to be zero at the desired speed v_c.
- **speed-capacity relationship**: the g-BM model reproduces this relationship because $\text{SD}[a(v(t))]$ is a decreasing function of the speed. In contrast, the BM model cannot because $\text{SD}[a(v(t))]$ is assumed constant.

Based on these observations, we conjecture that the acceleration error $\text{SD}[a(v(t))]$ has to be nonzero at v_c and be a decreasing function of the speed. Here, we impose that

$SD[a(v(t))]$ vanish at mv_c and the stochastic differential equation for the proposed acceleration model becomes:

$$\begin{cases} d\xi(t) = v(t)dt, \xi(0) = 0, \\ dv(t) = (v_c - v(t))\beta dt + (mv_c - v(t))\sigma dW(t), v(0) = v_0, \end{cases}$$

and has analytical solution. For the car-following implementation we simply generate $\xi(t)$ as a normal random variable. To eliminate irrelevant parameters from this analysis, we can formulate the key variables of this model in dimensionless form. One possibility is the following transformations:

$$\tilde{\sigma}^2 = \sigma^2/\beta$$

4 Car-Following Behaviors in Mixed Traffic Flow

Ding et al. [42] used three Tesla vehicles with SAE level-2 autonomous driving function to carry out an experiment on an open road and captured the significant heterogeneity for different operating scenarios: (1) AV follows AV (2) AV follows HV (3) HV follows AV (4) HV follows HV. In this chapter, we discuss the car-following features for these four scenarios.

4.1 Car-Following Behaviors of AVs

The data analysed in this chapter were collected from the empirical experiments conducted in [43].

The experiments used a three-vehicle platoon to produce a set of car-following scenarios, where the lead vehicle was a human-driven vehicle followed by two identical adaptive cruise control vehicles.

A high-accuracy GPS device was used to collect the location and velocity data.

The mean location and velocity errors of the GPS device were respectively 0.89 m and 0.10 m/s based on our validation.

During the experiments, the lead driver was instructed to produce driving cycles that consist of steady and non-steady traffic conditions. The driver first travelled at a stable speed, then conducted a deceleration-acceleration process, and finally resumed the initial stable speed. Consecutive driving cycles were separated by extensive stabilization periods. The periods with stable speed produce steady traffic conditions while the deceleration-acceleration represents a disturbance. The parameters of the driving cycles, including the initial stable speed and oscillation amplitude, were varied to reproduce different traffic scenarios.

Three adaptive cruise control car models were tested in the experiments. For each system, two headway settings, headway 1 (small) and headway 3 (medium), were tested. Additionally, Civic and Prius have different engine modes, which were separately tested. For each car model, a total of 96 driving cycles were conducted under three speed levels (32 in each speed level) in each headway setting and engine mode. The equilibrium speed range was from 15 m/s (35 mph) to 29 m/s (65 mph).

The parameter estimation results for different ACC groups are as shown in Table 1.

From the results, we can summarize the following features of the car-following features of AVs:

Prefer a large gap

In the proposed model, the two parameters in the congestion term δ and τ are the time shift and space shift in the Newell trajectory, which is equivalent to the jam spacing and the time gap at equilibrium. The gap between the vehicle and its leader would increase if these two parameter values increase.

One can find that: (i) when an ACC car uses headway 3 setting, the gap to its leader would be much larger than that with headway 1 setting, according to the increase of δ and τ for all 4 models (model 3, civic normal, civic sport, pirus normal). (ii) civic car-following behavior is more aggressive than model 3 and pirus normal, according to the smaller δ and τ values. (iii) all ACC models tested in this chapter tend to maintain a larger gap compared to human drivers. For reference, typical values for

Table 1 Parameter values for different ACC groups

ACC group	μ_δ	μ_τ	u	β	m	$\tilde{\sigma}$	ρ	σ_δ	σ_τ	α	ρ_0
Model 3 headway 1	14.37	0.4	97.6	150.0	1.4	0.11	-1	3.5	0.09	0.01	-0.85
Model 3 headway 3	17.92	0.59	84.1	150.2	7.8	0.01	-0.95	2.6	0.07	0.06	-0.75
Civic normal headway 1	5.15	0.96	96.4	150	1.7	0.09	-1	2.8	0.05	0.1	-0.79
Civic normal headway 3	6.87	1.8	92.5	151.1	2.1	0.06	-0.18	2.0	0.08	0.19	-0.82
Civic sport headway 1	7.15	0.9	103.2	150	1.2	0.18	-0.93	3.5	0.15	0.11	-0.88
Civic sport headway 3	10.53	1.5	92.8	150.0	1.6	0.09	-0.8	3.5	0.05	0.02	-0.43
Pirus normal headway 1	16.26	0.43	87.3	150.8	3.0	0.03	-0.01	2.9	0.08	0.07	-0.86
Pirus normal headway 3	18	1.24	81.9	150	10	0.01	1	3.5	0.08	0.17	-0.73

car-following behaviors are $\delta = 7$ m and $\tau = 0.4$ s [44], which are much smaller than those of ACC models.

Are more stable than human drivers

According to previous studies, the driver acceleration process is a stochastic process, whose error process could be a Brownian motion or a geometric Brownian motion [29, 35, 45]. We have shown that the product of m and $\tilde{\sigma}$ controls the acceleration process, which determines the (i) the traffic oscillation period and amplitude; (ii) the average speed and the discharge rate at the potential bottleneck; (iii) the concave growth pattern of the platoon oscillation. For simplicity, we use the product of the mean value of m and the mean value of $\tilde{\sigma}$ in the table. For reference, the estimated values for human drivers are $m\,\tilde{\sigma} = 0.24$ on a Chinese freeway, $m\,\tilde{\sigma} = 0.21$ on the Peachtree street in Atlanta [45] and $m\,\tilde{\sigma} = 0.38$ on the US101 freeway [46]. From the results, we can find that for most of the ACC models studied in this section, the acceleration process is more stable compared to human drivers. The only mode that has a similar $m\,\tilde{\sigma}$ value to human drivers is civic sport mode headway1.

Are insensitive to roadway grade

In the proposed model, the relationship between vehicle acceleration and roadway grade is shown in:

$$E[a(v)] = (u - v(t))\beta - \alpha g \max\{0, G\},$$

where u is the free-flow speed, i.e. the desired speed on a flat road segment, $g = 9.81$ m/s^2 is the acceleration of gravity, G is the roadway grade expressed as a decimal and α is a dimensionless parameter. Notice that in the literature $\alpha = 1$ which is consistent with the assumption that the acceleration due to gravity in the direction of movement, $g \max\{0, G\}$, is subtracted, in its entirety, from the acceleration the driver would impose to the vehicle on a flat segment, $(v_c - v(t))\beta$. Here, the parameter α is added to relax this strong assumption in the literature: $0 < \alpha < 1$ indicates they compensate for the upgrade, i.e. that they press the gas pedal harder than they would on a flat segment, while $\alpha > 1$ implies a softer than usual pressing. Values of $\alpha < 0$ are unlikely as it would indicate that acceleration increases with the upgrade.

The estimation results in Table 1 show that α is an insignificant parameter (|t-stat| < 2) for most of the ACC models. The parameter α is significant only for pirus normal mode headway 3 with $\alpha = 0.17$. Given the estimated free-flow speed, inverse relaxation time β and an upgrade of 5%, the desired speed would decrease from $u = 81.91$ km/h to $v_c = u - \frac{\alpha g \max\{0, G\}}{\beta} = 74.71$ km/h, which is a decrease less than 10%.

In general, we can say that the ACC models tested in this section are insensitive to roadway grade.

4.2   Car-Following Behaviors of HVs

The car-following behavior has been studied for a long time. The estimated parameter
values for traditional traffic flow can be found in the literature [45, 46]. In mixed
traffic flow, the key challenge is to find the impact of AVs on HVs. Toward that end,
we focus on the difference of acceleration, Δa_j, between the follower j and the
leader $j - 1$.

In this section, we apply MLE to estimate the standard deviation σ of Δa_j for
two different scenarios:

- Scenario A: an HV is following an AV
- Scenario B: an HV is following an HV.

Two datasets are used in the experiment. Dataset I is collected from a three-vehicle
platoon experiment [47]. In the experiment, the lead vehicle travels at a given speed
profile while the other two vehicles follow their leader naturally. However, the second
vehicle, which is the leader of the third vehicle, is an AV in Scenario A and is an HV
in Scenario B. The location, speed, acceleration, spacing to the leader, relative speed
to the leader, and other useful information are recorded for the third vehicle. There
are 5 different lead vehicle speed profiles and 9 different drivers for the third vehicle.
In this chapter, we use the data from 8 drivers because the data for the other driver is
not complete. Dataset II is collected from a car-following experiment performed on
a circular road [48]. There are three experiments. In each experiment, there is one
University of Arizona self-driving capable Cognitive and Autonomous Test (CAT)
Vehicle, which can be transitioned between manual mode and autonomous mode.
The CAT vehicle is first set to manual mode and then set to autonomous mode after
a certain amount of time. In this chapter, we treat CAT in manual mode as an HV
and CAT in autonomous mode as an AV. All vehicle trajectories are recorded with a
GPS at a 1/30 s interval.

For a single data point $\mathbf{x}$ from the sample, the log-likelihood function is:

$$f(\mathbf{x}; \sigma) = \frac{e^{-\frac{x^2}{2\sigma^2}}}{\sqrt{2\pi}\,\sigma}$$

By maximizing the sum of the log-likelihood function of data points, one can get
the MLE estimate of σ. The results can be found in Tables 2 and 3.

Table 2 Mean estimate of σ for different drivers and LLR test results from Dataset I

	σ following AV	σ following HV	LLR test statistic	Significant difference?
1	0.32	0.38	21.2	Yes
2	0.60	0.74	34.0	Yes
3	0.36	0.48	62.5	Yes
4	0.21	0.34	191.2	Yes
5	0.54	0.86	154.7	Yes
6	0.30	0.34	13.2	Yes
7	0.45	0.43	1.8	No
8	0.34	0.41	26.0	Yes

Table 3 Mean estimate of σ for different drivers and LLR test results from Dataset II

	σ following AV	σ following HV	LLR test statistic	Significant difference?
1	0.43	0.61	17.8	Yes
2	0.45	0.74	34.9	Yes
3	0.45	0.64	19.1	Yes

A likelihood-ratio test can be used to compare the goodness-of-fit of different model specifications.

For example, if we have two models with number of parameters n_1 and n_2 ($n_1 > n_2$), respectively, the likelihood-ratio test statistic is $\Lambda(\mathbf{x}) = 2[l\left(\mathbf{x}; \widehat{\Theta_1}\right) - l\left(\mathbf{x}; \widehat{\Theta_2}\right)$, which follows a chi-square distribution with $n_1 - n_2$ degrees of freedom. In this study, we have two models for each driver:

Model A: σ is different between following AVs and HVs, which has $n_1 = 2$ parameters: σ_{AV} and σ_{HV}

Model B: σ following AVs is the same as following HVs, which has $n_2 = 1$ parameter: σ

We define the hypothesis H_0: model A and model B are the same. Given a significance level of 0.01, the critical value is 6.6. That means if the LLR test statistic is greater than 6.6, we reject H_0 and conclude that parameter σ is significantly different for different leader types at a significance level of 0.01.

The LLR test results are shown in Tables 2 and 3. For 10 out of 11 drivers in two different datasets, we find that parameter σ shows a significant difference between following AVs and HVs. In other words, AVs in traffic flow have a significant impact on the car-following behavior of HVs. Δa_j has a significantly smaller standard deviation when following an AV.

5   Vehicle Platoon Simulation for Mixed Traffic Flow

In this section, we use the proposed model to simulate a 50-vehicle-platoon mixed with AV and HV. Based on our finding, we assume when a vehicle follows an AV, σ = 0.4 and when it follows an HV, σ = 0.6. We assume that the AVs are human-like, which means that the proposed HV car-following model can be applied to AVs.

In the simulation experiment, the lead vehicle is an HV and travels at a constant speed 10 m/s starting from x = 0 m and the initial spacing for every vehicle is 17 m. The initial speed of each vehicle is 10 m/s. The AV penetration rates in experiments range from 0 to 100% with a 5% interval and the AVs are distributed randomly in the platoon. Each simulation lasts for 200 s. A typical time–space diagram of the simulation is shown in Fig. 1.

To study the impact of the AV penetration rate on traffic flow, we choose the flow, average SD of vehicle speeds, Time Integrated Time-to-collision (TIT) as performance indexes. The flow is measured at x = 1000 m by recording the passing time of each vehicle. The vehicle speed is recorded at each time step and we can get the SD of speed for each vehicle. Then we take the average of them as the "average SD of vehicle speeds". TIT is an index to evaluate traffic safety and a decrease in TIT means an improvement in traffic safety. Its calculation method is reported in [49].

The simulation results in Fig. 2 show that traffic safety risk decreases with the growth of the AV penetration rate, indicated by the decreasing average SD of vehicle speeds and TIT, without undermining traffic throughput.

Fig. 1 A sample time space diagram for vehicle platoon simulation

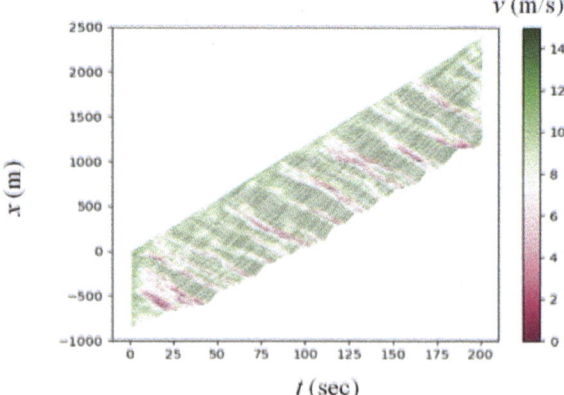

Fig. 2 Relationship between
traffic flow indexes with AV
PR

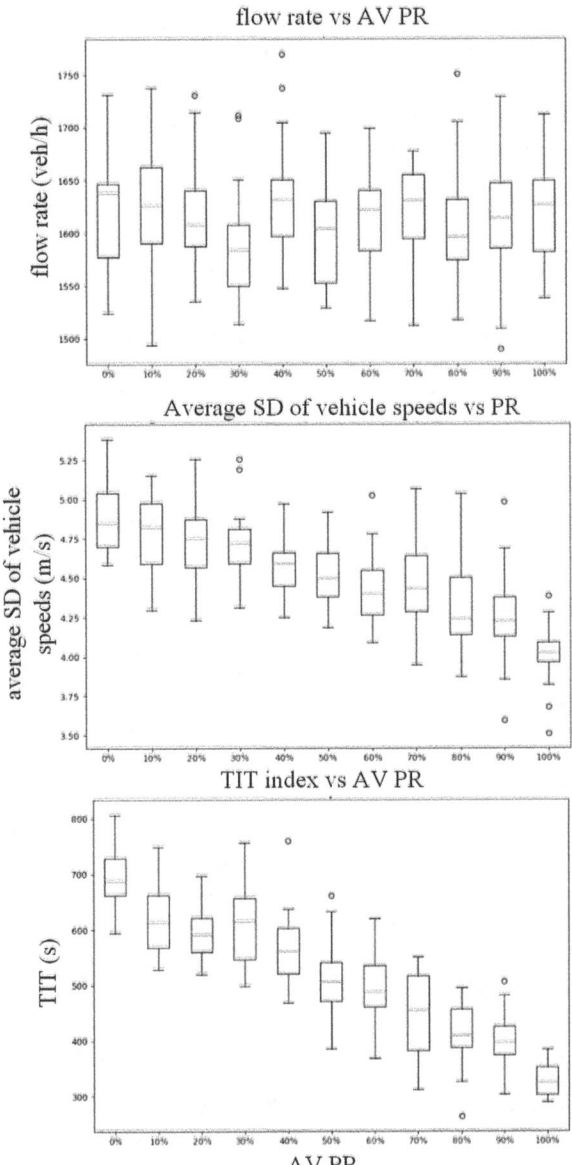

6   Conclusion

In this chapter, we show the recent research on how AVs impact the transportation
system and summarize several research gaps. To fill the gap, we first develop a
stochastic car-following model to capture the traffic instabilities in traffic flow. Then

we analyse the difference between the car-following behavior between AVs and HVs and how AVs impact the behavior of a single HV. Finally, we study how AV PR impact traffic safety and efficiency with a vehicle platoon simulation. The results show that traffic safety risk decreases with the growth of the AV penetration rate, without undermining traffic throughput.

References

1. G. Gunter, C. Janssen, W. Barbour, R. Stern, D. Work, Model- based string stability of adaptive cruise control systems using field data. IEEE Transa. Intell. Veh. **5**(1), 90–99 (2020)
2. J. Bilbao-Ubillos, The costs of urban congestion: estimation of welfare losses arising from congestion on cross-town link roads. Transp. Res. Part A: Policy Pract. **42**(8), 1098–1108 (2008)
3. B. Van Arem, C.J. Van Driel, R. Visser, The impact of cooperative adaptive cruise control on traffic-flow characteristics. IEEE Trans. Intell. Transp. Syst. **7**(4), 429–436 (2006)
4. A. Talebpour, H.S. Mahmassani, Influence of connected and autonomous vehicles on traffic flow stability and throughput. Transp. Res. Part C: Emerging Technol. **71**, 143–163 (2016)
5. H.S. Mahmassani, 50th anniversary invited article—autonomous vehicles and connected vehicle systems: flow and operations considerations. Transp. Sci. **50**(4), 1140–1162 (2016)
6. D. Chen, S. Ahn, M. Chitturi, D. A. Noyce, Towards vehicle automation: roadway capacity formulation for traffic mixed with regular and automated vehicles. Transp. Res. Part B: Methodol. **100**(C), 196–221 (2017)
7. D. Chen, A. Srivastava, S. Ahn, T. Li, Traffic dynamics under speed disturbance in mixed traffic with automated and non-automated vehicles. Transp. Res. Procedia **38**, 709–729 (2019), J. Transp. Traffic Theory
8. J. Zhou, F. Zhu, Modeling the fundamental diagram of mixed human-driven and connected automated vehicles. Transp. Res. Part C: Emerging Technol. **115**, 102614 (2020)
9. X. Li, Trade-off between safety, mobility and stability in automated vehicle following control: an analytical method. Research Gate Preprint, 07 (2020)
10. X. Shi, X. Li, Empirical study on car-following characteristics of commercial automated vehicles with different headway settings. Transp. Res. Part C: Emerging Technol. **128**, 103134 (2021)
11. L.A. Pipes, An operational analysis of traffic dynamics. J. Appl. Phys. **24**(3), 274–281 (1953)
12. R.E. Chandler, R. Herman, E.W. Montroll, Traffic dynamics: studies in car following. Oper. Res. **6**(2), 165–184 (1958)
13. R. Herman, E.W. Montroll, R.B. Potts, R.W. Rothery, Traffic dynamics: analysis of stability in car following. Oper. Res. **7**(1), 86–106 (1959)
14. D.C. Gazis, R. Herman, R.W. Rothery, Nonlinear follow-the- leader models of traffic flow. Oper. Res. **9**(4), 545–567 (1961)
15. R. Michaels, Perceptual factors in car following, in *Proceedings of the 2nd International Symposium on the Theory of Road Traffic Flow* (1963), pp. 44–59
16. M. Treiber, A. Hennecke, D. Helbing, *Congested Traffic States in Empirical Observations and Microscopic Simulations* (2000), pp. 1805–1824
17. G.F. Newell, A simplified car-following theory: a lower order model. Transp. Res. Part B **36**(3), 195–205 (2002)
18. M. Dougherty, A review of neural networks applied to transport. Transp. Res. Part C: Emerging Technol. **3**(4), 247–260 (1995)
19. R. Wiedemann, U. Reiter, Microscopic traffic simulation: the simulation system MISSION, background and actual state. Project ICARUS (V1052) Final Report, vol. 2 (1992)

20. M. Bando, K. Hasebe, A. Nakayama, A. Shibata, Y. Sugiyama, Dynamical model of traffic congestion and numerical simulation. Phys. Rev. E Stat. Nonlinear Soft Matter Phys. **51**(2), 1035–1042 (1995)
21. R. Jiang, Q. Wu, Z. Zhu, Full velocity difference model for a car-following theory. Phys. Rev. E Stat. Nonlinear Soft Matter Phys. **64**(1), 1–4 (2001)
22. J.A. Laval, L. Leclercq, A mechanism to describe the formation and propagation of stop-and-go waves in congested freeway traffic. Phil. Trans. R. Soc. A **368**(1928), 4519–4541 (2010)
23. D. Chen, J.A. Laval, S. Ahn, Z. Zheng, Microscopic traffic hysteresis in traffic oscillations: a behavioral perspective. Transp. Res. Part B **46**(10), 1440–1453 (2012)
24. M. Mauch, M.J. Cassidy, Freeway traffic oscillations: observations and predictions, in *15th International Symposium on Transportation and Traffic Theory*, ed. by M. Taylor (Pergamon-Elsevier, Oxford, U.K., 2002)
25. Z. Zheng, S. Ahn, D. Chen, J. Laval, Applications of wavelet transform for analysis of freeway traffic: bottlenecks, transient traffic, and traffic oscillations. Transp. Res. Part B: Methodol. **45**(2), 372–384 (2011)
26. Y. Sugiyama, M. Fukui, M. Kikuchi, K. Hasebe, A. Nakayama, K. Nishinari, S.-I. Tadaki, S. Yukawa, Traffic jams without bottlenecks-experimental evidence for the physical mechanism of the formation of a jam. New J. Phys. **10**, 033001 (2008)
27. R. Wilson, J. Ward, Car-following models: fifty years of linear stability analysis - a mathematical perspective. Transp. Plan. Technol. **34**(1), 3–18 (2011)
28. M. Treiber, A. Kesting, Validation of traffic flow models with respect to the spatiotemporal evolution of congested traffic patterns. Transp. Res. Part C: Emerging Technol. **21**(1), 31–41 (2012)
29. J.A. Laval, C.S. Toth, Y. Zhou, A parsimonious model for the formation of oscillations in car-following models. Transp. Res. Part B: Methodol. **70**, 228–238 (2014)
30. F.L. Hall, K. Agyemang-Duah, Freeway capacity drop and the definition of capacity. Transp. Res. Rec. **1320**, 91–98 (1991)
31. L. Leclercq, J.A. Laval, N. Chiabaut, Capacity drops at merges: an endogenous model. Procedia Soc. Behav. Sci. **17**(0), 12–26 (2011), *Papers selected for the 19th International Symposium on Transportation and Traffic Theory*
32. D. Chen, S. Ahn, Capacity-drop at extended bottlenecks: merge, diverge, and weave. Transp. Res. Part B: Methodol. **108**, 1–20 (2018)
33. S. Oh, H. Yeo, Impact of stop-and-go waves and lane changes on discharge rate in recovery flow. Transp. Res. Part B: Methodol. **77**, 88–102 (2015)
34. G. Wong, S. Wong, A multi-class traffic flow model: an extension of LWR model with heterogeneous drivers. Transp. Res. Part A **36**, 827–841 (2002)
35. K. Yuan, J. Laval, V.L. Knoop, R. Jiang, S. Hoogendoorn, A geometric Brownian motion car-following model: towards a better understanding of capacity drop. Transp. B (2018)
36. T. Xu, J. Laval, Parameter estimation of a stochastic microscopic car-following model, in *97th Annual Meeting of Transportation Research Board* (Washington, 2018)
37. R. Jiang, M.-B. Hu, H. M. Zhang, Z.-Y. Gao, B. Jia, Q.-S. Wu, B. Wang, M. Yang, Traffic experiment reveals the nature of car-following. PLoS ONE **9**(4), 1–9 (2014)
38. J. Tian, R. Jiang, B. Jia, Z. Gao, S. Ma, Empirical analysis and simulation of the concave growth pattern of traffic oscillations. Transp. Res. Part B: Methodol. **93**, 338–354 (2016)
39. M. Treiber, A. Kesting, The intelligent driver model with stochasticity -new insights into traffic flow oscillations. Transp. Res. Procedia **23**(Supp C), 174–187 (2017), *Papers Selected for the 22nd International Symposium on Transportation and Traffic Theory*, Chicago, Illinois, USA, 24–26 July, 2017
40. S. Nadarajah, S. Kotz, Exact distribution of the max/min of two Gaussian random variables, in *IEEE Transactions on Very Large Scale Integration (VLSI) Systems* (2008), pp. 210–212
41. S. Ahn, M. Cassidy, J.A. Laval, Verification of a simplified car- following theory. Transp. Res. Part B **38**(5), 431–440 (2003)
42. S. Ding, X. Chen, Z. Fu, F. Peng, An extended car-following model in connected and autonomous vehicle environment: perspective from the cooperation between drivers. J. Adv. Transp. **2021**, 1–17 (2021)

43. T. Li, D. Chen, H. Zhou, J. Laval, Y. Xie, Car-following behavior characteristics of adaptive cruise control vehicles based on empirical experiments. Transp. Res. Part B: Methodol. **147**, 67–91 (2021)
44. T. Xu, J. Laval, Analysis of a two-regime stochastic car-following model: explaining capacity drop and oscillation instabilities. Transp. Res. Rec. **2673**(10), 610–619 (2019)
45. T. Xu, J. Laval, Statistical inference for two-regime stochastic car-following models. Transp. Res. Part B **134**, 210–228 (2020)
46. T. Xu, J. Laval, Driver reactions to uphill grades: inference from a stochastic car-following model. Transp. Res. Rec. **2674**(11), 343–351 (2020)
47. Y. Rahmati, M.K. Hosseini, A. Talebpour, B. Swain, C. Nelson, Influence of autonomous vehicles on car-following behavior of human drivers. Transp. Res. Rec. **2673**(12), 367–379 (2019)
48. R.E. Stern, S. Cui, M.L. Delle Monache, R. Bhadani, M. Bunting, M. Churchill, N. Hamilton, H. Pohlmann, F. Wu, B. Piccoli, et al., Dissipation of stop-and-go waves via control of autonomous vehicles: field experiments. Transp. Res. Part C: Emerging Technol. **89**, 205–221
49. M.M. Minderhoud, P.H. Bovy, Extended time-to-collision measures for road traffic safety assessment. Accid. Anal. Prev. **33**(1), 89–97 (2001)

Chapter 2
ITS Traffic Management with Connected Vehicles: An Overview

Kan Wu and Yongdong Zhu

Abbreviations

CV	Connected vehicle
ITS	Intelligent transportation system
VOC	Volume-to-capacity
OD	Origin and destination
DTA	Dynamic traffic assignment
V2X	Vehicle-to-everything
MST	Minimum spanning tree
IIC	Incipient infinite cluster

1 Introduction

Traffic congestion burdens the urban's quality of life and will likely grow substantially if current trends continue. Tackling traffic congestions could be at least related to multiple UN's Sustainable Development Goals on urban mobility and climate change [1, 2]. Urban road transport is a dynamic and complex system, making it hard to identify and ameliorate bottleneck congestions. Intelligent transportation system (ITS) can reduce traffic congestion at least possible cost with a range of technologies.

K. Wu · Y. Zhu (✉)
Zhejiang Lab, Interdisciplinary Innovation Research Institute, Hangzhou, China
e-mail: zhuyd@zhejianglab.com

K. Wu
e-mail: kanwu@zhejianglab.com

© The Author(s), under exclusive license to Springer Nature Singapore Pte Ltd. 2023   17
Y. Zhu et al. (eds.), *Communication, Computation and Perception Technologies for Internet of Vehicles*, https://doi.org/10.1007/978-981-99-5439-1_2

Take urban congestion mitigation as an example, the ITS traffic management follows an iterative process as shown in Fig. 1. The congestion mitigation cycle often starts with traffic sensing and bottleneck identification. The precisely and timely identification of bottleneck road link set is pre-required for diagnosing cause of congestions. The countermeasures are prepared to address the causes of traffic congestion, which usually involves optimization of traffic control/guidance and bottleneck link or intersection capacities. After implementation of the countermeasures, a systematic evaluation of performance is often recommended, which requires continuously traffic sensing and monitoring and further countermeasure may be required if certain level of service is not satisfied. This formulates a congestion mitigation cycle that promote sustainable development of urban transport.

Traditional ITS method uses stationary and mobile sensors to monitoring traffic state, and optimize traffic at local level. In this chapter, we focus on ITS management under connected environment and provides a comprehensive review on emerging studies. For traffic sensing, this chapter explores the potential of CV-based mobile sensing in improvement of spatial coverage and temporal frequency. For traffic management, this chapter further reviews studies on CV-based the real-time traffic control, route guidance, and bottleneck capacity optimization. Moreover, this chapter further discusses the impact of CV penetration rate and robustness of CV-based ITS

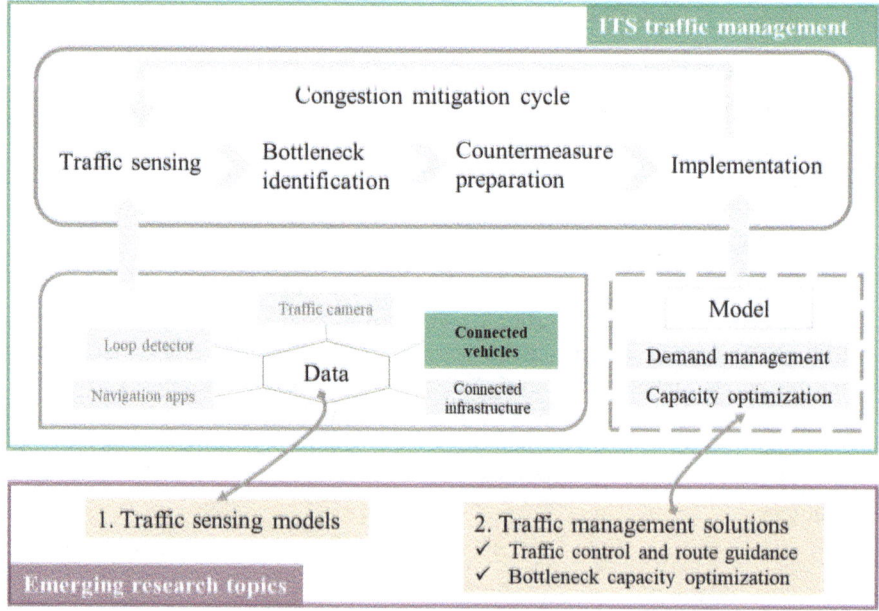

Fig. 1 ITS traffic management for congestion mitigation. The top section represents the congestion mitigation cycle, the bottom section shows the emerging research topics that will discuss in Sects. 2 and 3. With data collected under connected environment, congestion countermeasures can be designed and evaluated iteratively, the congestion mitigation cycle thus can be developed

solutions. With the comprehensive reviews, this chapter can help guide future traffic management policy design with incorporation of data-based solutions.

2 Traffic Sensing with Connected Vehicles

2.1 The Evolution of ITS Data Collection Techniques

Monitoring urban traffic both spatially and temporally is a challenging task. Existing sensing approaches to urban traffic, which generally fall into three main groups, have limitations. At one extreme, stationary sensor such as loop detectors and traffic surveillance cameras, collect traffic data over long periods of time, but with limited spatial coverage. Stationary sensors are pervasively used in the first-generation ITS since 1990s. At the other extreme, airborne sensors such as satellite and drone scan wide areas, but only during certain time window [3].

The third group, mobile sensors offer good coverage in both space and time. These sensors are mounted on "crowd-sourced" urban vehicles such as private cars, taxis, buses, or trucks, which allows the sensors to scan the road network traversed by their hosts. The mobile sensors can be used to estimate and monitor traffic state of a city with great ease and sound accuracy. However, the accuracy and immediacy of mobile traffic sensing is limited by the low penetration rate and update frequency (due to delay in data transmission and processing).

CV can be viewed as a type of mobile sensor (similar to floating cars), however, CVs are different from conventional mobile sensors when considering the data update frequency and spatial error. The temporal frequency (usually 1 Hz) is higher than conventional mobile sensors such as taxi data (secondly to minutely), and the spatial errors are usually in centimeter for CVs. Hence, it is possible to use CV data to obtain lane-based speed profiles and may also derive the trajectories of neighboring non-connected vehicles [4]. Constrained by temporal frequency and spatial errors, the existing mobile sensor data may not able to be used for real time traffic control, where data collected by CVs can be useful (since arrival information of CVs can be obtained or predicted in advance).

When compared with existing data collection approaches (stationary sensors, public travel records, and passive data collection), CVs approach offers higher level of spatial reach (same as the drive-by approach). However, in our eyes, the CVs approach is not a complete replacement to them, but a rather great enhancement as an additional data source. CVs can fill the gap of traffic flow estimation where stationary sensors are absent, and CVs data can be incorporated with public travel records and cellphone data to provide a more accurate estimation of urban travel demand which considers travels of different modes (Fig. 2).

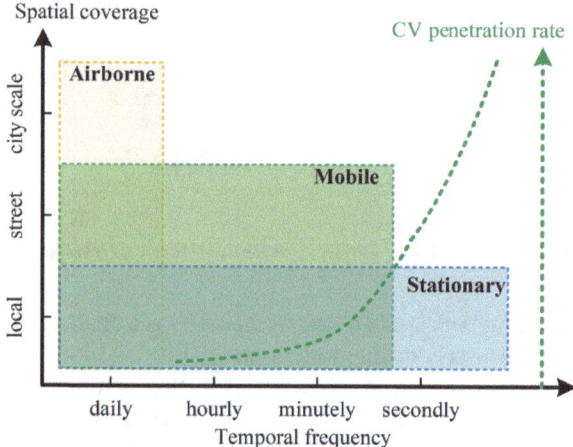

Fig. 2 Comparison of different ITS traffic sensing methods. Airborne sensors, such as drone, satellites, provide good spatial coverage, but it is often costly when collect sub-daily data. Stationary sensors collect data with high frequency, but have limited spatial coverage. Mobile sensors such as taxis offers some advantages of both methods, CV-based mobile sensing have further improvement potential in both spatial coverage and temporal frequency as penetration rate increase. This figure is reproduced from [3]

2.2 Traffic State Estimation

The precise and timely estimation of traffic states is critical for monitoring traffic state and identifying traffic congestions. The commonly used traffic state metrics are presented in Table 1. Note that some metrics are used to identify traffic congestions, while others are useful for designing countermeasures to congestions. For example, speed and travel time are commonly used for measure traffic system performance, flow and volume-to-capacity ratio are often used in designing traffic control algorithms and improving traffic system efficiency. In this section, we review a set of studies on estimating the traffic state under connected environment. The required data is speed and positioning measurements from CVs in combination with conventional traffic sensors (for example, stationary loop detectors). Hence, in real applications, the traffic state estimation is a mixture of models, often including both data assimilation and fusion techniques.

Traffic state metrics are generally correlated or even interconvertible, for example, measuring traffic density directly in the field is difficult: it requires a vantage point for photographing, videotaping, or observing significant lengths of highway. Density can be computed, however, from the average travel speed and flow rate, which are measured more easily. Moreover, the well-known Bureau of Public Roads (BPR) function relates the travel time to flow/demand volume (see Eq. 1) [6].

$$T_e = t_e \left[1 + \beta \left(\frac{x_e}{c_e} \right)^\alpha \right] \tag{1}$$

Table 1 Definitions of traffic state metrics (Reproduced from HCM 2010 [5])

Parameter	Definition
Average travel speed, average travel time	The length of the highway segment divided by the average travel time of all vehicles traversing the segment, including all stopped delay times (Equal to space mean speed). Average travel time can be computed as road segment length divide average travel speed
Flow rate, volume	The equivalent hourly rate at which vehicles or other roadway users pass over a given point or section of a lane or roadway during a given time interval of less than 1 h, usually 15 min
Capacity	The maximum potential hourly flow rate at which vehicles reasonably can be expected to traverse a point or a uniform section of a lane or roadway during a given period, under prevailing roadway condition
Density	The number of vehicles occupying a given length of a lane or roadway at a particular instant
Queue length	The distance between the upstream and downstream ends of the queue. (Queue is defined as a line of vehicles, bicycles, or persons waiting to be served due to traffic control, a bottleneck, or other causes.)
Volume to capacity ratio (VOC)	The ratio of volume (or flow) to capacity for a system element

where T_e is the travel time of road segment, t_e is the free flow travel time of a road segment e, x_e is the traffic flow of the road segment, c_e is the traffic capacity of the road segment, α, β are the model parameters (the recommended values are $\alpha = 4, \beta = 0.15$ in application).

In the following subsections, we focus on two most commonly studied traffic metrics: speed and flow, and explore the potential of using CV data to improve their estimation accuracy and immediacy. Note that other metrics can be also estimated directly or indirectly (i.e., convert using flow and speed data) with CV data.

2.2.1 Traffic Speed Estimation

Traffic speed estimation (or prediction) is one of most studied ITS topics [7]. The accurate estimation of traffic speed is essential for identifying traffic congestions. Existing speed estimation methods are mostly based on floating car data, where such data are usually collected by a navigation app such as Google Map. Floating car data normally stored timestamped geo-localization of vehicles. Link-based algorithms are proposed to estimate the speed at the road-segment, where they only use vehicles (geo-localization and timestamp) at the beginning and at the end of the road-segments [8]. But such an estimation is too coarse (producing the so-called space-mean speed) to reflect the true situation. It does not consider vehicle speed variation (remarkable change in a short time), and leads to great bias on insufficient samples.

Under connected environment, the availability of CV trajectories can improve traffic speed estimation. By assuming that the connected vehicles have the same distribution of speed as regular vehicles, the speed can be estimated as an average of the speeds of the connected vehicles. As the penetration rate of CV increases, the road segment speed profile can be updated at increased temporal frequency and a finer spatial resolution. The existing studies on fusion of CV and stationary sensor data for traffic state estimation often made the following essential assumptions [9, 10]:

- The connected vehicles continuously report their position, including information about their current road segment and speed, with a static frequency (for example, 1 Hz);
- The connected vehicles are assumed to have the same speed distribution as the nonconnected vehicles.

2.2.2 Traffic Flow Estimation

The accurate estimation of spatial–temporal change in traffic flow (or demand volume) is often required for traffic congestion management, since we need to measure the volume-to-capacity ratio to diagnose the causes of congestions.

The classical traffic flow estimation is model-based and follows a top-down proce-dure. The upper-level estimates the origin-demand (OD) demand (often use data from travel surveys); the lower-level model is used to load the demand volume to the network G, this problem is often known as dynamic traffic assignment (DTA) problem [11]. The dynamic OD trips resulting from off-line demand estimation are usually the starting point of the on-line demand estimation and prediction. This procedure involves sequentially updating dynamic OD matrices, previously adjusted off-line, in order to take into account the real-time variability of traffic conditions. Acquiring OD volumes is the key to estimate traffic network capacity afterwards. Generally, there are two approaches: the space–time consumption method and the optimization method. As with the former method, the road resource is assumed to be limited during a certain time interval, and therefore, the capacity is obtained with consideration of passing lane length, actual running time, time headway, and capacity reduction due to signals [12, 13]. In contrast, in the optimization method, the extreme value of largest flow is determined with given capacity and feasible flow through linear programming and max-flow mini-cut theorem [14].

Loop detectors are the primary data sources for collecting traffic flow data. But generally they are sparsely located and make it impossible to traffic volume profile for the city-wide network. In order to solve this issue, the traditional approach requires using additional data such as speed or trajectory of taxis (or similarly probe vehicles). By analyzing the speed pattern at the road-segment level, and assume roads with similar speed pattern may follow similar volume pattern, it is possible to fix missing flow data in the entire network via clustering road segments [15]. To estimate traffic flow, the CVs work in a similar manner as taxis or probe vehicles, but with better accuracy. In fact, given timestamped geo-locations and speed of CVs, it is possible to understand the traffic flow specific to CVs. Then a penetration rate can be applied to weight this value to retrieve traffic flow for all vehicles. This method could reduce estimation error since speed pattern and volume pattern might not be the same for some road segments.

With trajectories of connected vehicles available, it is now possible to retrieve information of origin–destination zones of individual trips as well as on their route choices and route travel times. This is a fundamental improvement for the calibration of traffic models and more importantly for the estimation of the OD travel demand (rather than adopting result from survey) [16]. For estimating capacity, due to the nearly instant wireless communication among CVs or between CVs and road side infrastructures, key vehicle states such as time headway, or arrivals of vehicles can be perceived in a timely manner, which allows for estimating traffic network demand and capacity at a high temporal frequency.

3 Traffic Management with Connected Vehicles

How to identify the bottleneck has very important theoretical significance and practical application worthiness in optimizing traffic organization and traffic control. From a transportation network perspective, the links with the maximum flow can be easily congested and become the congestion bottleneck. Therefore, in this paper, we attempt to study the bottleneck and flow distribution rules from a network perspective. Furthermore, a statistical-based method to identify the bottleneck is proposed. The formation of congestion bottleneck is due to several factors, which can be classified into two categories [17, 18].

- *Temporary obstruction.* Congestion in abnormal condition is commonly caused by unusual events, e.g. anomalistic car driving behavior of drivers, road work, accidents and crashes, special events, etc. There is little rule for the occurrence of this type of congestions that may have a great duration. Effective measures should be taken to disperse the congestion traffic in time; otherwise, a serious traffic jam may break out and the traffic may be destroyed both locally and network-wide.
- *Permanent capacity constraint in the network.* In general, the formation of bottleneck caused by the factor of this category is due to bad roadway line type or the width becoming narrow, which reduces the level of service by lowering speeds. Besides, indistinct roadway symbols could also cause bottlenecks.

In this section, we focus on using connected vehicle techniques for road traffic congestion mitigation. Depending on the optimization approaches, the countermeasures to urban traffic congestions can be roughly categorized into following two groups:

- *Demand management solutions.* Adaptive traffic management aims to ensure that the given roadway capacity is adapt to traffic demand via real-time traffic control or guidance methods, for example, regional traffic restriction (or congestion pricing), traffic signal control and route guidance are typical traffic management approaches;

- *Capacity optimization solutions.* Bottleneck capacity optimization seeks to maximize roadway design capacity at critical locations where traffic congestions occur recurrently, which often consists of a sequential decision-making process: (1) bottleneck identification, and (2) bottleneck amelioration.

3.1 Adaptive Traffic Control and Guidance

Traffic control and guidance are commonly used in urban area to mitigate traffic congestions. Since CV provides precise and timely traffic state information, real-time traffic control or route guidance could be more adaptive to change in traffic, hence, the traffic system efficiency could be improved.

3.1.1 Traffic Control

The commonly used traffic signal control can be categorized as feedback and prediction-based [19]. Conventional feedback-based traffic signal control collects traffic volume information via inductive loop detectors that usually installed tens of meters upstream to the stop lines. Models have been developed to describe traffic flow states which often fails to well present the variability in traffic demand [13, 20]. CV data provides a remedy for such problems [21, 22]. For example, data from CVs can be used to estimate the volume of each approach per signal cycle, and then calculated the optimum cycle length using the Webster's formula [23, 24]. The green time was allocated to produce equal degrees of saturation on each link. The preliminary signal plan for the next cycle was generated during the current cycle and adjusted to meet practical limitations like the minimum and maximum cycle lengths and pedestrian minimum green times. Simulation-based experiment showed that the system could reduce traffic delay and fuel consumption compared with traditional pre-timed traffic signals.

In contrast, prediction-based traffic signal control aims to schedule the signal timing plans to allow the platoons or individual vehicles to pass the intersection without severe interruptions, which can increase the overall traffic efficiency. This requires identification of the platoons and predicting their arrival times in advance. Although the idea of prediction-based traffic signal control has been proposed for several decades [19, 25], it became realistic only after CV techniques was introduced since it is now possible to properly identify platoons and predict their arrivals [26–30]. For example, a platoon-based "oldest arrival first" traffic signal control method is proposed to reduce delays at a single intersection [31]. After collecting real-time speed and position information of sample vehicles through V2X communication, they grouped the vehicular traffic into approximately equal-sized platoons. The grouped platoons could be scheduled by solving the reduced job scheduling problem. Compared with traditional feedback-based methods, the proposed method could significantly reduce delays when the traffic inflow rates are not large. The experimental results also showed that the proposed method did not perform well under low penetration rates, since the arrival rate cannot be accurately estimated under low penetration conditions. In addition, another group of research focus on control of individual vehicles, which often requires prediction of the arrival time of every vehicle and predict traffic conditions in a forward time horizon. Hence, this group of prediction-based control requires high to full penetration of CVs [19].

Other traffic control methods also benefit from CV data. For example, utilizing the Macroscopic Fundamental Diagram (MFD) as the traffic performance indicator, researchers designed feedback or prediction (i.e., model predictive control or MPC) based perimeter controllers to optimize traffic state at both local and network levels, where CV serve as only or supplementary data source [32, 33]. The studies found that low penetration rates of CV lead to strong noises in the controller, hence, the stochasticity in traffic state estimation need to be considered in design of robust controllers. In addition, CV-based ramp metering, transit priority and congestion pricing solutions have also been found in recent literatures [29, 34, 35].

3.1.2 Route Guidance

Route guidance in urban road network promises to improve the capacity of urban traffic management and mitigate traffic congestion in mega-cities [36]. The route guidance of urban traffic is essentially the shortest path problem in graph theory [37, 38]. Route guidance methods have been studied from the system's perspective and/or the user's perspective. User-oriented route guidance methods that focus on a user's preferences have long been studied. This may lead to traffic congestion because the optimal path for the user is the path that only maximizes the user's preferences. Route guidance methods for finding the system-optimal traffic distribution have also been studied [39]. It is well-known that a system-oriented guidance system sometimes seems unfair because certain users' preferences are sacrificed for the sake of the system's interests. Thus, in a system-oriented guidance system, when the unfairness reaches a certain level, the user may not follow the route guidance, thereby violating the original intention of the system optimization. By restraining users' selfish behavior or avoiding congested roads, a compromise can be achieved between individual and global benefits [6, 39].

Currently, the commonly used traffic guidance methods are navigation map and roadside variable message signs. The map navigation apps provide route choice suggestions for drivers using estimated speed by floating car data, the likeability of finding shortest-path highly depends on the penetration rate of floating cars, making it hard to avoid congested road segments with low traffic volume or penetration rate. Variable message signs provide site information about congestions caused by road work, weather condition and special events, however, they cannot prevent congestions from the very beginning. CV technique can provides more timely and accurate guidance information, which can help drivers to avoid congested roadway segments at an early stage.

Some CV-based route guidance can be found in recent literatures [36–38]. For example, some studies assume that connected vehicles (CVs) follow system optimum (SO), and non-CVs pursue the user equilibrium (UE) pattern [40]. The mixed UE–SO problem is formulated as a nonlinear complementarity problem. Although these studies also analyze users' vehicle choices, they assume an aggregate class of CVs. In addition, with connected infrastructure, a hybrid framework of destination choice (i.e., parking choice) and route choice can be further considered.

3.2 Bottleneck Capacity Optimization

The above-mentioned traffic management strategies require no (or minor) changes to transport infrastructures, however, it cannot address congestions imposed by capacity constraints at critical bottlenecks. In urban area, the volume of traffic flow fluctuates unevenly across complex road networks while simultaneously being hindered by some form of congestion or overload. In this section, we focus on studies on CV data in decision-making on bottleneck identification and amelioration. Urban transport infrastructure system is a weighted network, whose weakest links can be of particularly significance. For instance, the speed of traffic flow between two nodes is limited by the weak link with the smallest capacity (i.e., bottleneck). Furthermore, weak links can affect the overall network integrity [41]. Improving the infrastructure networks via protection or enhancement of a minimal set of links is receiving intense research interest [42, 43].

Minimum spanning tree (MST) is used to identify of the weakest links set in a weighted network [18]. Here, the weight is represented by the zero-flow cost (or free-flow travel time) of each link. The MST is widely used in different fields, such as the design and operation of communication networks, economic networks, the traveling salesman problem, and optimal traffic flow. By avoiding the strong links and preferentially following the weakest ones, the MST selects the lowest weight backbone of the network. Spanning trees show the robust features, such as scale-free degree distribution and distortive or neutral degree correlation. The benefit of CV in identifying bottleneck link set can be twofold: first, with real-time traffic speed collected by CVs, the weight of traffic network can be estimated more precisely; second, the flow distribution within traffic networks collected by CVs can be used to further adjust or calibrate the MST model.

To further find the most critical component supporting heavy traffic flow in MST, percolation-based method is used to obtain the incipient infinite cluster (IIC) or infinite cluster at criticality. The set of edges (nodes) inside the IIC is typically used by transport paths more often than other edges (nodes) in the MST, which can be understood as a critical percolation cluster [28]. Moreover, some recent studies used percolation-based approach to find out the bottleneck links directly from the road network (see Fig. 3). The above-mentioned studies investigate the statistical dynamics behaviors on static traffic networks. However, the real traffic is a dynamic

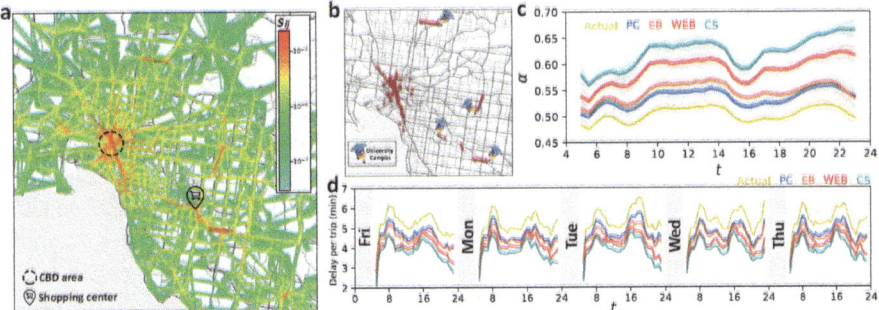

Fig. 3 Percolation-based bottleneck identification and amelioration on Melbourne's public transport network. a Spatial distribution of link criticality. **b** Top 100 weekday bottlenecks identified based on link criticality. **c, d** The impact of amelioration on bottlenecks identified by different approaches, i.e., criticality score (CS), edge betweenness centrality (EB), demand-weighted edge betweenness centrality (WEB), and percolation criticality (PC). This figure is reproduced from [44]

and complex network, a hybrid framework of model and data-based could be developed to identify bottlenecks with the use of CV data. Note bottleneck identification and amelioration is an iterative process, the traffic sensing system with CVs is essential in decision-making and evaluation.

4 Open Issues and Future Research Directions

4.1 The Impact of CV Penetration Rate

When speed measurements from CVs are available, the speed can be estimated by calculating an average of the speeds of the CVs. This requires that the CVs have the same distribution of speeds as regular vehicles. Otherwise, the speed estimate would be biased. Many interesting topics are identified for future research in CV-based traffic sensing. By including data assimilation techniques [45], the performance of the CV-based traffic sensing can be improved. By using a more advanced method to estimate the penetration rate, the accuracy of the traffic sensing can be further improved. For example, the dynamics of the penetration rate can be modeled using a macroscopic model as in [46, 47].

The use of the proposed method in traffic management systems is also a topic for future research. The penetration rate of CVs can significantly influence the performances of the afore-mentioned traffic management methods. Many existing studies assumed 100% penetrations of CVs so that the full information of all vehicles can be used and/or all vehicles can be controlled to better design the traffic control methods [48, 49]. The main advantage of assuming 100% penetration is that we can avoid estimating the information of unequipped vehicles, which can significantly reduce the complexity of the resulting model and estimation errors. Although it is expected that the penetration rate of CVs may dramatically increase in the future, there is still a long way to achieve such a goal of high CVs penetration. Therefore, it is practical and important to consider different levels of CVs penetration, i.e., to consider mixed traffic flow with both vehicles with and without connectivity, when designing CV-based traffic control methods in practice.

A few existing studies had begun to discuss the relationship between CV penetration rates and various performances of the control algorithms [50, 51]. For example, some previous studies found that the offline optimization algorithm over a 3-h window could perform well with a CV penetration as low as 1%, while the online optimization with 15-min windows requires at least 5% CV penetration [52]. A simulation-based studies revealed that a 40% CV penetration rate could prevent all types of accidents [51]. Some other studies showed that 100% CV penetration could improve the fuel efficiency under any traffic volume; in the mixed flow scenario, the fuel-saving benefits could be only achieved even when the traffic volume is low [53]. These studies could serve as a good starting point, while more efforts need to be conducted. For example, more traffic scenario studies are urgently needed. In addition, current works usually adopted simulation models to study the impact of penetration, theoretical models and real-world tests, which could better verify the impact of penetration of CVs, should also be thoroughly studied.

4.2 Robustness of ITS Solutions Under Connected Environment

The successful applications of the afore-mentioned ITS traffic management solutions highly rely on proper implementations of communication technologies. First, the reliability of V2X communications is one of the critical factors that may influence the performance of CV-based traffic control. The delayed/missed vehicle information and the position errors may lead to failures of pre-selected signal timing plans or even result in traffic accidents [54]. However, we believe that robust planning and stochastic programming is still be required, unless the V2X communications become robust [55]. Since real-time traffic control requires transmitting information with low communication delay, many researchers believe that 5G communication should be one of the supporting backbones for the communications of the next generation traffic management systems.

Second, the cybersecurity issues of V2X communication technologies and traffic management systems should also be carefully investigated. Cybersecurity should address at least two major issues: privacy protection of individual users, and the security of vehicles and traffic management systems. Existing research on this topic is relatively sparse. Privacy protection is mainly related to the collection and use of CV data. In a recent study, the vulnerability of actuated and adaptive traffic signal control systems under connected environment is tested [56]. Falsified data were sent from four typical elements including signal controllers, vehicle detectors, roadside units, and onboard units, the experimental results showed that some attacks could significantly increase congestion. Compared to the systems without CV-based signal control, the attacks could significantly reduce the mobility of CV-based signal control systems by up to 23.4% [56]. It is expected that with the wide deployment of CV technologies and the implementation of CV-based traffic control, the cybersecurity of vehicles and traffic control systems will become increasingly critical. This calls for comprehensive investigations of this important topic in the future.

Third, traffic sensing and management systems need to be continuously updated to keep the pace with CV technologies. With big data collected by CVs, it is an emerging research need on improvement of the computational capability of traffic sensing and management systems.

5 Conclusions

ITS can reduce traffic congestion at least possible cost with a range of technologies. The congestion mitigation cycle often starts with traffic sensing and bottleneck identification and follows an iterative process of traffic management and evaluation. In this chapter, we focus on ITS management under connected environment and provides a comprehensive review on emerging studies. First, this chapter explores the potential of CV techniques in improvement of spatial coverage and temporal frequency of traffic state sensing. Second, this chapter further reviews studies on CV-based the real-time traffic control, route guidance, and bottleneck capacity optimization. Moreover, this chapter further discusses the impact of CV penetration rate and robustness of CV-based ITS solutions.

References

1. D. Rolnick et al., Tackling climate change with machine learning. ACM Comput. Surv. (CSUR) **55**(2), 1–96 (2022)
2. V.M. Taghvaee et al., Sustainable development goals: transportation, health and public policy. Rev. Econ. Polit. Sci. (2021)
3. K.P. O'Keeffe, A. Anjomshoaa, S.H. Strogatz, P. Santi, C. Ratti, Quantifying the sensing power of vehicle fleets. Proc. Natl. Acad. Sci. **116**(26), 12752–12757 (2019)
4. X. Xie, H. van Lint, A. Verbraeck, A generic data assimilation framework for vehicle trajectory reconstruction on signalized urban arterials using particle filters. Transp. Res. Part C: Emerging Technol. **92**, 364–391 (2018)
5. H.C. Manual, HCM2010, in *Transportation Research Board, National Research Council, Washington, DC*, vol. 1207 (2010)
6. H. Youn, M.T. Gastner, H. Jeong, Price of anarchy in transportation networks: efficiency and optimality control. Phys. Rev. Lett. **101**(12), 128701 (2008)
7. W. Jiang, J. Luo, Graph neural network for traffic forecasting: a survey. Expert Syst. Appl. 117921 (2022)

8. X. Li, W. Shu, M. Li, H.-Y. Huang, P.-E. Luo, M.-Y. Wu, Performance evaluation of vehicle-based mobile sensor networks for traffic monitoring. IEEE Trans. Veh. Technol. **58**(4), 1647–1653 (2008)
9. E.F. Grumert. A. Tapani, Traffic state estimation using connected vehicles and stationary detectors. J. Adv. Transp. **2018** (2018)
10. S.M. Khan, K.C. Dey, M. Chowdhury, Real-time traffic state estimation with connected vehicles. IEEE Trans. Intell. Transp. Syst. **18**(7), 1687–1699 (2017)
11. C. Antoniou et al., Towards a generic benchmarking platform for origin–destination flows estimation/updating algorithms: design, demonstration and validation. Transp. Res. Part C: Emerging Technol. **66**, 79–98 (2016)
12. J.I.A. Shunping, P. Hongqin, L.I.U. Shuang, Urban traffic state estimation considering resident travel characteristics and road network capacity. J. Transp. Syst. Eng. Inf. Technol. **11**(5), 81–85 (2011)
13. J. Zheng, H.X. Liu, Estimating traffic volumes for signalized intersections using connected vehicle data. Transp. Res. Part C: Emerging Technol. **79**, 347–362 (2017)
14. L. Cheng, W. Wang, J.Y. Wang, Y. Shao, Y. Lu, Urban road network capacity, traffic planning and traffic management. J. Highw. Transp. Res. Dev. **22**(7), 118–122 (2005)
15. C. Meng, X. Yi, L. Su, J. Gao, Y. Zheng, City-wide traffic volume inference with loop detector data and taxi trajectories, in *Proceedings of the 25th ACM SIGSPATIAL International Conference on Advances in Geographic Information Systems* (2017), pp. 1–10
16. Y. Cao, K. Tang, J. Sun, Y. Ji, Day-to-day dynamic origin–destination flow estimation using connected vehicle trajectories and automatic vehicle identification data. Transp. Res. Part C: Emerging Technol. **129**, 103241 (2021)
17. J. Long, Z. Gao, H. Ren, A. Lian, Urban traffic congestion propagation and bottleneck identification. Sci. China Ser. F: Inf. Sci. **51**(7), 948–964 (2008)
18. J. Wu, Z. Gao, H. Sun, Topological-based bottleneck analysis and improvement strategies for traffic networks. Sci. China Ser. E: Technol. Sci. **52**(10), 2814–2822 (2009)
19. Q. Guo, L. Li, X.J. Ban, Urban traffic signal control with connected and automated vehicles: a survey. Transp. Res. Part C: Emerging Technol. **101**, 313–334 (2019)
20. I. Yun, B. Park, Stochastic optimization for coordinated actuated traffic signal systems. J. Transp. Eng. **138**(7), 819–829 (2012)
21. W. Wu, J. Zhang, A. Luo, J. Cao, Distributed mutual exclusion algorithms for intersection traffic control. IEEE Trans. Parallel Distrib. Syst. **26**(1), 65–74 (2014)
22. M.B. Younes, A. Boukerche, Intelligent traffic light controlling algorithms using vehicular networks. IEEE Trans. Veh. Technol. **65**(8), 5887–5899 (2015)
23. V. Gradinescu, C. Gorgorin, R. Diaconescu, V. Cristea, L. Iftode, Adaptive traffic lights using car-to-car communication, in *2007 IEEE 65th Vehicular Technology Conference-VTC2007-Spring* (2007), pp. 21–25
24. A.H. Chow, R. Sha, Y. Li, Adaptive control strategies for urban network traffic via a decentralized approach with user-optimal routing. IEEE Trans. Intell. Transp. Syst. **21**(4), 1697–1704 (2019)
25. Q. He, K.L. Head, J. Ding, Multi-modal traffic signal control with priority, signal actuation and coordination. Transp. Res. Part C: Emerging Technol. **46**, 65–82 (2014). https://doi.org/10.1016/j.trc.2014.05.001
26. X.J. Liang, S.I. Guler, V.V. Gayah, An equitable traffic signal control scheme at isolated signalized intersections using connected vehicle technology. Transp. Res. Part C: Emerging Technol. **110**, 81–97 (2020)
27. X. Liang, S.I. Guler, V.V. Gayah, Joint optimization of signal phasing and timing and vehicle speed guidance in a connected and autonomous vehicle environment. Transp. Res. Rec. **2673**(4), 70–83 (2019)
28. X. Liang, S.I. Guler, V.V. Gayah, Signal timing optimization with connected vehicle technology: platooning to improve computational efficiency. Transp. Res. Rec. **2672**(18), 81–92 (2018)
29. K. Wu, M. Lu, S.I. Guler, Modeling and optimizing bus transit priority along an arterial: a moving bottleneck approach. Transp. Res. Part C: Emerging Technol. **121**, 102873 (2020)

30. K. Wu, *Bus Transit Priority: Modeling and Evaluating Performance* (The Pennsylvania State University, 2019)
31. K. Pandit, D. Ghosal, H.M. Zhang, C.-N. Chuah, Adaptive traffic signal control with vehicular ad hoc networks. IEEE Trans. Veh. Technol. **62**(4), 1459–1471 (2013)
32. K. Yang, M. Menendez, N. Zheng, Heterogeneity aware urban traffic control in a connected vehicle environment: a joint framework for congestion pricing and perimeter control. Transp. Res. Part C: Emerging Technol. **105**, 439–455 (2019)
33. K. Yang, N. Zheng, M. Menendez, Multi-scale perimeter control approach in a connected-vehicle environment. Transp. Res. Procedia **23**, 101–120 (2017)
34. F. Tajdari, C. Roncoli, M. Papageorgiou, Feedback-based ramp metering and lane-changing control with connected and automated vehicles. IEEE Trans. Intell. Transp. Syst. (2020)
35. Y. Yang, Z. Liu, Y. Kong, X. Miao, J. Guo, Coordinated ramp metering under the connected vehicle environment, in *CICTP 2017: Transportation Reform and Change—Equity, Inclusiveness, Sharing, and Innovation* (American Society of Civil Engineers Reston, VA, 2018), pp. 774–783
36. T. Xie, Y. Liu, Impact of connected and autonomous vehicle technology on market penetration and route choices. Transp. Res. Part C: Emerging Technol. **139**, 103646 (2022)
37. X. Jiang, Y. Ji, M. Du, W. Deng, A study of driver's route choice behavior based on evolutionary game theory. Comput. Intell. Neurosci. **2014** (2014)
38. F. Yu-qin, L. Jun-qiang, X. Zhong-Yu, Z. Gui-e, H. Yi, Route choice model considering generalized travel cost based on game theory. Math. Probl. Eng. **2013** (2013)
39. S. Çolak, A. Lima, M.C. González, Understanding congested travel in urban areas. Nat. Commun. **7**(1), 1–8 (2016)
40. S.A. Bagloee, M. Sarvi, M. Patriksson, A. Rajabifard, A mixed user-equilibrium and system-optimal traffic flow for connected vehicles stated as a complementarity problem. Comput. Aided Civil Infrastruct. Eng. **32**(7), 562–580 (2017)
41. Z. Wu, L.A. Braunstein, S. Havlin, H.E. Stanley, Transport in weighted networks: partition into superhighways and roads. Phys. Rev. Lett. **96**(14), 148702 (2006)
42. A. Halu, A. Scala, A. Khiyami, M.C. González, Data-driven modeling of solar-powered urban microgrids. Sci. Adv. **2**(1), e1500700 (2016)
43. Y. Yang, T. Nishikawa, A.E. Motter, Small vulnerable sets determine large network cascades in power grids. Science **358**(6365), eaan3184 (2017)
44. H. Hamedmoghadam, M. Jalili, H.L. Vu, L. Stone, Percolation of heterogeneous flows uncovers the bottlenecks of infrastructure networks. Nat. Commun. **12**(1), 1–10 (2021)
45. T. Seo, T. Kusakabe, Probe vehicle-based traffic state estimation method with spacing information and conservation law. Transp. Res. Part C: Emerging Technol. **59**, 391–403 (2015)
46. V. Astarita, R.L. Bertini, S. d'Elia, G. Guido, Motorway traffic parameter estimation from mobile phone counts. Eur. J. Oper. Res. **175**(3), 1435–1446 (2006)
47. N. Bekiaris-Liberis, C. Roncoli, M. Papageorgiou, Highway traffic state estimation with mixed connected and conventional vehicles. IEEE Trans. Intell. Transp. Syst. **17**(12), 3484–3497 (2016)
48. W. Li, X.J. Ban, Traffic signal timing optimization in connected vehicles environment, in *2017 IEEE Intelligent Vehicles Symposium (IV)* (2017), pp. 1330–1335
49. S.M.A.B.A. Islam, A. Hajbabaie, Distributed coordinated signal timing optimization in connected transportation networks. Transp. Res. Part C: Emerging Technol. **80**, 272–285 (2017). https://doi.org/10.1016/j.trc.2017.04.017
50. B. Beak, K.L. Head, Y. Feng, Adaptive coordination based on connected vehicle technology. Transp. Res. Rec. **2619**(1), 1–12 (2017)
51. A. Validi, T. Ludwig, A. Hussein, C. Olaverri-Monreal, Examining the impact on road safety of different penetration rates of vehicle-to-vehicle communication and adaptive cruise control. IEEE Intell. Transp. Syst. Mag. **10**(4), 24–34 (2018)
52. C.M. Day, D.M. Bullock, Detector-free signal offset optimization with limited connected vehicle market penetration: proof-of-concept study. Transp. Res. Rec. **2558**(1), 54–65 (2016)

53. J. Rios-Torres, A.A. Malikopoulos, Impact of partial penetrations of connected and automated vehicles on fuel consumption and traffic flow. IEEE Trans. Intell. Veh. **3**(4), 453–462 (2018)
54. Y. Meng, L. Li, F.-Y. Wang, K. Li, Z. Li, Analysis of cooperative driving strategies for nonsignalized intersections. IEEE Trans. Veh. Technol. **67**(4), 2900–2911 (2017)
55. Y. Tong, L. Zhao, L. Li, Y. Zhang, Stochastic programming model for oversaturated intersection signal timing. Transp. Res. Part C: Emerging Technol. **58**, 474–486 (2015)
56. C. Yu, Y. Feng, H.X. Liu, W. Ma, X. Yang, Integrated optimization of traffic signals and vehicle trajectories at isolated urban intersections. Transp. Res. Part B: Methodol. **112**, 89–112 (2018)

Chapter 3
Evolution of Wireless Communication Technology for V2X Assisted Autonomous Driving

Kainan Zhu and Yongdong Zhu

Abbreviations

3GPP	Third Generation Partnership Project
5G	5Th Generation Mobile Communication Technology
5GAA	5G Automotive Association
8DPSK	Differential 8-Phase Shift Keying
ARIB	Association of Radio Industries and Businesses
ARQ	Automatic Repeat-reQuest
BPSK	Binary Phase-Shift Keying
BS	Base Station
BSM	Basic Service Message
C-V2X	Cellular Vehicle-to-Everything
CA	Carrier Aggregation
CAM	Cooperative Awareness Message
CAV	Connected Autonomous Vehicles
CEN	European Committee for Standardization
CP	Cyclic Prefix
CSMA/CA	Carrier Sense Multiple Access with Collision Avoidance
D2D	Device-to-Device
DENM	Decentralized Environmental Notification Message
DFT	Discrete Fourier Transform
DMRS	DeModulation Reference Signal
DQPSK	Differential Quadrature Phase Shift Keying

K. Zhu · Y. Zhu (✉)
Zhejiang Lab, Interdisciplinary Innovation Research Institute, Hangzhou, China
e-mail: zhuyd@zhejianglab.com

K. Zhu
e-mail: zhukainan@zhejianglab.com

DSRC	Dedicated Short-Range Communications
E-UTRAN	Evolved Universal Terrestrial Radio Access Network
ECP	Extended Cyclic Prefix
eMBMS	Enhanced Multimedia Broadcast Multicast Service
eNodeB	Evolved Node B
ETSI	European Telecommunications Standards Institute
FCC	Federal Communications Commission
FEC	Forward Error Correcting
FM	Frequency Modulation
FR1	Frequency Range 1
FR2	Frequency Range 2
GFSK	Gaussian Frequency-Shift Keying
GLOSA	Green Light Optimal Speed Advisory
GNSS	Global Navigation Satellite System
GPS	Global Positioning System
GSM	Global System for Mobile Communication
IEEE	Institute of Electrical and Electronics Engineers
IMU	Inertial Measurement Unit
ISI	Inter-Symbol Interference
ISM	Industrial, Scientific and Medical
ISO	International Standards Organization
ITS	Intelligent Transportation Systems
ITS-S	Intelligent Transportation Systems-Station
LiDAR	Laser Imaging, Detection, and Ranging
LoS	Line-of-Sight
LTE	Long Term Evolution
MAC	Medium Access Control
MCS	Modulation and Coding Scheme
MIMO	Multiple-Input Multiple-Output
MNO	Mobile Network Operators
MU-MIMO	Multi-User Multiple-Input Multiple-Output
NCP	Normal Cyclic Prefix
NLOS	Non-Line-of-Sight
NR	New Radio
O-QPSK	Offset Quadrature Phase Shift Keying
OBU	On Board Unit
ODD	Operational Design Domain
OFDM	Orthogonal Frequency Division Multiplexing
OFDMA	Orthogonal Frequency Division Multiple Access
PANs	Personal Area Networks
PHY	Physical Layer
POS	Personal Operating Space
ProSe	Proximity Services
PSCCH	Physical Sidelink Control Channel
PSFCH	Physical Sidelink Feedback Channel

PSSCH	Physical Sidelink Shared Channel
QAM	Quadrature Amplitude Modulation
QoS	Quality of Service
QPSK	Differential Quadrature Phase Shift Keying
RB	Resource Block
RDS	Radio Data System
RFID	Radio Frequency IDentification
RSU	Road Side Unit
SA	System Aspects
SAE	Society of Automotive Engineers International
SAP	Service Access Point
SC-FDMA	Single-Carrier Frequency-Division Multiple Access
SCI	Sidelink Control Information
SCS	Subcarrier Spacing
SDO	Standard Development Organizations
SLRs	Service Level Requirements
SPS	SemiPersistent Scheduling
TB	Transport Block
TCMA	Tiered Contention Multiple Access
UHF	Ultra-High Frequency
UMTS	Universal Mobile Telecommunication Systems
V2I	Vehicle-to-Infrastructure
V2N	Vehicle-to-Network
V2P	Vehicle-to-Pedestrian
V2V	Vehicle-to-Vehicle
V2X	Vehicle-to-Everything
VANET	Vehicular Ad Hoc Network
VRU	Vulnerable Road User
WAVE	Wireless Access in Vehicular Environments
Wi-Fi	Wireless Fidelity

1 Introduction

The outbreak of the coronavirus pandemic across the globe has impacted all areas of people's daily life. In terms of daily traffic behavior, research has observed a notable shift from public transportation towards private vehicles in different cities worldwide [1–4] due to factors such as the promotion of social distancing [5], hygiene considerations [6], etc. Private vehicle travel contributes substantially [7] to traffic congestion, environmental degradation, pollution, and energy inefficiency, only to name a few. Recently, the idea of smart urban mobility [8], which includes innovations such as Intelligent Transportation Systems (ITS) and autonomous driving vehicles, has been introduced to realize smart and sustainable urban transportation systems and to address the adverse effects of private vehicle travel on cities, societies and the environment [9].

The development of ITS dates back to as early as the beginning of the 1970s [10]. Intending to improve transport efficiency and safety, ITS utilizes advanced control, computing, and information and communication technologies to accomplish transport implementations such as optimal traffic signal control, autonomous vehicle control, and traffic flow control. Nowadays, ITS has been widely adopted in cities and the thriving of diverse ITS sources has contributed to the amount of the generated ITS data trending from the Trillion-byte level to Petabyte level [10]. Consequently, the ever-demanding need for data processing and communication in ITS calls for the assistance of emerging technologies such as 5G (5th Generation Mobile Communication Technology) [11], big data analytics [10], blockchain [12], artificial intelligence [13, 14] and edge cloud computing [15] to provide better performance and possibly to realize a wide range of new mobility services.

Autonomous driving envisages vehicles to perceive the environment through on-board sensors or through cooperative information exchange with other transport entities, such that human effort can be liberated and promises comprising improved traffic safety and efficiency can be realized. It is projected that more than 50% of all vehicles will be capable of autonomous driving by 2050 [16], and the resulting social benefits could reach nearly 800 billion US dollars [17]. Recent advances in the development of autonomous driving have witnessed a trend of reliance on parallel development in ITS. Wireless communication technology is a major enabling technology in the development of both ITS and autonomous driving. As in commonly referred Connected Autonomous Vehicles (CAV) [8], vehicles can be inter-connected with pedestrians and road infrastructure through cooperative wireless communication technologies, such that vehicles can comprehend and respond to broader network information and helps foster a smarter and safer road traffic system.

1.1 Motivation

With the ongoing pursuit of efficiency, safety and intelligence in the transportation industry, and the rapid development of information technology, autonomous driving has become an important interdisciplinary subject among transportation, communication, artificial intelligence and other related research areas. In order to identify vehicle automation levels, the Society of Automotive Engineers International (SAE) has developed a 6-level classification system [18], with level 0 being no autonomous and level 5 indicating fully autonomous, see Table 1 for features in each SAE level of driving automation.

Existing autonomous driving technologies can be divided into two major directions: standalone (or ego-only) autonomous driving and networked autonomous driving [17]. Standalone autonomous driving relies entirely on the input of on-board sensors, such as LiDAR (Laser imaging, Detection, and Ranging), IMU (Inertial Measurement Unit), millimeter-wave radar, and cameras to actualize environmental

Table 1 SAE level of driving automation

SAE level of driving automation	Description	Features
0	No driving automation	Human driver performs the entire driving task
1	Driver assistance	Driving automation system executes either the longitudinal or the lateral vehicle motion control subtask, and the human driver supervises the driving automation system and intervenes as necessary for safe operation
2	Partial driving automation	Driving automation system executes both the longitudinal and the lateral vehicle motion control subtask, and the human driver supervises the driving automation system and intervenes as necessary for safe operation
3	Conditional driving automation	Driving automation system performs the entire driving task, and the human driver intervenes upon system request
4	High driving automation	Driving automation system performs the entire driving task within its operational design domain (ODD) without the intervention of the human driver
5	Full driving automation	Driving automation system performs the entire driving task under all driver-manageable on-road conditions without the intervention of the human driver

perception, and to predict environmental changes and generate driving plans and control decisions through artificial intelligence technology. However, standalone autonomous driving is confronted with limitations including unreliability, infeasibility and inefficiency [19]. For example, cameras are vulnerable to severe weather conditions and sight blockage situations. Complex road conditions like intersections may also degrade the performance of the sensors. Moreover, the cost of LiDAR is far from affordable to general consumers. In addition, on-board sensors have blind spot problems and are incapable of acquiring road traffic conditions a few blocks away. Therefore, standalone autonomous driving can only alleviate human effort in driving and it is far from realizing a smarter and safer road traffic system.

To cope with the complex road traffic situations, networked autonomous driving integrates Vehicle-to-Everything (V2X) technology based on standalone autonomous driving to make up for sensor deficiencies. V2X technology enables autonomous driving vehicles equipped with the on-board unit (OBU) to communicate with pedestrians, nearby OBU equipped autonomous driving vehicles, road infrastructures including road side unit (RSU) and base station (BS) and other entities through wireless communication technology. In this way, autonomous driving vehicles can obtain real-time information like pre-collision warnings, and take corresponding measures promptly, thereby reducing traffic accidents and improving the safety of autonomous driving. Meanwhile, government sectors can use the V2X technology to obtain traffic information in real-time and to tackle urban traffic congestion problems through methods such as vehicle speed guidance and promoting the dissemination of public information.

Currently, there is a plethora of wireless communication technologies in V2X assisted autonomous driving. While dedicated short-range communications (DSRC) and Cellular-V2X (C-V2X) are the two major candidates for V2X, several other wireless communication technologies, including Bluetooth, ZigBee and Wireless Fidelity (Wi-Fi) have also been considered for V2X applications. Each of the above-mentioned wireless communication technologies has features that make it potentially promising for V2X, each also has some drawbacks. Therefore, it is important to shed light on the characteristics of each wireless access technology candidate for V2X and thus contribute to designing an efficient, smart and sustainable transportation system.

1.2 Main Contributions

The main contributions to this chapter are as follows:

(1) This chapter introduces wireless communication technologies for V2X assisted autonomous driving from a history evolution's point of view;
(2) This chapter introduces automation levels defined for autonomous driving, and elaborates typical use cases and requirements of V2X communication defined by different organizations, and presents a typical V2X system architecture;
(3) This chapter discusses the physical layer features of different potential wireless communication technologies for V2X, including Bluetooth, ZigBee, Wi-Fi, DSRC, and C-V2X, which is composed of LTE-V2X and NR-V2X. This chapter also highlights their pros and cons in providing communication services for V2X assisted autonomous driving, and summarizes some research efforts in this regard;
(4) This chapter shed light on standardization activities toward V2X communications to better coordinate various V2X use cases with stringent network requirements.

1.3 Structural Organization

Section 2 presents a brief history of wireless communication technologies for V2X. Section 3 introduces different V2X use cases and requirements defined by the Third Generation Partnership Project (3GPP) and 5GAA (5G Automotive Association). Section 4 elaborates in detail on the characteristics of different potential wireless communication technologies that can be applied in V2X assisted autonomous driving, with an emphasis on the physical layer aspect. Section 5 outlines the standardization activities toward cellular based V2X technology. Finally, Sect. 6 summarizes the whole chapter.

2 A Brief History of Wireless Communication Technology for V2X

V2X, or in some context referred to as connected vehicles, performs the coordination between vehicles and their environment through wireless communication technology [20], thus enabling Vehicle-to-Infrastructure (V2I), Vehicle-to-Vehicle (V2V), Vehicle-to-Network (V2N) and Vehicle-to-Pedestrian (V2P) communications. An early prototype [21] of modern V2X application has been around in 1926 when a patent "Radio warning systems for use on vehicles" [22] proposed for two independent vehicles to achieve peer-to-peer radio communication by equipping on-vehicle devices was granted. While this prototype is far from the V2X application today, it severs very similar demands, that is to improve vehicular safety through communication among vehicles, and this design also motivated the development and the consolidation of the later vehicular ad hoc network (VANET) technology [21].

With the development of Frequency Modulation (FM) radio broadcast technology, in 1984, the radio data system (RDS) became the first communication protocol standard for broadcasting digital information from the infrastructure to vehicles [23]. However, owing to the unidirectional nature of FM radio broadcast technology, vehicles in the RDS setting remained in the role of passive receivers of information from a centralized source.

The first vehicular bidirectional communication system took place in the 1980s when radio frequency identification (RFID) technology was adopted in tolling systems. The infrastructure can request information from RFID tags carried by vehicles passing through stationary beacons and RFID tags would respond with their identification, hence expediting payment processing and automatic access control [20]. Later in the 1990s, Philips invented the 5.8 GHz dedicated DSRC system [21], in which short communications sessions between road-side infrastructure and passing vehicles became possible within a short-range (2.5–10 m) of the stationary beacons. However, in such a communication architecture, the communication bandwidth and range were limited, and inter-vehicle communication was not possible and the concept of the vehicular network was still yet to come.

In the subsequent years, early efforts were devoted to the realization of inter-vehicle communications. The International Standards Organization (ISO) established technical committees focusing on ITS in 1992, which developed the architecture and infrastructure for general ITS communications systems [20]. The technical committees proposed the evolutionary idea of separating the application and communication concerns, and by adding network management and on-board communication functions, the architecture would be compatible with any standardized wireless media. In such a paradigm of design, only the "Service Access Point (SAP)" for each wireless medium that determines the controlling of the opening, management and closing of each communication session needs to be defined [21]. This architecture was later evolved into the concept of ITS-station (ITS-S, which could occur in the form of

Fig. 1 Simplified ITS-S reference architecture

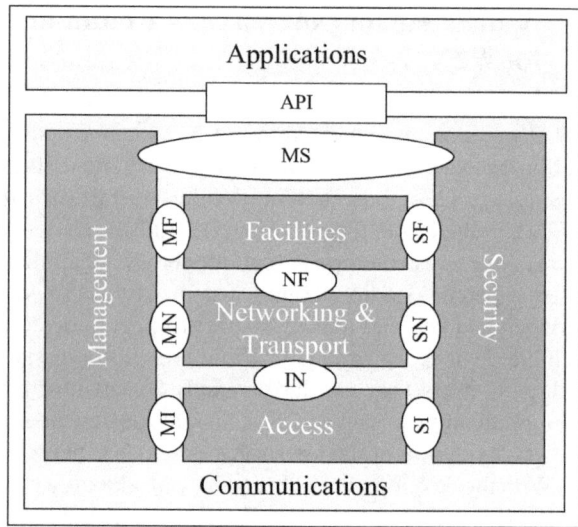

vehicle ITS-S, roadside infrastructure ITS-S, data center ITS-S, etc.) reference architecture by the ISO Standards, see Fig. 1. From this point of view, all ITS communications can be regarded as peer-to-peer relationships, which could involve ITS-stations in peer-to-peer, broadcast, or unicast communications.

The first true VANET architecture was developed around the early 2000s, in which each node can be treated as an ITS-S that can communicate with any other compatible node to construct an ad-hoc network. The underlining communication technology is DSRC with standardization activities undertaken by organizations including SAE in the USA, the European Committee for Standardization (CEN) and European Telecommunications Standards Institute (ETSI) in Europe, the Association of Radio Industries and Businesses (ARIB) in Japan, and ISO. In parallel with the development of standards in DSRC, numerous real-world implementations and field tests were piloted during the following decade with a primary focus on the application of driver-assist warning services [20]. For example, BMW exhibited a skid alert system between vehicles depending on the VANET technology at the ITS world congress in 2005 [21]. Mercedes and GM demonstrated practical applications of collision warning and passing car warning based on VANET at the same venue. These experiments validated the feasibility of DSRC, which operates in line-of-sight (LoS) or near LoS communication environment, and is based on the IEEE 802.11p standard and the IEEE 1609 WAVE (Wireless Access in Vehicular Environments) architecture [20] to provide pervasive inter-vehicle communication.

With the flourishing of research and development activities toward V2X by automotive companies, inconsistencies among the various Standard Development Organizations (SDO) emerged in different world regions. ETSI in Europe adopted the comprehensive cooperative awareness message (CAM) and decentralized environmental notification message (DENM), while IEEE (Institute of Electrical and Electronics Engineers) in the USA developed the counterpart called basic service message (BSM) [20]. Nevertheless, the ongoing research and development effort in DSRC-based communication led to the roll-outs of initial DSRC systems. In 2015, Toyota released its first V2I and V2V communication-enabled vehicles to the mass market.

Apart from DSRC technology, the mobile cellular network had also advanced significantly from early analogue communication to all-digital communication [21], which had undertaken an evolution from Global System for Mobile Communication (GSM) in 1991, to Universal Mobile Telecommunication Systems (UMTS) in 2001, to Evolved Universal Terrestrial Radio Access Network (E-UTRAN) in 2012, and to the latest New Radio (NR) network in 2018. While early generations of mobile cellular networks failed to provide stringent communication requirements as needed by V2X communication, later generations including E-UTRAN, or acknowledged in Europe and by the 3GPP as Long Term Evolution (LTE), LTE-Advanced and NR promise to deliver both uplink and downlink communications with high throughput, low latency and high reliability, and thus become viable wireless communication technology candidates for V2X applications.

The mobile cellular network based V2X communication takes advantage of the existing widely deployed infrastructure, and offers a larger cell coverage range, higher resilience to interference and better non-line-of-sight (NLOS) capabilities [24]. However, its centralized operation architecture poses a significant challenge in supporting low-latency V2V communications [25]. On the contrary, although DSRC technology operates in a distributed manner, the high mobility environment in V2X communication can jeopardize its communication performance, especially in high vehicle density scenarios. Some early field test results revealed that DSRC, LTE or LTE-Advanced technology alone falls short in accommodating V2X applications [20, 25]. Therefore, some researchers devoted efforts to exploiting the feasibility of interworking of DSRC and the mobile cellular network to guarantee communication reliability and efficiency [25].

Invented by Ericsson in 1994 [21], Bluetooth is a short-range wireless technology that employs Ultra-High Frequency (UHF) radio wave and operates in the Industrial, Scientific and Medical (ISM) band from 2.4 to 2.485 GHz. Relying on Bluetooth technology, data can be exchanged between fixed and mobile devices over short distances and forming personal area networks (PANs). Typical usage of Bluetooth within vehicles is to support hands-free use of the cell phone. Nowadays, automotive manufacturers are pressurized by consumers to provide ITS service provisioning such as infotainment through the Bluetooth link.

On the other hand, consumers also demand Wi-Fi technology, especially the latest evolutions compatible with the 802.11n and 802.11 ac standards, which poses significant regulatory pressure of shared use of spectrum such as the 5.9 GHz band which is allocated only for use by vehicular applications.

3   Use Case and Requirements of V2X Communications

V2X use cases are collections of traffic efficiency, safety and infotainment services
[26]. ETSI in Europe and the Department of Transportation in the USA have already
specified key functional and performance requirements for safety based on CAM or
DENM broadcast. V2X use cases have also been specified by 3GPP Services and
System Aspects (SA) Working Group 1 and have been further elaborated by the
5GAA Working Group 1.

3.1   3GPP

3GPP SA studied service requirements for V2X services in two stages, with the first
stage being LTE-V2X which supports basic active safety and traffic management use
cases, and the second stage being 5G NR-V2X which supports advanced use cases
and higher automation levels [27].

In the first stage study of V2X use cases by 3GPP SA Working Group 1, four
categories of V2X applications are defined based on LTE-V2X [28], which are V2V,
V2I, V2N and V2P, respectively, see Fig. 2. V2V enables nearby vehicles to exchange
V2V application related information that includes the position and dynamic status
information of vehicles, such that safety warnings can be achieved. V2I deals with
the information interaction between vehicles to RSU, the disseminating information could be traffic light status, road sign information, etc. V2N mainly focuses
on the communication between vehicles and application servers or communication
BSs, such that vehicles can have access to a broader V2X application and thereby
improving traffic efficiency. V2P concerns the exchange of information between
vehicles to surrounding pedestrians, such that safety warnings to both vehicles and
pedestrians can be provided. Based on these four categories of V2X applications,
3GPP SA Working Group 1 also defined 27 V2X use cases, see Table 2.

3GPP has also provided detailed LTE-V2X technical parameters for seven typical
scenarios, see Table 3.

5G NR-V2X can be regarded as a complement to LTE-V2X, in which advanced
use cases are defined, as described in detail in [29]. These advanced use cases are
divided into four groups including vehicles platooning, advanced driving, extended
sensors and remote driving, see Fig. 3. Vehicles platooning is the process of the
dynamic formation and management of groups of vehicles in platoons. Information
exchange within vehicles in the platoon occurs periodically to ensure proper platoon
operations and thereby increase traffic safety and efficiency. In the advanced driving
use case, vehicles disseminate information obtained from their local sensors alongside their driving intentions to surrounding vehicles, such that their trajectories or
maneuvers could be coordinated, and semi-automated or fully-automated driving can
be enabled. Extended sensors aim to improve the perception capability of vehicles
by enabling them to acquire information from adjacent vehicles, RSU, pedestrians,

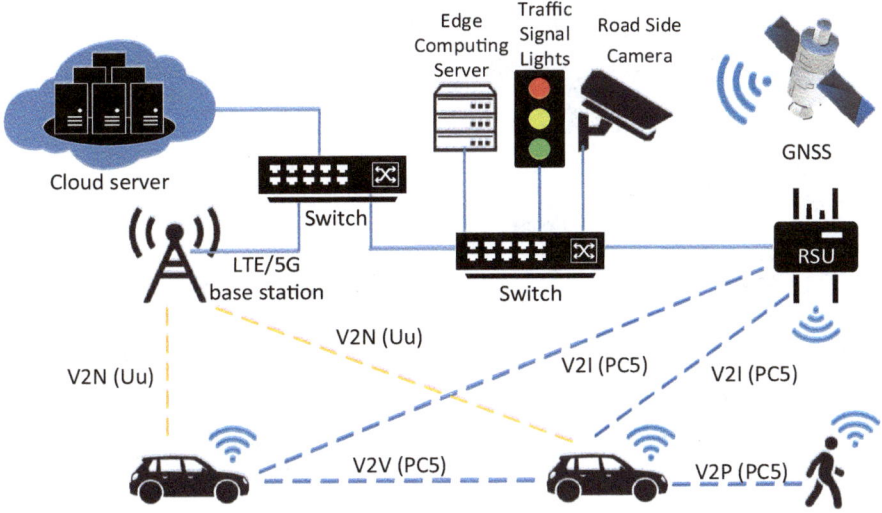

Fig. 2 V2X system architecture

etc., such that services like pre-collision warnings could be provided. Remote driving provides a remote driver or a V2X application the ability to (tele)operate vehicles, thereby allowing vehicles to operate withstanding hazardous environments or in cases where the driver or passengers are without driving abilities (e.g. injured). 3GPP has also provided detailed 5G NR-V2X technical parameters for the aforementioned four scenarios, as summarized in the range of values for each type of parameter in Table 4.

3.2 5GAA

5GAA combined the use cases that are defined by 3GPP with new innovative use cases, and has grouped these use cases into seven groups as described below [30, 31]. It is worth noting that a specific use case could fit into different groups.

(1) *Safety*: This group involves use cases that provide enhanced safety for vehicles and drivers. Typical use cases in this group are emergency braking, collision warning, see-through, etc., that could apply either to autonomous vehicles or to human drivers.
(2) *Vehicle operations management*: This group includes use cases that are provided by vehicle manufacturers to improve the operational and management services. Typical use cases in this group are sensors monitoring, software updates, remote support, etc.
(3) *Convenience*: This group includes use cases that could provide value and convenience to the driver or passengers in the vehicles. Typical use cases in this group

Table 2 27 user cases defined by 3GPP based on LTE-V2X [28]

Category	Use case	Category	Use case
V2V	Forward collision warning	V2I	V2X by UE-type RSU
	Control loss warning		Automated parking system
	V2V use case for emergency vehicle warning		Curve speed warning
	V2V emergency stop use case		V2X road safety service via infrastructure
	Cooperative adaptive cruise control		V2I emergency stop use case
	V2X message transfer under Mobile Network Operators (MNO) control		Road safety services
	Pre-crash sensing warning		Queue warning
	V2X in areas outside network coverage	V2N	V2N traffic flow optimization
	Wrong way driving warning		V2N use case to provide overview to road traffic participants and interested parties
	Privacy in the V2V communication environment		Enhancing positional precision for traffic participants
V2P	Warning to pedestrian against pedestrian collision		Remote diagnosis and just in time repair notification
	Pedestrian road safety via V2P awareness messages	V2X	Use case for V2X access when roaming
			Mixed use traffic management
	Vulnerable road user safety		V2X minimum Quality of Service (QoS)

are infotainment, assisted and cooperative navigation, and autonomous smart parking.

(4) *Autonomous driving*: This group includes use cases that are relevant for autonomous vehicles (equivalent to SAE level 4 and 5). Typical use cases in this group are tele-operation, updating and downloading of dynamic maps. This group may also apply to advanced driving, remote driving, and extended sensors that are defined by 3GPP.

(5) *Platooning*: Use cases in this group are the same as the 3GPP vehicles platooning use cases. Examples are collecting and establishing a platoon, determining position in the platoon, dissolving a platoon, managing distance within the platoon, leaving a platoon, controlling of the platoon in a steady state, requesting passing through a platoon, etc.

(6) *Traffic efficiency and environmental friendliness*: This group includes use cases that provide enhanced value to infrastructure or city providers in regions where the vehicles operate. Typical use cases in this group are Green Light Optimal Speed Advisory (GLOSA), traffic jam information, and routing advice.

Table 3 LTE-V2X services technical parameters [28]

Scenario	Effective distance (m)	Absolute speed of a UE supporting V2X services (km/h)	Relative speed between 2 UEs supporting V2X services (km/h)	Maximum tolerable latency (ms)	Minimum radio layer message reception reliability at effective distance (%)	Example Cumulative transmission reliability (%)
Suburban/major road	200	50	100	100	90	99
Freeway/motorway	320	160	280	100	80	96
Autobahn	320	280	280	100	80	96
NLOS/urban	150	50	100	100	90	99
Urban intersection	50	50	100	100	95	–
Campus/shopping area	50	30	30	100	90	99
Imminent crash	20	80	160	20	95	–

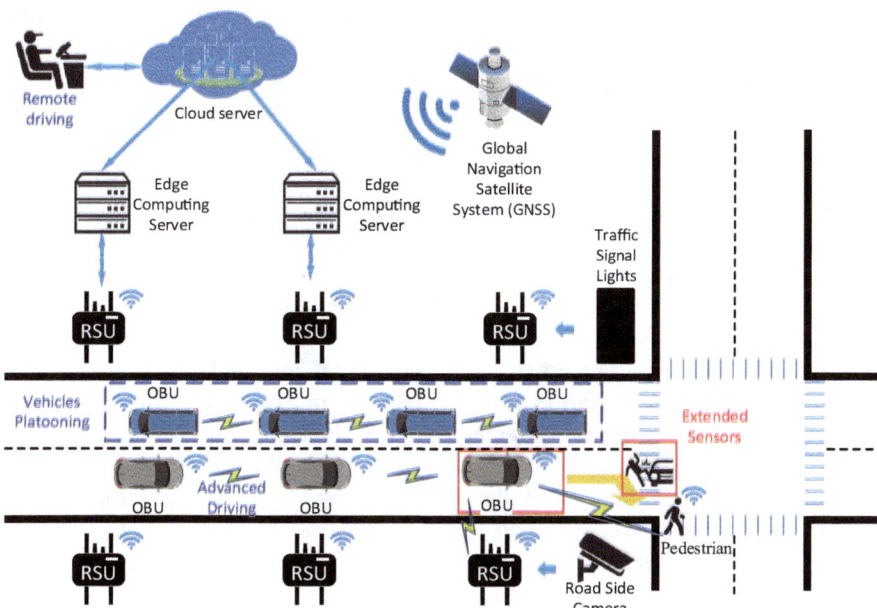

Fig. 3 Illustration on advanced V2X use cases

Table 4 A 5G NR-V2X services technical parameters [29]

Use case	Payload (bytes)	Transmission rate (message/sec)	Maximum end-to-end latency (ms)	Reliability (%)	Date rate (Mbps)	Required communication range (meters)
Vehicles platooning	50–6000	2–50	10–25	90–99.99	≤65	80–350
Advanced driving	SL: 300–12,000 UL: 450	SL: 10–100 UL: 50	10–100	90–99.999	SL: 10–50 UL:0.25–10 DL: 50	360–700
Extended sensors	1600	10	3–100	90–99.999	10–1000	50–1000
Remote driving	–	–	5	99.999	UL: 25 DL: 1	–

DL: Downlink; SL: Sidelink; UL: Uplink

(7) *Society and community*: This group includes use cases that provide enhanced value to the society and public. Typical use cases in this group are vulnerable road user (VRU) protection, emergency answering points, Emergency vehicle approaching, traffic light priority, patient monitoring, crash report, etc.

For each use case, 5GAA has provided a use case description, then has further defined multiple use case scenarios within this specific use case. These use scenarios within the same use case can differ in terms of road configuration/environment, actors involved, service flows, etc. Road environments refer to the typical locations where V2X use cases took place, such as intersections, urban and rural streets, high-speed roads, and parking lots, etc. Use cases are procedures at a high level for the execution of an application in a certain situation with a specific purpose. Use case scenarios are derived for different situations that may imply different specific requirements based on the high-level use case description. It should be noted that every road environment can be associated with one or more use cases, and every use case is mapped to at least one road environment and at least one use case scenario [30, 31].

While 3GPP focuses mainly on the requirements from the network aspect, 5GAA complements these requirements from the application level and defines the requirements of each use case scenario within a use case with the concept of service level requirements (SLRs). Typical SLRs include range, information requested/generated, service level latency, service level reliability, velocity, vehicle density, positioning, and interoperability/regulatory/standardization required [30, 31], which consist of both automotive-centric requirements (e.g., service level reliability, positioning) and system level requirements (e.g., vehicle density). A detailed description of the SLRs for all the use cases defined by 5GAA can be found in [31].

4 Wireless Communication Technology for V2X

The driving process of networked autonomous driving vehicles depends partially or wholly on the autopilot system, leading to the demand for high capacity, low latency, high reliability, and secure wireless communication technologies. Therefore, networked autonomous driving integrates V2X technology to cope with the stringent network requirements in various V2X use cases. Wireless communication technologies for V2X assisted autonomous driving are of paramount importance as they can greatly improve the perception, planning and control of autonomous driving. This section outlines characteristics of wireless communication technologies for V2X assisted autonomous driving, with an emphasis on the physical layer aspects.

4.1 Bluetooth

IEEE 802.15.1, or commonly known as Bluetooth, provides a low-cost, low-power consumption, and open short-range wireless communication solution for V2X. It has already been widely adopted in intra-vehicle infotainment systems, where external devices like cell phones can be connected directly to the in-car infotainment system, benefiting various services such as phone calls, messaging and global positioning system (GPS) navigation to be delivered directly through the in-car interface [20].

The modulation schemes in Bluetooth [19] include GFSK (Gaussian Frequency-Shift Keying), π/4-DQPSK (Differential quadrature Phase Shift Keying) and 8DPSK (Differential 8-Phase Shift Keying). The achievable data rate of Bluetooth varies from 1 to 24 Mb/s. For example, Bluetooth 4.0 provides a low-energy mode that adopts GFSK with a data rate of 1 Mb/s at the power consumption of 0.5 mW. To accommodate the highly dynamic nature of inter-vehicle communication networks, some literature has studied the feasibility of Bluetooth to connect high-speed moving vehicles in ad-hoc networks [32], and has studied the application of data protection methods including 1/3FEC (Forward Error Correcting), 2/3 FEC and ARQ (Automatic Repeat-reQuest) on the effectiveness in improving communication reliability of Bluetooth for mobile vehicles [33]. It is also proposed to utilize Bluetooth as a communication technology to control the vehicle platoon space [34].

Unfortunately, while the theoretical maximum range of Bluetooth is 100 m, its sphere in which Bluetooth devices forming networks is confined by the Personal Operating Space (POS) to 10 m [35]. This restriction coupled with the fact that Bluetooth technology requires time to discover the device and establish a connection, i.e., it lacks mobility support features to maintain a short-range ad-hoc network in a highly dynamic V2X environment, makes Bluetooth an unpromising wireless communication technology for generalized V2X communications.

4.2 ZigBee

Similar to Bluetooth technology, ZigBee also provides a low cost, low power consumption, short-range, high scalability, and low latency wireless communication solution for V2X based on IEEE 802.15.4 standard. Its operational range is between 10 to 100 m [36]. ZigBee operates primarily in the 2.4 GHz frequency band, except for operating in the 868 MHz frequency band in Europe and 915 MHz in America and Australia. The 2.4 GHz operating frequency band adopts O-QPSK (Offset Quadrature Phase Shift Keying) and can achieve a data rate of 250 kb/s [19, 37]. The 868 MHz and 915 MHz operating frequency bands both adopt BPSK (Binary Phase-Shift Keying) and can achieve data rates of 20 Kb/s and 40 Kb/s, respectively [19, 37].

Research in [38] concerned V2I in dense urban areas, where a large amount of infrastructure equipment could consume significant power consumption, and proposed to use ZigBee protocol with low power consumption as a viable solution for vehicle identification through V2I communication link. Lei et al. [39] studied the feasibility of utilizing ZigBee technology for forward collision warning system in low-speed scenarios. Zhang et al. [40] developed a localization and distance measurement mechanism for vehicles traveling on the freeway based on ZigBee technology. Dong et al. [41] proposed to deliver real-time traffic signal light information by ZigBee technology to vehicles, such that dynamic speed guidance could be advised and consequently realize energy/emission savings for vehicles.

In summary, although the transmission data rate of ZigBee is rather limited, it has the advantage of low cost, low power consumption and low latency as compared to Bluetooth technology. Meanwhile, it allows the simulation of complicated protocols and scenarios [36]. Therefore, ZigBee technology may find its place in V2X use cases with low data rate requirements or in the simulation research of V2X assisted autonomous driving.

4.3 Wi-Fi

Wi-Fi technology supports high-capacity wireless communication at the frequency bands of 2.4 and 5.8 GHz. The widely commercialized Wi-Fi technologies are based on the IEEE 802.11n standard and IEEE 802.11ac standard, both adopting MIMO (Multiple-Input Multiple-Output) and OFDM (Orthogonal Frequency Division Multiplexing) modulation schemes [37]. IEEE 802.11n can support BPSK, QPSK (Differential quadrature Phase Shift Keying), 16-QAM (Quadrature Amplitude Modulation) and 64-QAM, with a transmission rate of up to 600 Mb/s. IEEE 802.11ac can support all the modulation types as supported by IEEE 802.11n with the addition of 256-QAM. IEEE 802.11ac can reach a transmission rate of up to 1 Gb/s. Their later generation, called IEEE 802.11ax, adopts OFDM, OFDMA (Orthogonal Frequency Division Multiple Access) and MU-MIMO (Multi-User Multiple-Input

Multiple-Output) technology. Also, apart from all the modulation types supported by IEEE 802.11ac, IEEE 802.11ax also supports 1024-QAM, with a transmission rate of up to 9.6 Gb/s.

Both delay and throughput analysis of Wi-Fi networks have shown satisfactory results in accommodating V2X applications [42, 43]. Cheng et al. [44] and Han et al. [45] have studied the opportunistic Wi-Fi offloading problem in a vehicular environment, where vehicles opportunistically transmit to Wi-Fi access points along the road while driving through the coverage areas, and the trade-off between the task completion delay and the offloading efficiency, and between the task completion delay and the cost are pursued respectively. The application of Wi-Fi networks in V2V communication is studied in [46] for traffic information dissemination. Moreover, Wi-Fi network is proposed in [47] to serve as a medium for V2V content sharing for mining vehicles in the tunnel.

Wi-Fi technology can be utilized for information dissemination and Internet access in V2X assisted autonomous driving. Owing to the low-cost advantage of deploying Wi-Fi networks, it would even be feasible to deploy an array of Wi-Fi devices along the roadside and to improve the network performance [48].

4.4 DSRC

DSRC is designed specifically for automotive applications, and can be considered as an upgraded technology for Wi-Fi [19]. DSRC technology is developed based on three sets of standards. The first set of the standards is IEEE 802.11p, which regulates physical layer (PHY) and medium access control (MAC) layer specifications [19]. The second set of the standards is SAE J2735 and SAE J2945, which define the content and structure of the information [49]. Finally, the third set of the standards is IEEE 1609 WAVE, which as a supplement to the IEEE 802.11p standards, defines the interfaces and features of the V2X communication stack above the PHY and MAC layer [20], including the overall architecture, security management, routing, multi-channel operation and communications, etc.

DSRC adopts OFDM technology and supports modulation types including BPSK, QPSK, 16-QAM and 64-QAM. It supports high data rate transmission with a maximum data rate of 54 Mb/s as well as low latency communication. The operating frequency band of DSRC varies in different world regions. For example, 70 MHz (5855–5925 MHz) frequency band is allocated to general V2X communications in Europe, with 30 MHz (5875–5905 MHz) specially devoted to traffic safety applications [20]. The USA counterpart reserves 75 MHz (5850–5925 MHz) frequency band for V2X applications, with 10 MHz (5885–5895 MHz) frequency band dedicated to vehicle safety traffic, and another 10 MHz (5915–5925 MHz) frequency band reserved for public safety communication [20].

Vehicles with DSRC technology depend on the IEEE 802.11p standard to form a VANET. Vehicles periodically broadcast messages including their locations and trajectories or maneuvers. In order to avoid transmission collision [20], the IEEE

802.11p standard adopts the Tiered Contention Multiple Access (TCMA) mechanism, which is an extension of Carrier Sense Multiple Access with Collision Avoidance (CSMA/CA). Its working principle follows that each vehicle would listen for activity on a channel between transmissions. Once activity on a particular channel is detected, the vehicle should wait for a random back-off period which is confined by minimum and maximum durations before its next transmission [20].

To enable vehicles with only one radio to access both safety and service-related communications, the IEEE 1609 WAVE standard is designed with a multi-channel operation feature, by which single radio on vehicles could divide transmissions between the control channel (for safety-related messages) and the service channel [20].

Numerous research paper has been published on the study of DSRC technology using analytical models, extensive simulations, or field trials during the past decade [24]. DSRC technology is robust to Doppler Spreads and can support short-distance low-latency communication, which makes it a potential communication technology for V2V communications. However, it suffers from a poor scalability problem which is caused by high packet collision probability due to its random channel access mechanism, particularly under medium or high traffic density conditions. The poor scalability problem is further deteriorated by the high-speed movement of vehicles. In addition, RSUs are required to establish V2I communication, which incurs additional costs for infrastructure installation.

4.5 C-V2X

With the ongoing development of autonomous driving technology, the demand for long transmission distances and high transmission capacity also upsurges. DSRC technology may fail to deliver the stringent transmission requirements due to its short-range transmission and random channel access characteristics. Compared to DSRC technology, cellular based V2X technology has wider cell coverage and possesses a much larger bandwidth to consume. Also, cellular based V2X technology does not need the installation of dedicated road side devices. Currently, there are two major C-V2X technologies, namely LTE-V2X and NR-V2X, which will be introduced in the following.

4.5.1 LTE-V2X

LTE-V2X standard is defined by 3GPP in Release 14 for V2X communications through the LTE air interface, and is later refined in 3GPP Release 15 [27]. It operates in the 5.9 GHz frequency band.

For its physical layer aspects, LTE-V2X uses SC-FDMA (Single-Carrier Frequency-Division Multiple Access) and OFDMA modulation schemes and supports 10 and 20 MHz channels. In the frequency domain, each channel can be divided into 180 kHz resource blocks (RBs), which are the smallest unit of frequency resources that can be allocated to a vehicle [50]. Each RB also corresponds to 12 subcarriers of 15 kHz.

In the time domain, the channel is divided into subframes with 1 ms long, where each subframe contains 14 OFDM symbols with normal cyclic prefix. These OFDM symbols are organized such that 9 OFDM symbols are used for data transmission, 4 OFDM symbols are used to transmit demodulation reference signals (DMRSs) for channel estimation and mitigating the Doppler effect caused by vehicle movement, and the last symbol is reserved for timing adjustments and for vehicles to switch between transmission and reception across subframes [27].

LTE-V2X also defines subchannels as a collection of RBs in the same subframe, which are used for data and control information transmission. The number of RBs within each subchannel depends on the specific configuration and can vary from each other.

LTE-V2X has two radio interfaces to enable its functioning in both in-coverage and out-of-coverage conditions. The cellular interface, or Uu interface supports V2I communications, and the LTE sidelink, or PC5 interface supports V2V communications, see Fig. 2 for an illustration of Uu and PC5 interfaces.

(1) **LTE-Uu radio interface**

LTE-Uu is the traditional radio interface connecting the eNodeB (Evolved Node B) and user equipment. Vehicles communicating with other entities through the LTE-Uu radio interface need to transmit their message to the eNodeB in the uplink, then the same or a different eNodeB can transmit the message to the destination vehicle using unicast download or enhanced Multimedia Broadcast Multicast Service (eMBMS) [24]. Therefore, LTE-Uu has a major advantage in terms of the communication range as it can utilize the widespread cellular core network.

Compared to DSRC technology, where radio access is in a random fashion, data transmission in LTE-Uu is scheduled, and packet collisions and radio interference can be controlled. This enables the QoS guarantee in terms of data rate and transmission delay for various V2X use cases.

(2) **LTE-PC5 Radio Interface**

LTE-PC5 radio interface enables V2V communication based on direct LTE sidelink (or device-to-device communication), allowing vehicles in close proximity to communicate without messages passing through the eNodeB [48]. It should be noted that the PC5 radio interface can function with or without the presence of the eNodeB. Therefore, the LTE-PC5 radio interface can offload traffic from the cellular infrastructure, and better network throughput, spectrum efficiency and delay performance can be achieved.

Data in LTE-PC5 is transmitted in transport blocks (TBs) over physical sidelink shared channels (PSSCH). A TB contains a full packet to be transmitted and can occupy one or more subchannels depending on the packet size, the configuration of the number of RBs within each subchannel, and the adopted modulation and coding scheme (MCS). Typical MCSs used in LTE-V2X are QPSK, 16-QAM, 64-QAM and turbo coding [27]. Moreover, a vehicle that transmits a TB must also transmit its associated sidelink control information (SCI) messages, which are carried over physical sidelink control channels (PSCCH), and contains information detailing the MCS, allocation of RBs, and the reservation interval of resource for semipersistent scheduling (SPS) [50].

LTE sidelink was initially introduced in 3GPP Release 12 for public safety purposes [50], and has two resource allocation modes, namely, mode 1 and mode 2. Both resource allocation modes were intended to prolong the battery operational lifetime for mobile devices at the expense of increased latency. While the LTE-PC5 radio interface is based on LTE sidelink, its resource allocation modes are not suitable for V2X use cases as high reliable and low latency communication performance is pursued.

3GPP Release 14 introduces two new resource allocation modes intended specifically for the LTE-PC5 radio interface, namely mode 3 and mode 4 [50]. In mode 3, the PC5 radio resource is scheduled by the cellular network, or eNodeB. Therefore, mode 3 demands the communicating vehicles to be within the eNodeB coverage, resulting in a centralized resource allocation scheme. In mode 4, the vehicle autonomously schedules the resource for PC5 radio in a distributed manner, and can operate without the coverage of eNodeB.

In summary, thanks to the utilization of the existing cellular infrastructures, LTE-V2X can provide much wider coverage and can support QoS guarantee for various V2X use cases. Meanwhile, LTE-V2X can significantly improve network throughput, spectrum and energy efficiency, and communication delay performance by offloading traffic through the LTE-PC5 radio interface. In addition, LTE V2X can support the high-speed movement of vehicles [19]. Nevertheless, some research results show that LTE-V2X may fail to support the highest degree of automation, which remains to be enhanced by 5G NR-V2X technology.

4.5.2 NR-V2X

NR-V2X standard is developed by 3GPP in Release 15 with a focus on Uu features, and is later refined in 3GPP Release 16 to include sidelink communication aspects [27]. The physical layer structure for NR sidelink is based on the design of NR Uu. Both NR Uu and NR sidelink operates in the same frequency bands, namely Frequency Range 1 (FR1) and Frequency Range 2 (FR2). 3GPP TS 38.104 standard has defined FR1 to range from 410 to 7125 MHz, and has defined FR2 as between 24,250 and 52,600 MHz [51]. While NR sidelink supports both frequency ranges, the design of NR sidelink is primarily based on FR1 [27].

Signals transmitted on different carrier frequencies may experience different degrees of multipath fading and cause different degrees of frequency selective fading. In order to tackle this problem, 5G NR supports different values of subcarrier spacing (SCS), and deploys OFDM subcarriers with smaller subcarrier spacing to the spectrum with lower carrier frequencies, and deploys OFDM subcarriers with larger subcarrier spacing to on the spectrum of higher carrier frequencies [52, 53]. Moreover, multipath fading can also cause different degrees of Inter-Symbol Interference (ISI). 5G NR uses different cyclic prefix (CP) lengths for different values of SCS to reduce the impact of ISI [52]. According to the ratio of CP length to symbol duration, 5G NR supports two types of CP, namely normal cyclic prefix (NCP) and extended cyclic prefix (ECP).

To support diverse requirements and operating frequencies in FR1 and FR2, a scalable OFDM numerology is adopted in NR-V2X. Each OFDM numerology is associated with an SCS and a CP. 3GPP standards TS 38.104 [51] and TS 38.211 [54] defined the numerology for FR1 and FR2 as follows. For FR1, the 15, 30, and 60 kHz SCSs with NCP are allowed, and ECP can only be used together with 60 kHz SCS. For FR2, the 60 and 120 kHz SCSs with NCP are allowed, and ECP can only be used together with 60 kHz SCS.

In terms of the transmission waveform of 5G NR, the downlink transmission adopts the CP-OFDM technology, and the uplink transmission adopts both CP-OFDM and Discrete Fourier Transform (DFT)-spread-OFDM. For sidelink transmission waveforms, only CP-OFDM is supported [52].

In the time domain, resources in NR-V2X are composed by frame, subframe, slot and mini-slot [55], see Fig. 4. The time duration of each frame is 10 ms, and each frame can be divided into 10 subframes. Therefore, the time duration of each subframe is fixed at 1 ms. Each subframe contains one or more slots, the specific number depends on the adopted SCS value [54]. For the 15 kHz SCS, each subframe contains one slot, that is the time duration of each slot is 1 ms. For the 30 kHz, 60 kHz, and 120 kHz SCSs, each subframe contains 2, 4, and 8 slots, respectively, which corresponds to the time duration of each slot to be 0.5 ms, 0.25 ms, and 0.125 ms, respectively. With NCP, each slot contains 12, and 14 OFDM symbols in the case of ECP and NCP, respectively. To further reduce the uplink and downlink transmission latency, NR-V2X allows the transmission of a part of a slot at a time, which is known as the mini-slot transmission mechanism. According to the actual configuration of the system, a mini-slot contains 2, 4 or 7 OFDM symbols.

NR-V2X sidelink transmission does not support the mini-slot transmission mechanism, so the minimum unit of NR sidelink transmission in time domain resource scheduling is a slot. However, the partial slot transmission for sidelink is supported in case that only partial symbols in a slot are available for sidelink transmission and the remaining symbols are reserved or occupied for Uu transmission for the shared band operation [52].

In the frequency domain, resources in NR-V2X are composed by resource element, resource block, resource grid and bandwidth part [55], see Fig. 5. A resource element consists of a subcarrier on an OFDM symbol in the time domain, and is the smallest unit of network resources. A resource block contains 12 consecutive subcarriers with the same SCS. Therefore, the bandwidth of one resource block is determined by the SCS value of the subcarriers. A resource grid consists of multiple resource blocks with the same SCS. A bandwidth part consists of multiple consecutive resource blocks, and the carrier bandwidth in NR can be up to 400 MHz (including 275 resource blocks). However, user terminals in most mobile services may not

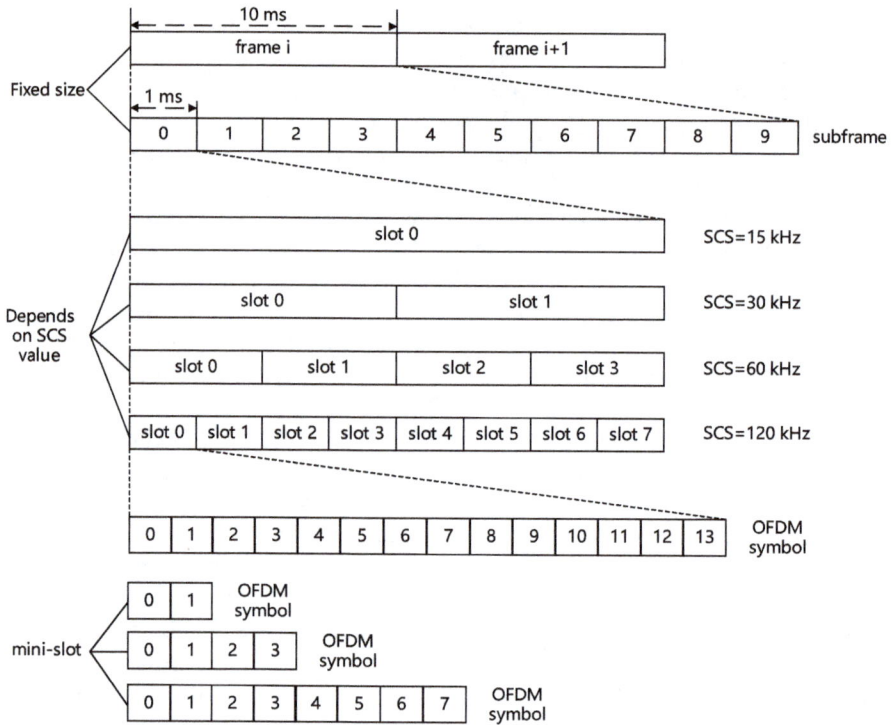

Fig. 4 NR time domain resource structure [52]

fully utilize the 275 resource blocks, and only part of the resource blocks are used for energy-saving purposes. For a user terminal, at most four bandwidth parts are configured on one carrier, and at most one bandwidth part can be activated at any time [52].

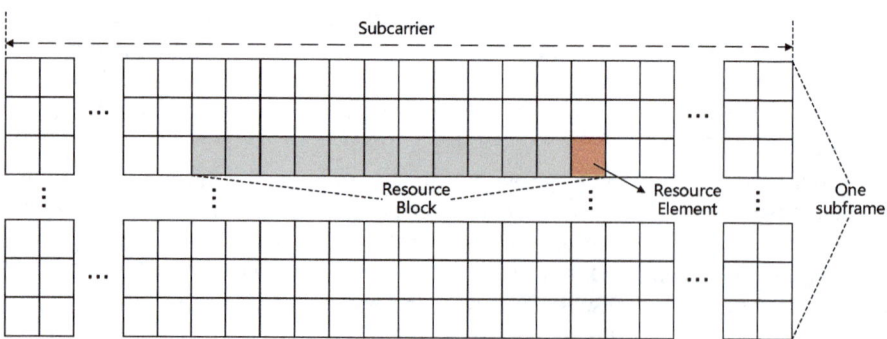

Fig. 5 NR frequency domain resource structure [52]

Table 5 Comparison between LTE-V2X and NR-V2X

Features	LTE-V2X	NR-V2X
Subcarrier spacing	15 kHz	15, 30, 60, 120 kHz
Latency	Less than 10 ms	Less than 1 ms
Reliability	95–99%	99.9–99.999%
Channel coding	Turbo	LDPC, Polar
Modulation	Up to 64-QAM	Up to 256-QAM
Cast type	Broadcast	Broadcast, multicast, unicast
Positioning accuracy	More than 1 m	m

NR sidelink transmission allows at most one sidelink bandwidth part to be configured on a carrier, and the smallest unit in frequency domain resource scheduling is a subchannel [52]. According to the actual configuration of the system, a subchannel contains 10, 15, 20, 25, 50, 75 or 100 consecutive resource blocks.

NR-V2X supports two sidelink resource allocation schemes, namely, mode 1 and mode 2 [56]. In mode 1, the BS configures and schedules sidelink resources to the user equipment through the Uu radio interface for transmission. In mode 2, the user equipment autonomously determines the sidelink resources (pre)configured by the BS or the network. Mode 2 can be further divided into 4 sub-modes. According to 3GPP TR 38.885 standard [56], in mode 2(a), user equipment autonomously selects sidelink resources for transmission; in mode 2(b), user equipment assists in sidelink resource selection for other user equipment; in mode 2(c), user equipment is configured with NR configured grant for sidelink transmission; in mode 2(d) user equipment schedules sidelink transmissions of other user equipment.

To better support the advanced V2X use cases, NR-V2X further supports three cast types [57]. Unicast allows direct communication between a pair of user equipment. Broadcast enables a single transmitter user equipment to send packets that can be received by all other user equipment within the radio communication range. Multicast is intended for a transmitter user equipment to send packets that can be received by a group of receiver user equipment fulfilling certain conditions. A comparison between LTE-V2X and NR-V2X [58] is provided in Table 5.

5 Standardization Activities for V2X Communications

The two major contender technologies for V2X communications are DSRC and C-V2X. While the Federal Communications Commission (FCC) allocated 75 MHz in the 5.9 GHz frequency band in 1999 for ITS services and DSRC, the deployment of DSRC is rather unsatisfactory. On November 20, 2020, FCC issued a Report and Order to reallocate the 5.9 GHz frequency spectrum [59], with the lower 45 MHz of the 5.9 GHz (5850–5895 MHz portion) to serve as an expansion of unlicensed mid-band spectrum operations, and the upper 30 MHz of the 5.9 GHz (5895–5925 MHz portion) reserved for ITS operation. The Report and Order further issued that ITS operations should transit from DSRC to C-V2X based technology following a transition period. Therefore, this section will only introduce standardization activities for C-V2X communication.

3GPP Release 12 [60] was the first standard to introduce device-to-device (D2D) communication for proximity services (ProSe) using cellular technologies. Sidelink communication in 3GPP Release 12 only supports broadcast mode, and although it can operate with or without the coverage of eNBs, it is still highly desired to access the network whenever eNBs are in presence [52]. To enhance sidelink communication capability, 3GPP Release 13 [61] introduced User Equipment-to-Network Relay, by which the remote user equipment outside the coverage of the eNBs is connected to the cellular network through the relay of the user equipment within the coverage of the eNBs, thus to realize the communication between the remote and the network.

Based on 3GPP Release 12 and 13, 3GPP developed LTE-V2X in its Release 14, in which sidelink communication is extended from D2D ProSe solely for public safety purposes to V2X message dissemination [62]. The detailed V2X messages transmitted between user equipment include CAM and DENM. 3GPP Release 14 also enhanced the PC5 interface and Uu interface functions, and the sidelink transmission communication technology has been redefined, including channel structure, resource scheduling and allocation methods, and the related radio features [63].

3GPP introduced enhanced V2X in Release 15 [64] with the purpose to support advanced V2X use cases including platooning, advanced driving, extended sensors, remote driving and vehicle QoS support. To further enhance the throughput and reduce the latency of V2X, 3GPP Release 15 also introduced carrier aggregation (CA), 64-QAM, transmission diversity and short transmission time interval mechanisms.

The 5G NR standard was developed in 3GPP Release 15 without sidelink features. 3GPP Release 16 became the first standard to support V2X communication based on the 5G NR radio interface including sidelink communications. In 3GPP Release 16, a new radio interface was designed, and apart from the support of broadcast, it introduced Physical Sidelink Feedback Channel (PSFCH) to support unicast and groupcast [56].

Aiming at better support all V2X requirements and use cases, 3GPP Release 17 focuses on the development of enhanced features including power saving for the battery-powered user equipment, reliability and latency improvement, enhancement on autonomous resource allocation, beamforming, uplink/downlink multicast, sidelink relaying and sidelink positioning and ranging [27, 57].

6 Conclusion

In this chapter, the history evolution of wireless communication technologies for V2X assisted autonomous driving was introduced. Automation levels for autonomous driving and typical use cases and the corresponding network requirements as well as

V2X system architecture were presented. This chapter also discussed the physical layer aspects of different potential wireless communication technologies for V2X, and highlighted their pros and cons in providing communication service for V2X assisted autonomous driving. Finally, this chapter also introduced standardization activities toward V2X communications to better coordinate various V2X use cases with stringent network requirements.

References

1. P. Bucsky, Modal share changes due to COVID-19: the case of Budapest. Transp. Res. Interdisc. Perspect. **8** (2020)
2. C. Eisenmann, C. Nobis, V. Kolarova, B. Lenz, C. Winkler, Transport mode use during the COVID-19 lockdown period in Germany: the car became more important, public transport lost ground. Transp. Policy **103**, 60–67 (2021)
3. M.J. Beck, D.A. Hensher, E. Wei, Slowly coming out of COVID-19 restrictions in Australia: implications for working from home and commuting trips by car and public transport. J. Transp. Geogr. **88** (2020)
4. J.A. Vallejo-Borda, R. Giesen, P. Basnak, J.P. Reyes, B.M. Lira, M.J. Beck, D.A. Hensher, J.D.D. Ortúzar, Characterising public transport shifting to active and private modes in South American capitals during the COVID-19 pandemic. Transp. Res. Part A: Policy Pract. **164**, 186–205 (2022)
5. J.D. Vos, The effect of COVID-19 and subsequent social distancing on travel behavior. Transp. Res. Interdisc. Perspect **5** (2020)
6. M.J. Beck, D.A. Hensher, Insights into the impact of COVID-19 on household travel and activities in Australia—the early days under restrictions. Transp. Policy **96**, 76–93 (2020)
7. D. Tarasi, T. Daras, S. Tournaki, T. Tsoutsos, Transportation in the Mediterranean during the COVID-19 pandemic era. Glob. Transitions **3**, 55–71 (2021)
8. L. Butler, T. Yigitcanlar, A. Paz, Smart urban mobility innovations: a comprehensive review and evaluation. IEEE Access **8**, 196034–196049 (2020)
9. F. Golbabaei, T. Yigitcanlar, J. Bunker, The role of shared autonomous vehicle systems in delivering smart urban mobility: a systematic review of the literature. Int. J. Sustain. Transp. **15**(10), 731–748 (2021)
10. L. Zhu, F.R. Yu, Y. Wang, B. Ning, T. Tang, Big data analytics in intelligent transportation systems: a survey. IEEE Trans. Intell. Transp. Syst. **20**(1), 383–398 (2019)
11. M. Yu, Construction of regional intelligent transportation system in smart city road network via 5G network. IEEE Trans. Intell. Transp. Syst. Early Access
12. M.B. Mollah et al., Blockchain for the internet of vehicles towards intelligent transportation systems: a survey. IEEE Internet Things J. **8**(6), 4157–4185 (2021)
13. A. Haydari, Y. Yılmaz, Deep reinforcement learning for intelligent transportation systems: a survey. IEEE Trans. Intell. Transp. Syst. **23**(1), 11–32 (2022)
14. Z. Lv, R. Lou, A.K. Singh, AI empowered communication systems for intelligent transportation systems. IEEE Trans. Intell. Transp. Syst. **22**(7), 4579–4587 (2021)
15. P. Arthurs, L. Gillam, P. Krause, N. Wang, K. Halder, A. Mouzakitis, A taxonomy and survey of edge cloud computing for intelligent transportation systems and connected vehicles. IEEE Trans. Intell. Transp. Syst. **23**(7), 6206–6221 (2022)
16. C.Y.D. Yang, K. Ozbay, X. Ban, Developments in connected and automated vehicles. J. Intell. Transp. Syst. **21**(4), 251–254 (2017)
17. E. Yurtsever, J. Lambert, A. Carballo, K. Takeda, A survey of autonomous driving: common practices and emerging technologies. IEEE Access **8**, 58443–58469 (2020)

18. *Taxonomy and Definitions for Terms Related to Driving Automation Systems for On-Road Motor Vehicles*, SAE Standard J3016, SAE International (2018)
19. J. Wang, J. Liu, N. Kato, Networking and communications in autonomous driving: a survey. In: IEEE Commun. Surv. Tutor. **21**(2), 1243–1274 (2019)
20. Z. MacHardy, A. Khan, K. Obana, S. Iwashina, V2X access technologies: regulation, research, and remaining challenges. IEEE Commun. Surv. Tutor. **20**(3), 1858–1877 (2018)
21. M. Annoni, B. Williams, The history of vehicular networks. Veh. Ad Hoc Netw. 3–21 (2015)
22. H. Flurscheim, Radio warning system for use on vehicles. US Patent 1612427 (1926)
23. D. Kopitz, B. Marks, *RDS: The Radio Data System* (Artech House, 1999)
24. G. Naik, B. Choudhury, J.-M. Park, IEEE 802.11bd & 5G NR V2X: evolution of radio access technologies for V2X communications. IEEE Access **7**, 70169–70184 (2019)
25. K. Abboud, H.A. Omar, W. Zhuang, Interworking of DSRC and cellular network technologies for V2X communications: a survey. IEEE Trans. Veh. Technol. **65**(12), 9457–9470 (2016)
26. M. Boban, A. Kousaridas, K. Manolakis, J. Eichinger, W. Xu, Connected roads of the future: use cases, requirements, and design considerations for vehicle-to-everything communications. IEEE Veh. Technol. Mag. **13**(3), 110–123 (2018)
27. M.H.C. Garcia et al., A tutorial on 5G NR V2X communications. IEEE Commun. Surv. Tutor. **23**(3), 1972–2026 (2021)
28. *Study on LTE support for Vehicle to Everything (V2X) services*, 3GPP TR 22.885 (2015)
29. *Service Requirements for Enhanced V2X Scenarios* (3GPP TS 22.186, 2022)
30. *C-V2X Use Cases: Methodology, Examples and Service Level Requirements* (5GAA White Paper, 2019)
31. *C-V2X Use Cases Volume II: Examples and Service Level Requirements* (5GAA White Paper, 2020)
32. P. Murphy, E. Welsh, J.P. Frantz, Using Bluetooth for short-term ad hoc connections between moving vehicles: a feasibility study, in *Proceedings of the IEEE 55th Vehicular Technology Conference* (vol. 1, 2002), pp. 414–418
33. T. Zheng, S. Wang, A.E. Kamel, Bluetooth communication reliability of mobile vehicles, in *Proceedings of the International Conference on Fluid Power and Mechatronics* (2011), pp. 873–877
34. S. Gillijns, M.L.R. de Arbulo Gubía, M. Engels, A fast simulation approach to assess the influence of bluetooth communication on distance control between vehicles, in *Proceedings of the IEEE 72nd Vehicular Technology Conference* (2010), pp. 1–5
35. I. C. S. L. M. S. Committee et al., Wireless LAN medium access control (MAC) and physical layer (PHY) specifications, *IEEE Standard 902.11-1997* (1997)
36. A. Maimaris, G. Papageorgiou, A review of Intelligent Transportation Systems from a communications technology perspective, in *Proceedings of the IEEE International Conference on Intelligent Transportation Systems (ITSC)* (2016), pp. 54–59
37. J.E. Aasri, M. Arioua, A. Zakriti, I. Ez-zazi, Modulator performance measurement in wireless sensor transmission chain, in *Proceedings of the International Conference on Wireless Networks and Mobile Communications (WINCOM)* (2017), pp. 1–5
38. R.A. Gheorghiu, M. Minea, Energy-efficient solution for vehicle prioritisation employing ZigBee V2I communications, in *Proceedings of the International Conference on Applied and Theoretical Electricity (ICATE)* (2016), pp. 1–6
39. Y. Lei, J. Wu, Study of applying ZigBee technology into forward collision warning system (FCWS) under low-speed circumstance, in *Proceedings of the 25th Wireless and Optical Communication Conference (WOCC)* (2016), pp. 1–4
40. K. Zhang, L. Zhang, F. Lu, Y. Zhao, Distance measurement algorithm for freeway vehicles based on Zigbee technology, in *Proceedings of the IEEE Advanced Information Technology, Electronic and Automation Control Conference (IAEAC)* (2017), pp. 2007–2010
41. C. Dong, X. Chen, H. Dong, K. Yang, J. Guo, Y. Bai, Research on intelligent vehicle infrastructure cooperative system based on Zigbee, in *Proceedings of the International Conference on Transportation Information and Safety (ICTIS)* (2019), pp. 1337–1343

42. W. Xu, H.A. Omar, W. Zhuang, X.S. Shen, Delay analysis of in-vehicle internet access via on-road WiFi access points. IEEE Access **5**, 2736–2746 (2017)
43. W. Xu, W. Shi, F. Lyu, H. Zhou, N. Cheng, X. Shen, Throughput analysis of vehicular internet access via roadside WiFi hotspot. IEEE Trans. Veh. Technol. **68**(4), 3980–3991 (2019)
44. N. Cheng, N. Lu, N. Zhang, X.S. Shen, J.W. Mark, Opportunistic WiFi offloading in vehicular environment: a queueing analysis, in *Proceedings of the IEEE Global Communications Conference* (2014), 211–216
45. D. Han, W. Chen, Y. Fang, Opportunistic WiFi offloading in a vehicular environment: an MDP approach, in *Proceedings of the ICC IEEE International Conference on Communications (ICC)* (2020), pp. 1–6
46. S. Goel, T. Imielinski, K. Ozbay, Ascertaining viability of WiFi based vehicle-to-vehicle network for traffic information dissemination, in *Proceedings of the IEEE Conference on Intelligent Transportation Systems* (2004), pp. 1086–1091
47. H. Viittala, S. Soderi, J. Saloranta, M. Hamalainen, J. Iinatti, An experimental evaluation of WiFi-based vehicle-to-vehicle (V2V) communication in a tunnel, in *Proceedings of the IEEE Vehicular Technology Conference (VTC Spring)* (2013), pp. 1–5
48. H. Zhou, W. Xu, J. Chen, W. Wang, Evolutionary V2X technologies toward the internet of vehicles: challenges and opportunities. Proc. IEEE **108**(2), 308–323 (2020)
49. J.B. Kenney, Dedicated short-range communications (DSRC) standards in the United States. Proc. IEEE **99**(7), 1162–1182 (2011)
50. R. Molina-Masegosa, J. Gozalvez, LTE-V for sidelink 5G V2X vehicular communications: a new 5G technology for short-range vehicle-to-everything communications. IEEE Veh. Technol. Mag. **12**(4), 30–39 (2017)
51. *NR; Base Station (BS) radio transmission and reception*, 3GPP TS 38.104, (2020)
52. S.-Y. Lien et al., 3GPP NR sidelink transmissions toward 5G V2X. IEEE Access **8**, 35368–35382 (2020)
53. S.-Y. Lien, S.-L. Shieh, Y. Huang, B. Su, Y.-L. Hsu, H.-Y. Wei, 5G new radio: waveform, frame structure, multiple access, and initial access. IEEE Commun. Mag. **55**(6), 64–71 (2017)
54. *NR; Physical channels and modulation*, 3GPP TS 38.211 (2020)
55. S. Ahmadi, *5G NR: Architecture, Technology, Implementation, and Operation of 3GPP New Radio Standards* (Academic Press, 2019)
56. *Study on NR Vehicle-to-Everything (V2X), (Release 16)*, V16.0.0: 3GPP TR 38.885 (2019)
57. M. Harounabadi, D.M. Soleymani, S. Bhadauria, M. Leyh, E. Roth-Mandutz, V2X in 3GPP standardization: NR sidelink in release-16 and beyond. IEEE Commun. Standards Mag. **5**(1), 12–21 (2021)
58. H. Bagheri et al., 5G NR-V2X: toward connected and cooperative autonomous driving. IEEE Commun. Standards Mag. **5**(1), 48–54 (2021)
59. Dedicated Short Range Communications (DSRC) Service, https://www.fcc.gov/wireless/bureau-divisions/mobility-division/dedicated-short-range-communications-dsrc-service
60. *Evolved Universal Terrestrial Radio Access (E-UTRA) and Evolved Universal Terrestrial Radio Access Network (E-UTRAN); Overall description; Stage 2 (Release 12)*, V12.10.0, 3GPP TS 36.300 (2016)
61. *Evolved Universal Terrestrial Radio Access (E-UTRA) and Evolved Universal Terrestrial Radio Access Network (E-UTRAN); Overall description; Stage 2 (Release 13)*, V13.14.0, 3GPP TS 36.300 (2020)
62. *Study on LTE-based V2X Services; (Release 14)*, V14.0.0: 3GPP TR 36.885 (2016)
63. *Evolved Universal Terrestrial Radio Access (E-UTRA); Physical channels and modulation, (Release 14)*, V14.15.0: 3GPP TS 36.211 (2020)
64. *Study on enhancement of 3GPP Support for 5G V2X Services (Release 15)*, V15.3.0: 3GPP TR 22.886 (2018)

Chapter 4
5G Meets V2X: Integration, Application, Standard and Industrialization

Junling Shi, Mao Xu, Min Jia, Liang Zhao, Tian Xiao, Cheng Wang, Miaowen Wen, Yuting Luan, and Lexi Xu

Abbreviations

5GAA	5G Automotive Association
ADAS	Advanced Driver Assistance Systems
AGV	Automated Guided Vehicle
AR	Augmented Reality
BS	Base Station

J. Shi · L. Zhao (✉)
School of Computer, Shenyang Aerospace University, Shenyang, China
e-mail: lzhao@sau.edu.cn

J. Shi
e-mail: jlshi@sau.edu.cn

M. Xu
School of Big Data and Information Industry, Chongqing City Management College, Chongqing, China
e-mail: 896699640@qq.com

M. Jia
School of Electronics and Information Engineering, Harbin Institute of Technology, Harbin, China
e-mail: jiamin@hit.edu.cn

T. Xiao
China Unicom Research Institute, Beijing, China
e-mail: xiaot6@chinaunicom.cn

C. Wang
School of Electronic Engineering, Beijing University of Posts and Telecommunications, Beijing, China
e-mail: wangcheng@bupt.edu.cn

M. Wen
School of Electronic and Information Engineering, South China University of Technology, Guangzhou, China
e-mail: eemwwen@scut.edu.cn

© The Author(s), under exclusive license to Springer Nature Singapore Pte Ltd. 2023
Y. Zhu et al. (eds.), *Communication, Computation and Perception Technologies for Internet of Vehicles*, https://doi.org/10.1007/978-981-99-5439-1_4

C-V2X	Cellular Vehicle to Everything
CAN	Controller Area Network
CVIS	Cooperative Vehicle Infrastructure System
DSRC	Dedicated Short-Range Communications
eMBB	Enhanced Mobile Broadband
FCC	Federal Communications Commission
FCW	Forwarding Collision Warning
GPS	Global Position System
GSA	Global mobile Suppliers Association
HMI	Human Machine Interface
IMT	International Mobile Telecommunications
IoV	Internet of Vehicles
ITS	Intelligent Transportation System
MEC	Mobile Edge Computing
MIIT	Ministry of Industry and Information Technology
mMTC	Massive Machine Type Communication
NFV	Network Functions Virtualization
OBU	On-Board Units
OEMs	Original Equipment Manufacturers
RSU	Road Side Unit
SDN	Software Defined Network
uRLLC	ultra-Reliable Low-Latency Communication
UPF	User Plane Function
V2C	Vehicle to Cloud
V2I	Vehicle to Infrastructure
V2N	Vehicle to Network
V2P	Vehicle to Pedestrian
V2V	Vehicle to Vehicle
V2X	Vehicle to Everything
VR	Virtual Reality

Y. Luan
China Railway Engineering Consulting Group Corporation, Beijing, China
e-mail: 176976816@qq.com

L. Xu (✉)
China Unicom Research Institute & Beijing University of Posts and Telecommunications, Beijing,
China
e-mail: davidlexi@hotmail.com

1 Introduction

In the past 40 years, the mobile communication experiences fast development from 1 to 5G. Recently, 5G mobile communication begins the large-scale deployment worldwide, and 5G has superior capability and performance. According to ITU, 5G supports three categories of scenarios, including Enhanced Mobile Broadband (eMBB), Massive Machine Type Communication (mMTC), Ultra-Reliable Low-Latency Communications (uRLLC) [1].

uRLLC is mainly designed for vertical industries which have high requirements in latency and reliability, and uRLLC can provide ultra-reliable and low latency communication services [2]. Therefore, 5G uRLLC is suitable for Vehicle to Everything (V2X) services. 5G applications for V2X have attracted the attentions from both the academics and industrials. The typical applications are V2X security applications, especially V2V-based autonomous driving applications, V2I-based autonomous driving applications, V2P-based autonomous driving applications. In addition, there are also 5G transportation efficiency applications, and V2X entertainment applications. Furthermore, Metaverse applications of V2X are also future direction of 5G V2X applications.

In order to accelerate the industrialization, 5G for V2X also starts the standardization process. Among which, Cellular Vehicle to Everything (C-V2X) is the milestone [3]. C-V2X technology and standard employs 4G/5G cellular communication to provide low delay and high reliability communication for V2X. Meanwhile, C-V2X technology and standard has taken many tests and demonstrations in many countries around the world, such as USA, China, UK, Australia, Italy, Germany, Japan, etc.

Main contributions to this chapter are as follows:

(1) This chapter introduces mobile communication evolution from 1 to 5G, and then introduces 5G key performance, as well as 5G three scenarios and standardization. Furthermore, this chapter analyses 5G uRLLC scenario, characteristics, and challenges. Then, the principle of 5G uRLLC integration with V2X is discussed.

(2) This chapter elaborates on 5G applications for V2X. Initially, this chapter introduces V2X security applications, including V2V-based, V2I-based, V2P-based autonomous driving applications. Furthermore, this chapter researches on 5G transportation efficiency applications, and V2X entertainment applications. Furthermore, Metaverse applications of V2X is introduced.

(3) This chapter discusses the standardization and industrialization of C-V2X, especially C-V2X Technology and standardization, spectrum allocation for Internet of Vehicles (IoV), C-V2X application trends and industrialization, important C-V2X tests and demonstrations around the world.

This chapter is organized as follows: Sect. 2 introduces 5G, and 5G uRLLC integration with V2X. Section 3 presents 5G applications for V2X, including V2X security applications, transportation efficiency applications, V2X entertainment applications, Metaverse applications. Section 4 introduces standardization and industrialization of C-V2X. Section 5 summarizes the whole chapter.

2 5G and Its Integration with V2X

2.1 5G Introduction

In the past 40 years, the mobile communication undergoes fast development, as described in Fig. 1. In 1980s, 1G mobile communication employs analog technology to provide voice services to telecom customers. In 1990s, 2G mobile communication are widely used, and the typical 2G mobile communication systems include GSM and CDMA-IS95. 2G mobile communication employs digital technology to provide telecom services. In addition to voice service, 2G also provides new services, especially text service, namely instant short message service. In 2G, its mainstream data rate is 384Kbps.

From 2000 to 2010, 3G mobile communication systems are deployed worldwide. The typical 3G mobile communication systems include WCDMA, CDMA2000, TD-SCDMA. Compared with 1G and 2G, 3G can provide a series of multimedia services, for example, rings, music, pictures, videos, etc. In 3G, its mainstream data rate can reach 21Mbps.

Since 2010, 4G mobile communication is widely used. The typical 4G mobile communication systems include TDD-LTE and FDD-LTE. Compared with 3G, 4G mobile communication can provide various mobile internet services, with the achievable user data rate of over 100Mbps.

Since 2020, 4G envisages many challenges, including future explosive growth of data traffic, massive device connections, the continuous emergence of new service and application scenarios. In order to address above-mentioned challenges, 5G mobile communication starts the commercialization worldwide. Compared with 3G and 4G, 5G can provide extremely promising network performance.

5G mobile communication employs many new technology, including massive MIMO, software defined network (SDN), network functions virtualization (NFV),

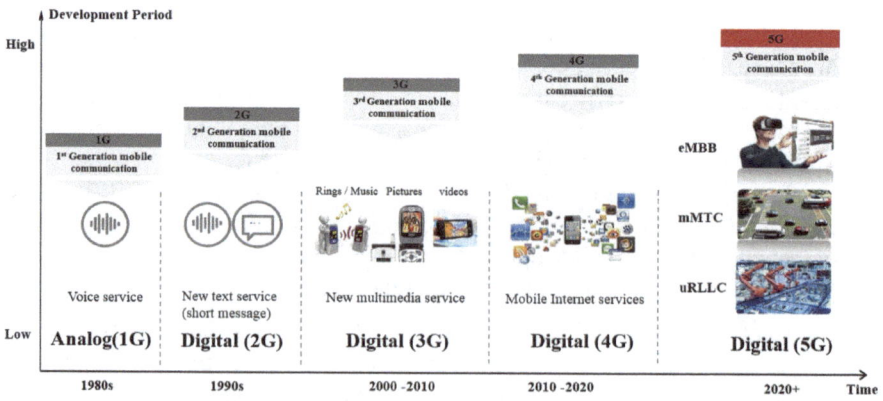

Fig. 1 1G to 5G mobile communication standard evolution

Fig. 2 Overview of 5G capability and performance

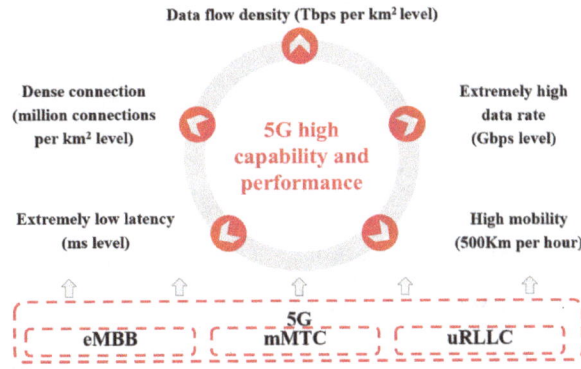

mobile edge computing (MEC), network slicing, etc. 5G mobile communication has superior capability and performance [1]. Specifically, 5G supports the achievable user data rate of over 1 Gbps, meanwhile, 5G supports extremely low latency, namely millisecond (ms) level of end-to-end latency. 5G also employs heterogeneous integrated network architecture and dense networking technology, hence, 5G supports dense connection of mobile terminal and NB-IoT, reaching one million connections per square kilometer [4]. In 5G mobile communication, the data flow density can reach tens of Tbps per square kilometer, meanwhile, the mobility capability can reach around 500 km per hour [1].

Among various performance indicators, there are three basic indicators of 5G, including the user experience rate, connection density, latency. In addition, 5G also significantly improves the efficiency of network deployment and operation, as shown in Fig. 2. Compared with 4G, the spectrum efficiency has been improved by 5 to 15 times, and energy efficiency and cost efficiency has been increased by around 100 times [1, 5].

2.2 5G Three Scenarios and Standardization

According to ITU, 5G application scenarios can be generally categorized into three scenarios, including eMBB (Enhanced Mobile Broadband), mMTC (Massive Machine Type Communication), uRLLC (Ultra-Reliable Low-Latency Communications) [6].

(1) eMBB

5G sets eMBB scenario, and 3GPP also designs eMBB related standard. Generally, 5G eMBB can be considered as the evolution of 4G, and most of 4G telecom users and traditional 4G services can be served/provided under 5G eMBB scenario. Compared with 4G, 5G eMBB will provide higher data rate, better telecom user perception, higher quality of service (QoS), as well as support new services (e.g., AR, VR).

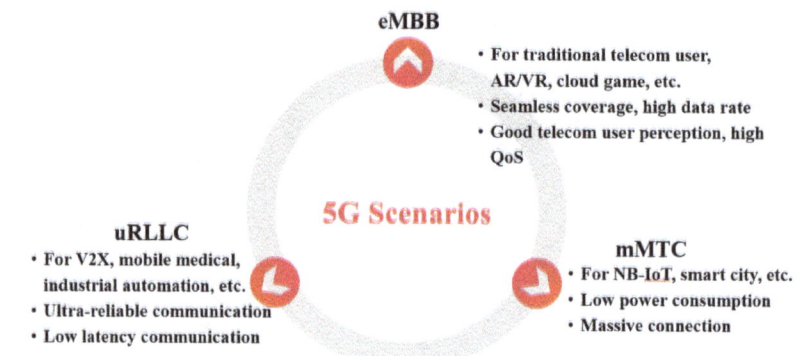

Fig. 3 Typical three scenarios of 5G

In eMBB scenario, 5G can provide three categories of capability:

High capacity capability: 5G can support mobile communication with high capacity in both outdoor and indoor densely populated areas (e.g., office buildings, city downtown, shopping mall, conference centers, stadiums, etc.).

Enhanced connectivity capability: 5G can support all-available mobile broadband access to telecom users, with seamless user perception.

High user mobility capability: 5G can provide mobile broadband services to users in trains, UAV, vehicles, etc.

(2) mMTC

5G sets mMTC scenario to support massive machine type communication as well as massive NB-IoT services, as shown in Fig. 3. The 5G characteristics include low power consumption, massive connection, low latency and high reliability. Therefore, above-mentioned characteristics are well adapted to NB-IoT services. NB-IoT services can be widely used for vertical industry applications.

Specifically, the low power consumption and massive connection characteristics can meet the requirements of environment sensing and data acquisition. For example, NB-IoT terminals are widely distributed with large numbers, hence, terminals have the features of small data packets, low power consumption, and massive connections [4]. NB-IoT terminals require 5G mobile networks to support high connection density, namely nearly 1 million connection per square kilometers. In addition, NB-IoT terminals also consume ultra-low power with low terminal-cost.

Therefore, 5G mMTC can be employed for smart cities, environmental monitoring, intelligent agriculture, forest fire monitoring, and other relevant applications.

(3) uRLLC

uRLLC is mainly designed for vertical industries, which have high requirements in latency and reliability, and uRLLC is capable of providing ultra-reliable and low latency communication services. 5G uRLLC scenario will be detailed introduced in the following sub-sections.

For the standardization, initially, ITU names 5G as IMT-2020. IMT-2020 indicates that 5G is designed for commercialization around 2020. ITU also pre-plan the key performance indicators of 5G.

3GPP starts the standardization process of 5G from 2016. In 3GPP, Release 15 is the first version of 5G standard [6]. Release 15 is done in 2018, whilst Release 15 mainly considers eMBB scenario. 3GPP Release 16 is the first complete and comprehensive 5G standard. Release 16 is done in 2020, and Release 16 standard relates to all three scenarios, including eMBB, mMTC, uRLLC.

3GPP Release 17 is done in 2022. Release 17 makes a series of enhancement on the basis of Release 16 standard, for example, coverage enhancement, terminal energy saving, spectrum expansion of massive MIMO, relay enhancement, uRLLC enhancement for 5G non-public network, NR-Light, etc. From 2022, 3GPP starts the research and standard towards Release 18.

2.3 5G URLLC Scenario, Characteristics, Challenges

uRLLC is one of the three major application scenarios in 5G [7]. With the improvement of the 5G network's low-latency and high-reliability communication guarantee capability, some businesses in the general consumer field and vertical industry will be potential bearers of the uRLLC network in the future. In the vertical industry, the goal is to replace the existing business wired and wireless solutions with the 5G uRLLC network in the early stage and to boost the upgrade of the industry by 5G uRLLC in the later stage.

5G network operators need to clarify the needs of different scenarios and services for 5G uRLLC network deployment and comprehensively use critical technologies, such as operational MEC and network slicing, in order to develop diverse network deployment solutions to suit different industries and applications.

2.3.1 Typical Service Scenarios and Requirements for 5G uRLLC

The key requirement metrics for typical uRLLC services in 3GPP are shown in Table 1 [7]. Intelligent industrial services also require packet determinism, high-precision synchronization, and delay jitter, etc.

uRLLC applications are mainly oriented to low-latency and ultra-high reliability business scenarios in limited areas, including augmented reality (AR), virtual reality (VR), intelligent grid, factory automation, and intelligent transportation with vehicle-infrastructure cooperation [2].

(1) AR/VR

AR utilizes the additional information generated by modern high-tech means with computers to enhance the real world perceived by users. VR is a new digital, artificial environment with multi-sensor integration generated by users with the aid of auxiliary

Table 1 Key requirements metrics for 5G uRLLC services

Application scenarios		AR/VR		Intelligent grid		Industrial automation	Intelligent transportation	
Reliability/%		99.999	99.9	99.9999	99.999	99.9999	99.999	99.999
Delay	E2E	1 ms@32Bytes	–	5 ms	15 ms	2 ms	5 ms	10 ms
	Air interface	4 ms@200Bytes	7 ms	2–3 ms	6–7 ms	1 ms	5 ms	7 ms
Packet size and traffic model		DL&UL periodic FTP traffic model with different arrival rates		DL&UL fixed packet size and interval period		Deterministic DL&UL flow model	Large-BW-based DL&UL periodic flow model	
network capacity assurance requirements		Critical	Normal	Extremely critical	Critical	Extremely critical	Critical	Critical

sensing devices. It naturally interacts with virtual world objects and each other to create the feeling and experience of being in the real environment.

In the consumption field, AR/VR can be applied to 360-degree panoramic live broadcasting, games, video entertainment, and social and immersive telecommunications. AR/VR can be widely used in education and training, medical care, industry, real estate, sports, tourism, etc. In order to provide users with a "real" experience, high resolution, and high-frequency updating rate, it is also necessary to keep the extremely low delay.

(2) **Intelligent Grid**

The rise of new energy brings new demand for information perception interaction and power distribution capability of the incoming power network. Many applications have strict requirements for intelligent grid scenarios on delay, communication service reliability, equipment availability, jitter, and certainty. Meanwhile, new sensors and actuators are deployed in the emerging intelligent grid, which can more effectively monitor and control the unstable state of the grid. When problems occur, rapid and automatic measures and diagnosis can address the interruption.

(3) **Industrial Automation**

Industrial automation is a hot application scenario with low delay and high reliability, which can significantly improve the factory's production efficiency, including Automated Guided Vehicle (AGV) and motion control. AGV is widely used in warehousing, logistics, ports, factories, and other scenarios to transport goods/packets automatically. The operator monitors the real-time status information reported by AGV through the data management platform and sends operation instructions. The platform interacts with the AGV trolley through a 5G low-delay network to instruct the AGV trolley to complete relevant actions.

Future factories can improve efficiency of future factories through communication among machines or among people and machines (e.g., printing presses, machine tools, packaging equipment). For remote control, data transmission's reliability must be very high so that any information and commands transmitted must be successfully received. Meanwhile, all data transmission needs to be completed within strict time requirements [8].

(4) **Intelligent Transportation**

Intelligent transportation mainly focuses on enhancing V2X scenarios. Specific applications include vehicle formation, advanced driving and extended sensors. The vehicle formation scheme can realize the automatic formation of multiple vehicles through the periodic information interaction between all vehicles in the formation, and then help follow the vehicles to achieve automatic driving.

In the advanced driving scenario, each vehicle and road side unit (RSU) will share the data obtained through local sensors with surrounding vehicles to achieve semi-automatic or fully automatic driving. Meanwhile, the remote driving of the target vehicle can be realized when the passengers cannot drive the vehicle by themselves.

The extended sensor exchanges raw/processed data or real-time video data collected by local sensors between vehicles, RSUs, and pedestrian devices. V2X application servers expand the detection range of vehicle sensors and enhance the vehicle's perception of the environment that its sensors cannot detect.

2.3.2 5G uRLLC Network Classification and Challenges

As shown in Fig. 4, to meet the requirements of different uRLLC services, 5G network can be divided into three levels according to the degree of the 5G network's guaranteed requirements for delay and reliability, as well as the degree of concentration of network coverage and service application scope [5].

Specific definitions are as follows:

- Level 1 network refers to the uRLLC business needs based on 5G wide coverage network, mainly for the personal consumption market, such as video games and AR/VR. The network can provide a basis for uRLLC services by adopting technical support schemes, including soft isolated network slice, QoS guarantee, air-interface delay enhancement scheme, and MEC. Level 1 network requires around 100 ms delay and 99.9% reliability.
- Level 2 network is mainly aimed at industrial users. The network is built according to the needs of industrial users and is generally deployed within the region, such as power distribution automation, differential protection, and remote control. For users in different industries, Level 2 network has corresponding enhancement capabilities in network security and high-precision time synchronization. Potential technical solutions include hard isolated network slice, QoS enhancement, low delay, high-reliability enhancement, redundant transmission, dedicated user plane function (UPF), and MEC. Level 2 network requires tens of milliseconds of delay and 99.9–99.999% reliability.

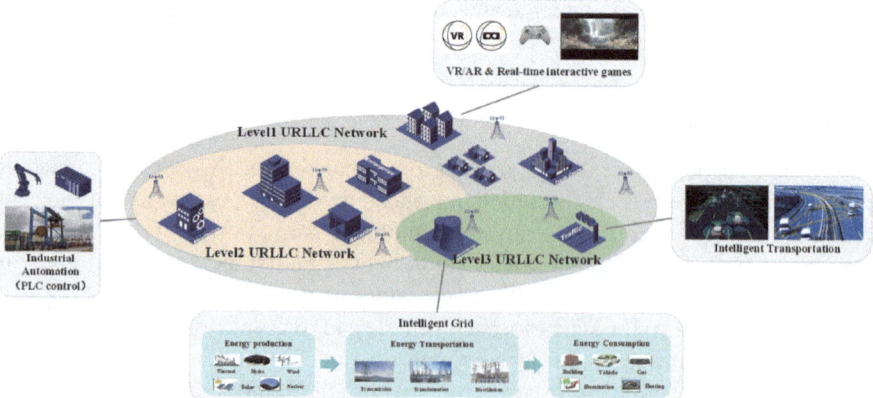

Fig. 4 Schematic diagram of 5G uRLLC layered system

- Level 3 network is targeted at users from specific industries. The network coverage is generally enterprise zone or factory levels, such as industrial site real-time control, motion control, and other businesses. The services carried out by the Level 3 network are complex and diverse, requiring high network performance. There are enhanced requirements for deterministic communication, high-precision time synchronization, security, and high-precision positioning. Potential technical solutions include QoS enhancement for uRLLC services, low delay and high-reliability enhancement, 5G and time-sensitive network integration, non-public network, redundant transmission, and private core network scheme. Level 3 network requires several milliseconds of delay and 99.999–99.9999% of reliability.

2.3.3 Challenges of 5G uRLLC Network Application Operations

5G uRLLC still faces a series of challenges for future application. In terms of business development, 5G uRLLC should gradually mine its business value in the vertical industry, define deployment scenarios and network requirements, and provide customized network solutions for vertical industry [9].

For the product implementation, with the freezing of uRLLC standards, it is necessary to clarify the uRLLC functional evolution roadmap in transforming the standards into products to promote the development of the industrial chain of network equipment, modules, and terminal products.

For the network operation and maintenance, uRLLC services require higher real-time monitoring of network indicators, such as service delay, reliability, and delay jitter. Moreover, it also needs the ability to realize network fault location and quickly recovery. Therefore, efficient operation of uRLLC is essential.

2.4 Principle of 5G uRLLC Integration with V2X

V2X technology is a new generation of information and communication technology that connects vehicles with everything [3]. V2X mainly includes the Vehicle to Vehicle (V2V) scenario, Vehicle to Pedestrian (V2P) scenario, Vehicle to Infrastructure (V2I) scenario, and Vehicle to Network (V2N) scenario. V2X organically links such elements as vehicles, pedestrians, infrastructure, and networks [10]. It can obtain more information for automatic driving of vehicles and promote the development of new business forms of vehicles and traffic services. It is of great significance for transportation management, especially traffic efficiency improvement, resources saving, pollution reduction, accident analysis, etc. [11].

From the perspective of standard, V2X can be divided into 802.11p/DSRC (Dedicated Short-Range Communications) based V2X and 3GPP-based V2X, namely, C-V2X (Cellular Vehicle to Everything). As shown in Fig. 5, C-V2X provides two communication interfaces: the Uu interface (cellular communication interface) and the PC5 interface (direct connection communication interface). When the terminal

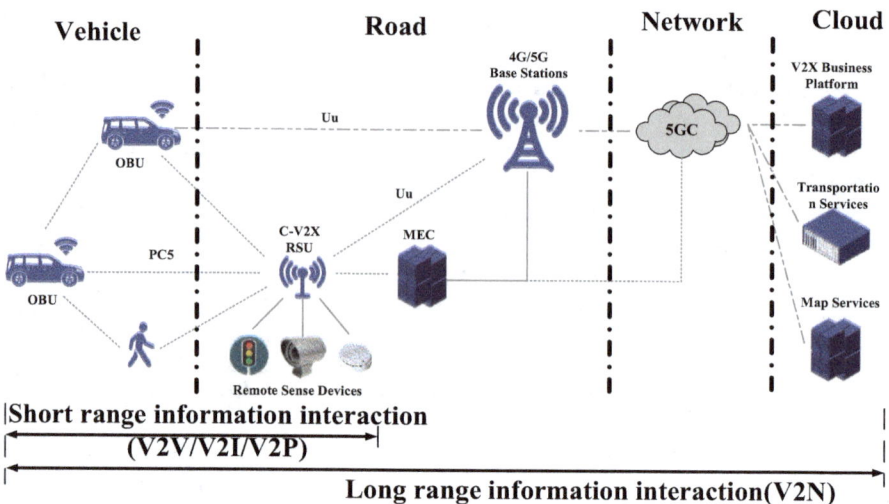

Fig. 5 Schematic diagram of C-V2X networking

equipment supporting C-V2X (e.g., vehicle terminals, smartphones, roadside units.) is within the cellular network's coverage, Uu interfaces can be used under the control of the cellular network to achieve reliable communication over a long distance in a broader range. When there is network coverage, the PC5 interface can be used for V2X communication. C-V2X combines the Uu interface and PC5 interface, supports each other, and is used for V2X service transmission to guarantee communication reliability [3].

In order to guarantee the real-time information interaction and transmission, the IoV (Internet of Vehicles) needs to control the communication delay within 10 ms. uRLLC technology can well address the problems of millisecond-level delay guarantee, network security, hard switching, and road condition prediction in IoV. Therefore, the combination of uRLLC technology and IoV is an inevitable trend in the future [12].

It should be noted that only 5G uRLLC may not fully support V2X applications. 5G uRLLC establishes signaling connections through the Uu interface for data transmission. It takes more than 100 ms from idle state to connected state, and user data needs to be forwarded through the base station (BS). The NR-V2X communication based on the PC5 interface is direct communication among vehicles, or among vehicles and infrastructure.

Moreover, 5G uRLLC services have specific requirements for the extensive bandwidth capability of eMBB based on low latency and high reliability. On one hand, the 5G uRLLC network has implemented low-latency enhancement technologies (e.g., uplink configuration transmission, mini slot) and high-reliability enhancement technologies (e.g., uRLLC modulation and coding strategies). On the other hand, the network sacrifices spectral efficiency, repeated transmission, and redundancy for eMBB performance. Therefore, only the integrated scheme based on 5G eMBB Uu+C-V2X PC5 can simultaneously realize extensive bandwidth, low delay, and high-reliability communication to support various applications of intelligent driving and intelligent transportation [13].

2.5   5G Applications for V2X

V2X is the key technology of Intelligent Transportation System (ITS), enabling vehicles to communicate with other entities via V2V, V2I, V2N and V2P [14, 15]. By the effective connection, it can strengthen driving safety, reduce congestion, improve transportation efficiency, and provide vehicle entertainment. In this section, V2X applications are generally divided into four categories. The first category is the security application, which commits itself to improve collision avoidance by perceiving the surrounding vehicles and pedestrians. The second category is the transportation efficiency application, which assists city transportation to reach a high level of effectiveness by rational scheduling of vehicles. The third category is the entertainment application. By using such kind of application, drivers and passengers can effectively spend their time when they are in vehicles, rather than losing time by idleness. The fourth category is the future application.

2.6   V2X Security Applications

DSRC has reserved a dedicated channel for security applications because security information has high priority among all V2X applications, meanwhile, security information is also the basis for many other V2X applications. Vehicles broadcast their original information periodically, thus the surrounding vehicles can perceive the existence of other vehicles. The information includes speed, coordinates, accelerated velocity, course angles, braking, etc. As the representative security application, autonomous driving can replace the driver to complete the driving mission and ensure safety. According to the different communication modes, autonomous driving applications can be further divided into V2V-based, V2I-based and V2P-based, which will be discussed as follows.

2.6.1   V2V-Based Autonomous Driving Applications

According to the real-time interaction of V2V information, applications can comprehensively judge whether the collision will happen by integrating the motion state information of the local vehicles with the received motion state information of the surrounding vehicles [3, 15]. Then, autonomous vehicles or drivers can take steps in advance after receiving the alert from the application. V2V-based autonomous

Fig. 6 Collision warning
flow diagram

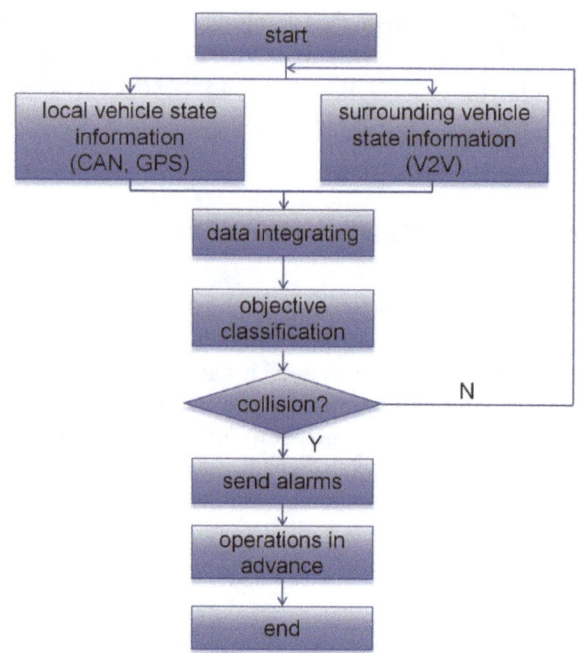

driving applications can avoid accidents caused by vehicle obstructions and dead
zone, including Forwarding Collision Warning (FCW), backward collision warning,
side collision warning and intersecting collision warning. No matter which collision,
it can be avoided by the in advance operations only if vehicles are not speeding.
Controller Area Network (CAN) and Global Position System (GPS) are used to
require the local vehicle state information, and Dedicated Short Range Communications (DSRC) is used to obtain the surrounding vehicle state information. Next,
the required data is integrated in the sensing layer by computing with the fusion
algorithm, and the final decisions are made based on the analyzed results, as shown
in Fig. 6 [14].

2.6.2 V2I-Based Autonomous Driving Applications

Infrastructures sense a wider range of environments, offering more comprehensive
information about autonomous driving compared to V2V mode. Autonomous vehicles can react automatically by combining the traffic information received from RSU
with the Advanced Driver Assistance Systems (ADAS). For instance, the countdown
interval of red lights and the recommended velocity will be presented on the Human
Machine Interface (HMI) when there is a traffic signal light in front. Vehicles will
decide to speed cut or up according to the obtained interval.

2.6.3   V2P-Based Autonomous Driving Applications

Portable equipment such as smartphones can be embedded with V2P units, which will send the V2P signal to vehicles. The V2P signal assists vehicles to simulate the positions and trajectories of humans and decides whether speed cut or stop, solving the problem that the camera and the radar cannot perceive the crossing the street of pedestrians. Even though this application cannot be popularized in a short period, it will become a very practical application.

2.7   Transportation Efficiency Applications

Recently, the smart city requires to employ a series of innovative services, which can provide information to all citizens about all aspects of city life via interactive and internet-based applications [15, 16]. In order to meet above requirements, V2X integrates the future mobile networks into vehicles to empower communication among vehicles, V2X is attractive to be integrated into smart city [15].

With the growth of 5G, C-V2X can enable vehicles to become smarter than before, which promotes the development of Cooperative Vehicle Infrastructure System (CVIS). A mature CVIS makes both vehicles and roads intelligent, of which the most central value is exchanging information.

2.8   V2X Entertainment Applications

Except for the security applications and transportation efficiency applications, vehicles can act as intelligent mobile spaces to provide immersive entertainment to drivers and passengers. It is certain that such entertainment applications should be supported by sound autonomous driving technologies. On the basis of advanced autonomous driving, the time on the road of the drivers can be freely controlled, which provides a huge opportunity for merchants to design entertainment products, such as the smart musical cockpit and the mobile workspace. Compared to other entertainment applications, for example, smartphone apps, V2X entertainment applications are different by the leak-proofness of the vehicle spaces, which is the necessary condition to ensure the immerse.

The smart cockpit actually is an AI-based man–machine interactional system. To realize a true sense of man–machine interaction, there should be four basic elements, i.e., hardware, interaction mode, onboard software and network [17]. To be specific, the hardware includes some input and output equipment, such as vehicle screens, voice boxes, and vehicle cameras, which can guarantee a high-quality experience for the consumers. In particular, there are already some merchants that have embedded hardware into vehicle seats or windows to make these parts become input or output equipment. Referring to the interaction mode, intelligent voice is the most significant technology compared to other interaction modes. Intelligent voice is mainly responsible for voice recognition and reply. By using it, vehicles can understand what consumers want to express and then conduct the mission. Up to now, many automobile manufacturers have developed their own intelligent voice products, and some of them choose to cooperate with the phonetic explaining merchant to accelerate their plans. Next is about the onboard software, which includes various developed applications and can be designed not only by the merchants but also individuals in the future. It is worth mentioning that many existing popular applications can be embedded in vehicles by extension. In addition, vehicle network is essential to realize all the networked entertainment applications. Benefiting from 5G, low latency and high-quality network quality become true, which speeds up the progress of V2X entertainment applications.

The smart musical cockpit is able to provide an immersive musical space for the consumers. In such a confined space as a vehicle, a VR-based entertainment system will provide a whole-scale music experience and make consumers relax and enjoy music. Another typical application is the mobile workspace, which is devised and customized for some specific people such as office clerks. The above entertainment application is a brand-new environment of the vehicle cockpit, where AI, cloud and edge computing and big data analysis all should play important roles, and the language-based user interaction, visual-based recognition, and image-based are core technologies in the whole development stage.

2.9   Future V2X Applications

The future V2X applications are representative by the metaverse and the integration of everything, which are discussed in the following two sub-sections.

2.9.1   Integration of Everything Applications

Current V2X mainly focuses on the interconnections between vehicles and the entities on or near the road. Furthermore, in the future, vehicles can also interconnect with the entities beyond the road, such as the charging piles, hospitals, parking lots, houses, and intelligent household electrical appliances, etc. Everything that exists in life will be connected to vehicles to make our lives more convenient, as shown in Fig. 7.

Automatic driving can assist vehicles to make the right decisions by the obtained information of the whole related objects. Only if the accurate information about the related objects can be obtained, the safe and efficient automatic driving can go into service. By that time, vehicles are not just intelligent transports, but butlers that take charge of full service of lives. However, to realize the real interconnection

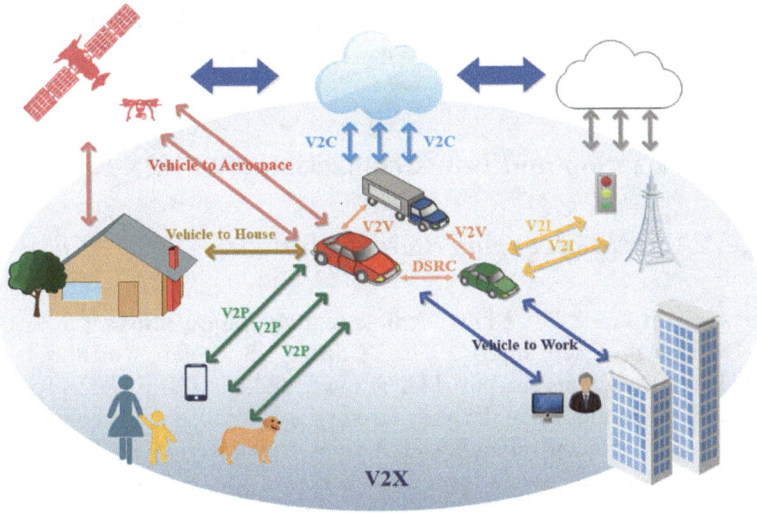

Fig. 7 Integration of everything in V2X

to everything is extremely difficult. Except for the high intelligence of the vehicles themselves, it also requires a rapid and safe network, and the intelligent and networked "everything". In fact, the interconnection of everything to vehicles has to wait for a long time since it depends on the overall development of science and technology.

2.9.2 Metaverse Applications

Metaverse is a 3D sharing digital platform with integrated and sound economic and social architecture. Metaverse has attracted the attentions from both the academics and industrials. Few recent years have witnessed an unprecedented explosion of the metaverse, mostly derived from 3D gaming, which is fueled by the improvement of both hardware and software to build the virtual world more solidly and creatively [18].

Metaverse is known as the most promising application in the future world and also an important application in V2X. To realize Metaverse in V2X, more hardware equipment such as smartphones, smart homes, and headphones, should be compatible with existing equipment. Simultaneously, accurate and rapid feedback and interaction are also necessary. As a consequence, the underlying display technologies like VR, AR are required to guarantee high-quality frames. Moreover, the network computing power technologies should be used to accelerate and accurate the feedback, including spatial orientation algorithms, VR fitting, built-in sensors, real-time network transmission, GPU servers, edge computing, etc. [19]. The faster technological change and more precisely integrated real-time scenario of metaverse will be droved by the V2X technologies and applications.

3   Standardization and Industrialization of C-V2X

3.1   C-V2X Technology and Standardization

The milestone of C-V2X (Cellular Vehicle to Everything) research and standard is from Dr. Shanzhi Chen. In 2011, Dr. Shanzhi Chen's team, which includes Dr. Chen and his PhD students and post-PhD fellows, studied the IEEE802.11p standard and then discovered that when the nodes are densely populated, the communication performance (e.g., latency, reliability, etc.) drops dramatically [20]. Meanwhile, the coverage and connectivity performance also becomes poor, and the deployment cost becomes high. At that time, the 4G TD-LTE national standard had essentially just been completed. A new technical approach was developed by Dr. Chen Shanzhi and the Datang Telecom. This approach involves designing IoV technology based on cellular mobile communication and attempting to be compatible with the physical layer technology of cellular mobile communication. In this way, the cellular mobile communication technology and network deployment can be utilized, and the IoV chips can benefit from the economies of scale of its mobile terminals to reduce costs. Moreover, technological innovation is required to address the issues of low delay and high reliability communication between vehicles and various traffic elements. This can help concurrent communication among multiple vehicle nodes. This can also help deal with the specific requirements and challenges of IoV communication, such high-speed mobility, complex interference environment, high frequency and periodic data transmission. As a result, it can bring both technical and business benefits.

Based on the in-depth research, C-V2X technology was developed. Dr. Shanzhi Chen and the Datang Telecom began conducting research in 2012. C-V2X is the world-leading wireless communication technology and standard for the IoV that integrates cellular communication and direct communication (i.e., LTE-V2X) [20]. LTE-V2X is the first version of C-V2X wireless communication technology. It provides significant technological advances, such as cellular and direct communication integration system architecture, wireless transmission, access control and resource scheduling methods, and synchronization mechanisms [3]. LTE-V2X essentially set the groundwork for C-V2X's cellular and direct fusion system architecture, as well as key technical principles of direct links. As shown in Fig. 8, it may adaptably support V2V, V2I, V2P, V2N, and vehicle-to-cloud (V2C) communication capabilities, as well as communication scenarios within and beyond the coverage of cellular networks.

Since 2015, Datang Telecom has collaborated with LG and Huawei to advance the development of international standards for LTE-V2X in 3GPP. C-V2X technology

Direct safety communication independent of cellular network

Low latency V2V, V2I, V2P operates in ITS bands(e.g. 5.9 GHz)

Direct PC5 interface

e.g. location, speed, local hazards

Network communications for complementary services

V2V operates in a mobile operator's licensed spectrum

Network Uu interface

e.g. accident 2 kilometer ahead

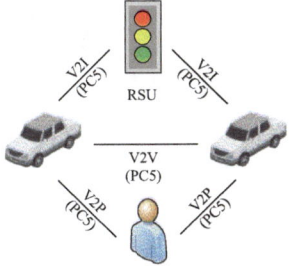
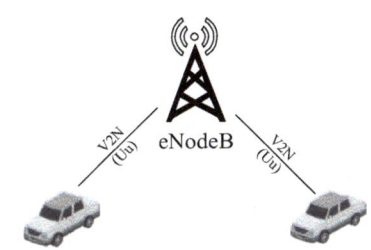

Fig. 8 C-V2X enables network independent communication

standardization in 3GPP is divided into two stages, including LTE-V2X and NR-V2X. Instead of replacing one another, LTE-V2X and NR-V2X enhance with one another. The standard design fully allows the coexistence of LTE-V2X and NR-V2X in the devices, since both LTE-V2X and NR-V2X fully consider the backward compatibility and forward compatibility [20].

- LTE-V2X: The PC5 interface enabling V2X short-distance direct communication, which can communicate directly with each other without relying on the 4G/5G BS (Base Station), is introduced into the cellular communication. Its aim is to support the communication requirements for basic road safety services and primarily implement the auxiliary driving function.
- NR-V2X: In order to serve advanced V2X business requirements, the enhancement of the PC5 interface and the Uu interface based on 5G NR is designed [3]. Figure 9 shows the relationship between 5G and C-V2X.

3.2   Global Spectrum Allocation for C-V2X

In October 1999, the Federal Communications Commission (FCC) of USA approved the allocation of 75 MHz in the 5.9 GHz band (5.850–5.925 GHz) as a dedicated spectrum to support the DSRC-based ITS services.

In July 2018, 5G Automotive Association (5GAA) released the White Paper on ITS spectrum utilization in the Asia Pacific Region, which highly recommended regulators and governments planning to allocate spectrum to ITS applications consider 5.9 GHz as the target ITS spectrum. It is also suggested that the C-V2X PC5 interface

C-V2X integrates direct communication (PC5 interface) and cellular communication (Uu interface). It is an effective V2X communication technology with continuous evolution that can achieve 5G NR C-V2X forward compatibility.

- PC5 interface provides low latency and high reliability in V2V/V2I communication, and currently supports commercial use on large-Scale LTE-V2X.
- Uu interface provides vehicle-to-cloud communication, and currently supports commercial use on large-scale 4G and 5G eMBB.

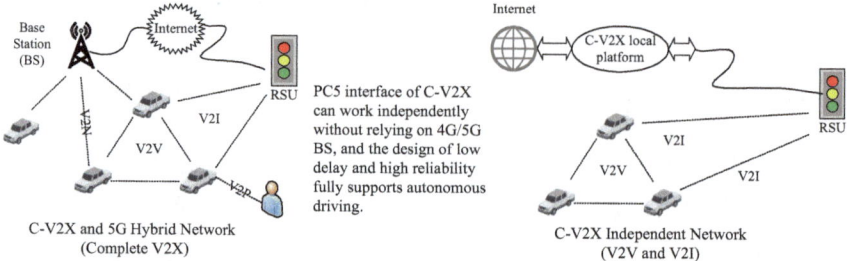

5G URLLC (low latency, high reliability) characteristics rely on 5G BS coverage and cannot solve problems, such as the high frequency and pairwise communication of V2V/V2R.

5G autonomous driving and 5G intelligent high-speed are just two combinations of professional words

Fig. 9 The relationship between 5G and C-V2X

should be used on dedicated ITS spectrum to provide safety-related services, and the C-V2X Uu interface can be used to supplement the licensed spectrum.

In November 2018, China allocates the dedicated spectrum for cellular IoV, when the Radio Management Bureau of Ministry of Industry and Information Technology (MIIT) assigned 20 MHz (5.905–5.925 GHz) as the dedicated spectrum for LTE-V2X direct communication, as shown in Fig. 10.

To seek a global or regional unified spectrum for ITS, World Radiocommunication Conference 2019 encouraged national authorities to use the 5.9 GHz band or a portion of it as a global or regional unified band when planning and deploying evolving ITS applications. In the future, the utilization of this frequency band for C-V2X technology will become a worldwide development trend.

In December 2019, the FCC reassigned the 75 MHz (5.850–5.925 GHz) spectrum originally allocated to DCRC and allocated the 20 MHz (5.905–5.925 GHz) dedicated spectrum to C-V2X technology, which is similar as that in China.

In November 2020, the FCC officially voted to revoke the whole 75 MHz (5.850–5.925 GHz) spectrum that had originally been allotted to DCRC and instead allocate it to Wi-Fi and C-V2X, in which 30 MHz (5.895–5.925 GHz) spectrum is dedicated to C-V2X technology to improve vehicle safety. The allocation of spectrum resources in USA reflects its support towards C-V2X technology.

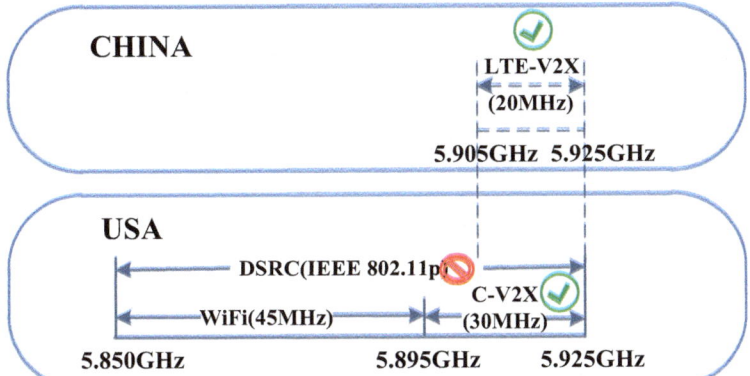

Note: The national spectrum resource allocation policy reflects the decision of its technology industry direction.

- **China: China allocates 20 MHz bandwidth of dedicated 5.9 GHz band for LTE-V2X direct communication in 2018.**
- **USA: US Federal Communications Commission (FCC) cancels spectrum that was already allocated to DSRC (IEEE 802.11p), while assigning 30MHz of it to C-V2X in 2020.**

Fig. 10 C-V2X spectrum allocation in China and the USA

3.3   C-V2X Technology Application Trends and Industrialization

3.3.1   USA

Recently, the FCC and the Society of Automotive Engineers (SAE) of USA support C-V2X experiments and industrialization. As C-V2X technology has been acknowledged by China and USA as the worldwide standard for vehicle-to-vehicle/vehicle-to-road wireless communication, several cities and government institutions in USA are proactively deploying C-V2X. In order to keep up with industry trends, USA automakers like Ford and GM are already deploying C-V2X. Ford, for example, has released several models with C-V2X capabilities, including new domestically made Explorer, Edge Plus, and Mustang Mach-E. The 2021 Buick GL8, was also equipped with the first batch of C-V2X technology.

The OmniAir Consortium is a significant testing and certification group in the field of IoV. The OmniAir Consortium, comprised of infrastructure owners/operators, enterprises deploying C-V2X technology, device manufacturers, and OEMs (original equipment manufacturers), has provided test and certification since 2012, starting with RFID charging and then progressing to DSRC. The OmniAir Consortium announced the official certification procedure for the V2X IoV industry, and launched the LTE-V2X OBU and RSU certification program in August 2021. Furthermore, the OmniAir Consortium promotes independent third-party test and certification.

OEMs and infrastructure owners/operators rely on the OmniAir certification program when making deployment decisions, and the program is also required as part of the IoV pilot in USA. Other countries (e.g., South Korea, Canada, etc.) are gradually adopting OmniAir certification.

Simultaneously, the FCC is developing a new set of rules to further promote the popularization of C-V2X. These new rules will provide regulatory clarity not only for OEMs making decisions and engineering plans, but also for infrastructure owners and operators investing in C-V2X.

3.3.2   China

In December 2015, the MIIT (Ministry of Industry and Information Technology) proposed the "Innovative Action Plan for the Development of IoV (2015–2020)," which called for the promotion of IoV research and standardization, as well as 5G-based IoV pilot and demonstration.

In June 2017, the International Mobile Telecommunications (IMT) -2020 (5G) promotion group established the C-V2X working group, with China Academy of Information and Communications Technology as the leaingd institution for organizing technical research, test verification, industry and application of LTE-V2X and 5G-V2X [3].

In March 2021, the Intelligent Connected Vehicle promotion group (ICV-2035) was established under the coordination of the National Manufacturing Power Leading Group's IoV special committee. The leading institution is from MIIT. There are six working groups, including regulatory platform, technical standards, test applications, operating systems, network security and industrial ecology, as well as an expert group from industry, academia, research and application.

Under the guidance of the Chinese government and the collaboration of the industry in the past decade, C-V2X has been in-depth researched in China and USA, which are the two major automobile and transportation countries. C-V2X also becomes the promosing international standard for the wireless communication technology of IoV [11]. Meanwhile, C-V2X has established a relatively complete industrial chain ecology that includes communication chips, communication modules, on-board units (OBU), RSU, test instruments, automobile manufacturing, operation services, test certification, high-precision positioning and map services, etc.

The MIIT, Ministry of Public Security, and Ministry of Transport, have collaborated to promote collaboration between ministries and provinces in China. Abovementioned ministries have supported the construction of IoV testing areas and pilot areas, deployed IoV infrastructure on a large-scale in multiple scenarios (e.g., cities, expressways), and promoted the implementation of IoV applications. At present, MIIT has approved and supported the establishment of national IoV pilot zones in Wuxi city, Tianjin city, Changsha city, and Chongqing city. C-V2X networks will be deployed in a variety of road situations, including expressways and urban roads. It aims to complete the transformation of IoV functions and the promotion of core system capabilities of transportation facilities in important areas and scenarios. It also aims to build an open, integrated, and innovative industrial ecology for C-V2X. There are also other Chinese cities preparing for C-V2X deployment, including Guangzhou city, Liuzhou city, Chengdu city, Hefei city, and Deqing city, etc.

On the roadside, by the end of 2020, four national pilot areas of IoV, including Wuxi city, Tianjin city, Changsha city, Chongqing city, have deployed 1,200 RSUs on more than 700 km of expressways and urban roads. There are also varied degrees of deployment in other cities across the China. As of June 2021, more than 1,100 RSUs have been deployed on urban roads, and all sections of the G1 Expressway are projected to be deployed in the near future. Meanwhile, China Academy of Information and Communications Technology is actively developing a state statistics platform for IoV infrastructure, which collects statistical data from the IoV demonstration areas, pilot areas, and the C-V2X infrastructure.

On the vehicle side, a large number of automakers have declared intentions for C-V2X mass production models, for example, FAW Hongqi, SAIC GM, SAIC Audi, Ford Motor, Great Wall Motor, etc.

3.3.3 C-V2X Tests and Demonstrations

In January 2022, the Global mobile Suppliers Association (GSA) released the statistics on the development of the global C-V2X industry chain [21]. There are some suppliers of C-V2X chips that conform to 3GPP Release14 standards, for example, Qualcomm, Autotalks, Huawei and Morningcore in China. In addition, 42 C-V2X vehicular modules support LTE or 5G, 25 C-V2X RSU and 31 C-V2X OBU [21].

Over the past few years, investments in the C-V2X have been steadily increasing as telecom network operators and government communications regulators plan and conduct trials, as well as develop and build test sites or required infrastructure. Since the beginning of 2020, the worldwide C-V2X industry has changed significantly, with an increasing number of automotive OEMs supporting C-V2X technology.

Operators also invest in and support C-V2X experiments, including the deployment of 4G LTE, LTE-Advanced, and 5G networks. There are over 26 global C-V2X operators, which were identified by GSA, involving in evaluation, development, test, or experiment, including AT&T (American Telephone and Telegraph), BT/EE (British Telecom/ Everything Everywhere), Bell Canada, China Mobile, China Unicom, Deutsche Telekom, HKT (Hong Kong Telecom), KDDI, KPN, KT, LG Uplus, NTT Docomo, Orange France, Proximus, SK Telecom, SoftBank Group, Telecom Italia, Spain Telecom, Telekom Malaysia, A1 Telekom Austria, Telstra Corporation, T-Mobile Czech Republic, Verizon Communications, Vodafone Germany, Vodafone UK, etc. Furthermore, more country/regional transportation and urban or regional agencies are expected to engage in C-V2X trials.

There are some C-V2X tests, which have significant impact:

In Australia, the Victoria government, the state government agency VicRoads, the Traffic Accident Commission, Telstra Corporation, and Lexus Australia are involved in the "Advanced Connected Vehicles Victoria" test.

In Europe, Audi, Ericsson, Qualcomm, SWARCO Group, and Technical University of Kaiserslautern completed the trilateral C-V2X test platform. In addition, the first live demonstration of C-V2X direct communication interoperability between motorcycles, automobiles, and roadside infrastructures was held. They also demonstrated cross-border C-V2X in France, Germany, and Luxembourg.

In Europe, 25 IoV relevant companies are working together to conduct the 5G-CARMEN project, which is a cross-border 5G, C-V2X, and C-ITS interoperability trial. The trial ranges from Bologna, Italy, to Munich, Germany.

In Germany, Deutsche Telekom, Vodafone, Huawei, Bosch, Continental AG, Fraunhofer ESK, Nokia, the ConVeX Consortium (Audi, Ericsson, Qualcomm, SWARCO Group, and Technical University of Kaiserslautern), and the Federal Ministry for Digital and Transport have collaborated to build a digital A9 highway test platform for testing connected vehicle applications, including C-V2X trials.

In Turin of Italy, several business partners, including Telecom Italia, Telefónica, BT/EE, Cisco, Capgemini, Harman, and Stellantis Group, jointly conducted interconnectivity experiments to investigate the impact of waiting times and the interoperability and roaming performance for C-V2X.

In Japan, C-V2X tests were conducted at JARI City and JARI Tsukuba tracks with the participation of Continental AG, Ericsson, Nissan, NTT DOCOMO, OKI, and Qualcomm.

In South Korea, SK Telecom, KT, Samsung, and LG Uplus K-City collaborated to develop a city specifically designed for automatic driving tests of the 5G network, and the C-V2X 5G trial was held.

In UK, Vodafone created a dedicated 5G network with MEC for C-V2X test at the HORIBA MIRA track and conducted C-V2X test in collaboration with HORIBA MIRA, which is a global supplier of vehicle engineering, research, and test services.

In Colorado of USA, the Colorado Department of Transportation, Qualcomm, Panasonic, Ford, Austrian transportation technology company Kapsch, and Spanish automotive systems/components company Ficosa are working together to promote the C-V2X. Kapsch provided 100 sets of C-V2X RSU and Ficosa provided 500 sets of C-V2X OBU. C-V2X was deployed on specific roads at the Panasonic "CityNOW" Smart City project head office in Denver, as well as along Interstate 70's mountain corridor.

In Atlanta, Boston, Dallas, Miami, New York, San Francisco and Washington of USA, C-V2X was jointly promoted by Verizon, Amazon, automotive operating system developer RenovoMotors, automotive technology companies Savari and LG Electronics. They conducted 5G MEC to support ADAS (Advanced Driving Assistance Services) and C-V2X trials in driver and pedestrian systems.

In addition, many other countries have conducted regional C-V2X trials to verify C-V2X performance.

4   Conclusion

In recent years, 5G mobile communication begins the commercialization worldwide, and 5G can provide promising communication and network performance. Among various 5G scenarios, 5G uRLLC scenario can provide ultra-reliable and low-latency communication. Hence, 5G uRLLC is suitable for V2X services. In this chapter, we briefly introduce 5G, then we focus on 5G uRLLC scenario and discuss its characteristics, challenges, and 5G uRLLC integration with V2X. Furthermore, this chapter elaborates on 5G applications for V2X, especially V2X security applications, transportation efficiency applications, V2X entertainment applications, Metaverse applications, etc. Furthermore, the standardization and industrialization of C-V2X is discussed.

References

1. IMT2020(5G) Advancing Group, *5G Vision and Requirements* (2014)
2. H. Chen, R. Abbas et al., Ultra-reliable low latency cellular networks: use cases, challenges and approaches. IEEE Commun. Mag. **56**(12), 119–125 (2018)
3. M. Fallgren, M. Dillinger, T. Mahmoodi, T. Svensson, *Cellular V2X for Connected Automated Driving* (Wiley, 2021)
4. G. Cao, J. Li, Y. Li, F. Li, Research advances of 5G network architecture standards. Mob. Commun. **41**(2), 32–37 (2017)
5. A.K. Bairagi, M.S. Munir, M. Alsenwi et al., Coexistence mechanism between eMBB and uRLLC in 5G wireless networks. IEEE Trans. Commun. **69**(3), 1736–1749 (2021)
6. X. Liu, F. N, P. Li, C. Liu, Introduction of 5G Standardization and analysis of network architecture, Telecom Eng. Tech. Stand. **30**(8), 44–49 (2017)
7. 3GPP TR 38.824 (v2.0.1), *Study on Physical Layer Enhancements for NR Ultra-Reliable and Low Latency Case (uRLLC)* (2019)
8. C. Pan, Z. Wang, Z. Zhou, X. Ren, Deep reinforcement learning-based URLLC-aware task offloading in collaborative vehicular networks. China Commun. **18**(7), 134–146 (2021)
9. D. Segura, E.J. Khatib, R. Barco, Dynamic packet duplication for industrial URLLC. Sensors **22**(2), 587–587 (2022)
10. H. Bagheri, M. Noor-A-Rahim, Z. Liu et al., 5G NR-V2X: toward connected and cooperative autonomous driving. IEEE Commun. Stand. Mag. **5**(1), 48–54 (2021)
11. S. Chen, S. Kang, A tutorial on 5G and the progress in China. Front. Inf. Technol. & Electron. Eng. (2018)
12. Y. Yoon, H. Seon, H. Kim, A defensive scheduling scheme to accommodate random selection devices in 5G NR V2X. IEEE Commun. Lett. **25**(6), 2068–2072 (2021)
13. J. Hu, X. Ren, X. Zhao, L. Zhao, S. Zheng, Y. Shi, Design and Evaluation of Synchronization Signals for NR-V2X Sidelink, in *2020 IEEE 91st Vehicular Technology Conference (VTC2020-Spring)* (2020), pp. 1–6
14. P. Wang, "Vehicle to Everything," China Machine Press, 2020, 536–538.
15. J. Shi, X. Wang, M. Huang, K. Li, S.l K. Das, Social-based routing scheme for fixed-line VANET. Comput. Netw. (2017)
16. M. Bonola, L. Bracciale, P. Loreti, R. Amici, Opportunistic communication in smart city: experimental insight with small-scale taxi fleets as data carriers. Ad Hoc Netw. **43**, 43–55 (2016)
17. https://www.zhihu.com/question/401566776

18. T. Huynh-The, Q. V. Pham, X. Q. Pham, T. T. Nguyen, Z. Han, D. S. Kin, *Artificial Intelligence for the Metaverse: A Survey* (Cornell University, 2022), pp.1–24
19. Y. Wang, Z. Su, N. Zhang, D. Liu, R. Xing, T. H. Luan, X. Shen, "A Survey on Metaverse: Fundamentals, Security, and Privacy," arXiv e-prints, pp.1–23, 2022.
20. S. Chen, J. Hu, Y. Shi, L. Zhao, W. Li, A vision of C-V2X: technologies, field testing, and challenges with Chinese development. IEEE Internet Things J. **7**(5), 3872–3881 (2020)
21. Global mobile Suppliers Association (GSA), *Global C-V2X Ecosystem: Status Update Executive Summary* (2022)

Chapter 5
Enabling Reconfigurable Intelligent Surface for V2X Communication Systems

Bin Yang and Yongdong Zhu

Abbreviations

RIS	Reconfigurable Intelligent Surface
V2X	Vehicle-to-Everything
IoVs	Internet of Vehicles
MEC	Mobile Edge Computing
AI	Artificial Intelligence
AWGN	Additive White Gaussian Noise
DNNs	Deep Neural Networks

1 Introduction

The future of transportation will be a comprehensive entry into the era of intelligent transportation systems. The intelligent transportation system is an extensive integration of advanced fifth-generation (5G) Internet of Vehicles (IoVs) technology, artificial intelligence (AI) technology and other techniques. Utilizing the above techniques, it comprehensively optimizes and improves the efficiency and reliability of transportation management. In the intelligent transportation system, vehicles, roads, edges, and cloud computing platforms are fully connected and intelligently interacted, which will further spawn a large number of new transportation services. For

B. Yang · Y. Zhu (✉)
Zhejiang Lab, Interdisciplinary Innovation Research Institute, Hangzhou, China
e-mail: zhuyd@zhejianglab.com

B. Yang
e-mail: binyang@zhejianglab.com

87

example, AI technology can process and analyze traffic in real-time, e.g., realizing highly reliable intelligent path planning, and effectively improving traffic safety [1].

In recent years, the reconfigurable intelligent surface (RIS) has attracted widespread attention for its potential of extending network coverage and enhancing spectrum efficiency, by intelligently adjusting the phase shifts of each passive element in the surface [2]. RISs are intelligently designed artificial planar structures with reconfigurable properties enabled via integrated electronic circuits, which can be programmed to reflect an impinging electromagnetic wave in a controlled manner. RISs are manufactured with low-profile, lightweight, cheap materials that can be shaped with conformal geometries, thus easing their deployment on a variety of environment surfaces, such as facades of buildings, walls, ceilings, etc. [3]. The signal propagation from transmitters to receivers can therefore be assisted by steering the RIS-reflected signals in directions that enhance the resulting signal quality, which in turn can be exploited to achieve substantially higher spectral efficiencies compared to current wireless systems.

Despite the widespread interest in applying RIS in wireless communication systems, the discussions concerning RIS-aided vehicle-to-everything (V2X) are still in their infancy [4]. By tuning the RISs' coefficients, it is possible to boost vehicle-to-infrastructure (V2I) capacity. Also, by appropriately placing RIS elements, the path loss between transmitters and receivers in higher-frequency bands can also be suppressed. In this way, benefits introduced by RISs can provide new degrees of freedom for advanced V2X applications, including autonomous driving.

1.1 Motivation

Today, the intelligent transportation brought by the 5G IoVs technology and AI also puts forward corresponding development requirements for the level of vehicle informatization and intelligence. For connected vehicles, their own computing power and communication range are limited. Thus, it is difficult to cope with the high computing needs of intelligent transportation services only by the limited computing resources of the vehicle itself. For transportation systems, due to factors such as uneven distribution of traffic flow on the road and intricate road information, the efficiency of IoV resources (communication bandwidth and computing resources) is low, and it is difficult to cope with the increasingly complex intelligent transportation system.

RIS can actively and intelligently control space electromagnetic waves through programming. This mechanism provides an interface between the electromagnetic world of RIS and the digital world of information science, which is very attractive for the development of wireless networks in the future. As mentioned above, on the one hand, RIS can actively enrich channel scattering conditions and enhance the multiplexing gain of wireless communication systems; on the other hand, RIS can realize signal propagation direction regulation and in-phase superposition in three-dimensional space, increase received signal strength, and improve transmission performance. Therefore, RIS has great potential for coverage enhancement and capacity improvement of future wireless networks, providing virtual line-of-sight (LoS) links, eliminating local coverage holes, serving cell edge users, solving inter-cell co-channel interference, and realizing intelligent reconfiguration wireless environment. As a result, RIS-assisted V2X is a promising approach for the future intelligent transportation systems.

1.2 Main Contributions

The main contributions to this chapter are as follows:

(1) This chapter introduces the literature review of RIS-assisted communications. Specifically, we give a short review of V2X technology and its standard evolution. Besides, we also introduce the fundamentals of RIS, including the structure and several potential application scenarios of RIS-assisted communication systems, including enhancement of network coverage, signal transmission robustness enhancement, secure transmission, energy harvesting/transmission and massive antenna transceiver.
(2) This chapter introduces modeling of the combination of V2X and RIS, including RIS-assisted vehicular communication network and RIS-assisted vehicular edge computing.
(3) This chapter discusses the future development trend and challenge of RIS-assisted V2X communication in intelligent transportation systems.

1.3 Structural Organization

Section 2 introduces V2X communication protocols and techniques evolution, and fundamentals of RIS. Section 3 presents the modeling and application of RIS-assisted communication systems. Section 4 shows the challenges and the modeling of RIS-assisted V2X communication systems. Section 5 discusses the future development trend and challenge of RIS-assisted V2X communication. Section 6 concludes the whole chapter.

2 Background

2.1 V2X Technology and Standard Evolution

It is anticipated that the near future will witness the proliferation of intelligent vehicles. Vehicular communication is the key component of intelligent transportation systems (ITSs) relying on platooning and autonomous driving [4]. There are currently two standards for V2X technology: one is the dedicated short-range communication (DSRC) standard, which was mainly researched and developed by the United States, Europe, and Japan since the 1990s. The other is C-V2X, which was launched by the third-generation partnership program (3GPP) in 2015 based on the cellular network vehicle wireless communication technology standard, including LTE-V2X standard and 5G V2X standard. We compare the key technical indicators of DSRC and C-V2X technology as follows (Table 1).

Based on 3GPP TS22.886, the 5G-V2X will support a total of 4 main areas, including platoon driving, advanced driving, remote driving, and sensor sharing, which will be subdivided into 22 scenarios, and will also provide comprehensive information for L1–L5 autonomous driving. The requirements are mainly reflected in three aspects, i.e., delay, bandwidth, and reliability. Some special application scenarios of the V2X require a delay of 5–10 ms, which can provide a stable bandwidth of 10–100 MHz, and the reliability should be greater than 99.999%. In the 5G-V2X system, the 5G network data transmission rate can theoretically reach 5 Gbps or even 100 Gbps, which will be 50 100 times compared with the 4G network, and the average delay will also be reduced to less than 10 ms. Besides, the 5G network can adopt network slicing technology to achieve flexible slice management of network data considering service priorities. Compared with the current DSRC, 5G technology has a larger transmission distance, i.e., 1000 m, which is 3 times the range supported by DSRC technology. The maximum moving speed that can be supported is 500 km/h, which is much higher than DSRC, i.e., 200 km/h. Therefore, key technologies specification of C-V2X such as high bandwidth, low latency, and slicing network can fully meet the needs of future V2X communication, which will undoubtedly accelerate the process of future autonomous driving.

Table 1 Comparison of key indicators between DSRC and C-V2X

Indicator	DSRC	C-V2X
Transmission range (m)	300–500	1000
Speed (km/h)	200	500
Transmission rate (Mbps)	27	500
Delay (ms)	50–100	50
Infrastructure deployment	Road side unit	Base station (BS)

2.2 Fundamentals and Introduction of RIS

RIS is an artificial 2D material with subwavelength size, usually composed of metals, dielectrics, and tunable elements, which can be equivalently characterized as RLC circuits. The physical properties of the electromagnetic unit can be adjusted, such as capacitive reactance, impedance, and inductive reactance, to change the radiation characteristics of the RIS, and realize unconventional physical phenomena such as irregular reflection, negative refraction, wave absorption, focusing and polarization conversion [5].

RIS generates the electromagnetic characteristics required for each electromagnetic unit by controlling the bias voltage of varactor diodes, PIN switches, MEMS switches, liquid crystals, graphene, etc. [6–10]. Through the integration of active controllers, RIS regulates the state of each electromagnetic unit, thus promoting the transformation of RIS from *static* to *dynamic*, and organically connecting the physical world and the digital world [11–14]. RIS can be controlled in real-time by designing and storing different digital coding sequences in advance and then completing the dynamic regulation of electromagnetic waves by switching the coding sequences, achieving single-beam reflection, multi-beam reflection, diffuse scattering, and transmission [15–18]. The electromagnetic unit using digital mode can directly process digital information. Combined with AI, the information can be perceived, understood, remembered, and learned [19–23]. As a result, RIS has great potential to become a new physical platform for regulating both electromagnetic waves and digital information, enabling the construction of intelligent electronic information systems with new architectures.

3 Modeling and Application of RIS-Assisted Communication System

3.1 Received Signal Modeling of RIS-Assisted Communication System

Application scenarios of RIS include non-line-of-sight (NLoS) enhancement, solving local voids, supporting edge users, realizing secure communication, reducing electromagnetic pollution, constructing passive IoTs, high-precision positioning, and integration of communication and sensing [9, 24–26]. To better realize the potential of RIS communication systems, realistic channel measurements, communication performance analysis, accurate channel estimation, flexible beamforming, and AI-enabled design are all critical [27–29]. Without loss of generality, in this chapter we consider a three-node communication system consisting of a transmitter, a receiver, and a RIS with a massive electromagnetic unit, which is illustrated in Fig. 1.

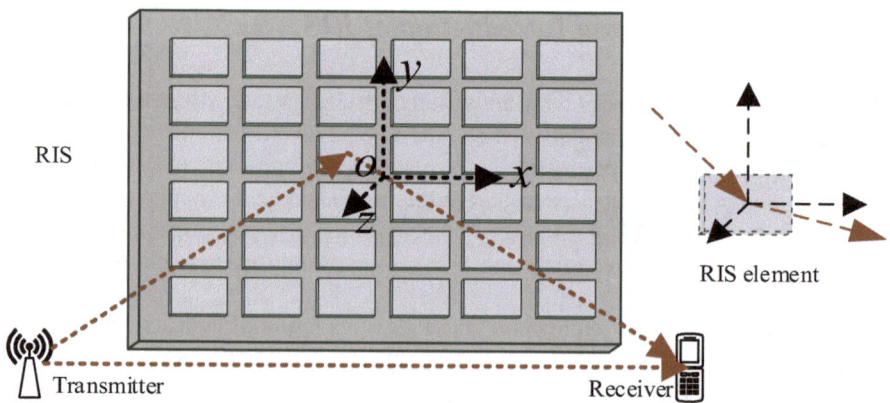

Fig. 1 RIS-assisted communication system

The received signal of the receiver y is

$$y = \sqrt{P}\left(h_{SR}^{T}\omega h_{RD} + g\right) \cdot s + n, \tag{1}$$

where P is the transmit power of the transmitter, s is the transmitted signal with unit energy, n is the zero-mean additive white Gaussian noise (AWGN) with variance N_0, ω is the diagonal matrix consisting of the reflection coefficients produced by each reflection element of the RIS, h_{SR} is a vector that contains the channel gains from the transmitter to each element of RIS and h_{RD} is the channel gains from each element of RIS to the receiver. Due to the quasi-passive nature of the RIS, the heat introduced by the radiation process noise can be ignored [30].

A possible optimization mechanism is to design the phase shift matrix of the RIS so that the reflected signals of the RIS are superimposed in phase at the user end to optimize the signal-to-noise ratio received by the user end, thereby increasing the transmission rate of the system [31–34]. Similarly, the model can be easily extended to multi-BS and multi-RIS scenarios. Thanks to the channel freedom provided by RIS and according to the needs of different scenarios, it is necessary to design the RIS control matrix in the future, which can further improve the transmission performance in various scenarios [35–37].

Generally, RIS can be divided into two categories, i.e., active RIS and passive RIS. The key differences between active and passive RIS are summarized as follows [38] (Table 2).

Table 2 Active versus passive RIS

Active RIS	Passive RIS
Reflects signal with amplification	Reflects signal without amplification
Can achieve high-capacity gain when the direct link is strong	Cannot achieve high-capacity gain when the direct link is strong
Introduces thermal noise	Introduces negligible thermal noise
Consumes power	Consumes negligible power
Has RF chains	No RF chains
Overcomes double fading	Suffers from double fading

3.2 Potential Application Scenarios of RIS-Assisted Communication System

3.2.1 Enhancement of Network Coverage

According to statistics, more than 80% of the traffic in the current 4G mobile network occurs in indoor scenarios. With the advent of the 5G era, various new services emerge in an endless stream. The industry predicts that more than 85% of mobile services will occur in indoor scenarios in the future.

Signal blockage by indoor walls and furniture results in more coverage holes and blind spots. RIS can be reconfigured for target users, which is beneficial for indoor coverage enhancement. As illustrated in Fig. 1, the signal experiences path loss and penetration loss due to refraction, reflection and diffusion, and the received signal at the target user is weak. However, the signal propagation can be reconstructed by RIS, so that the received signal to the target user can be enhanced.

Moreover, due to the large penetration loss, it has always been difficult to realize indoor coverage by using outdoor BSs. The RIS can be deployed on the glass surface of the building, which can effectively receive the signal transmitted by the BS and forward it to indoor users, thus resulting in an improvement of signal quality of indoor users.

3.2.2 Signal Transmission Robustness Enhancement

For high-frequency communication systems, i.e., millimeter wave (mmWave) communication and Terahertz communication systems, high beamforming gain is introduced to overcome the effect of path loss. However, the beams usually have narrow beamwidths and are easily blocked by buildings or trees on the street, which can have an impact on the robustness of the received signal. Through the ubiquitous deployment of RIS, more transmission paths can be brought, thereby enhancing the system transmission robustness. RIS generates more than one reflected beam aimed at different receiving panels of the receiver so that even if one beam is blocked, the other beam can still ensure reliable communication.

In addition, the RIS device can realize the manipulation of the parameters of the partial paths in the multipath channel. By manipulating the amplitude and phase of some paths, the multipath signals are superimposed in the forward direction at the receiving end, so as to suppress the multipath effect and improve the robustness of wireless data transmission. The phase of the RIS reflected beam is always the same as the phase of the direct path signal of the BS, so that the terminal received signal maintains the best quality.

3.2.3 Secure Transmission

The uncertainty and uncontrollability of the electromagnetic environment will bring about problems such as leakage of confidential information and complex interference. The existing external security mechanism results in mutual constraints between communication and security, which leads to low energy efficiency. Only by relying on properties of the electromagnetic environment to design endogenous security functions, we can solve future communication security problems. The design of wireless network safety faces passive adaptation to the electromagnetic environment, and the efficacy is close to the limit. Using RIS to build a unified integration of electromagnetic theory and information theory, a new paradigm of wireless security based on electromagnetic information theory can be constructed. RIS can realize real-time reconfiguration of electromagnetic environment and dynamic programmability of wireless channels, further enabling refined channel perception and channel customized generation, minimizing and eliminating the uncertainty and uncontrollability of electromagnetic environment, and improving wireless communication and security. Moreover, the 6G security vision driven by RIS should make use of the refined channel perception and customization capabilities of RIS to simultaneously approach the Shannon channel capacity and the one-time pad security capacity. As described in Fig. 2, the RIS device can be deployed near the eavesdropping user, and the signal reflected by the RIS device can be tuned to cancel the direct link signal between the BS and the eavesdropper received from the eavesdropping user, thereby effectively reducing information leakage.

3.2.4 Energy Harvesting and Transmission

Simultaneous wireless information and power transfer (SWIPT) is a new type of wireless communication. Different from traditional wireless communication that only transmits information, apart from transmitting traditional information wireless signals, SWIPT can transmit energy signals to wireless devices at the same time. Since the RIS reflector is a passive device, it can act as a quasi-relay, reflecting the signal mixed with information and energy. In this way, the quality of service for users

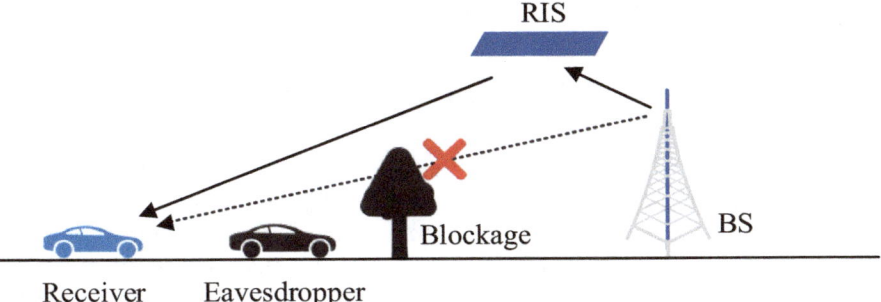

Fig. 2 Secure transmission in RIS-assisted networks

with weak transmission signals can be ensured. Combined with energy harvesting, it can effectively prolong the service life of users and improve energy efficiency.

In wireless power transfer (WPT), the power transmitter only provides energy to the user for charging, but does not transmit information. It is used in the fields of household electronic product charging, electric vehicles, and medical implants. Applying RIS to the WPT network can firstly provide stable and continuous energy for the passive reflective surface. Secondly, it can provide an energy supply for the target user by reflecting the RF energy on the reflective surface. This provides new solutions and feasible methods for green communications.

3.2.5 Massive Antenna Transceiver

RIS can be combined with massive MIMO antenna technology. A certain phase shift introduced by the surface can achieve focused beam emission in any direction. This type of antenna can overcome the problems of increased cost and power consumption caused by the increase in the number of transmitting and receiving antennas. It can reduce equipment costs while improving the spatial diversity gain of MIMO and the flexibility of focusing beams. In the future, beam scanning, polarization switching and beamforming have great application potential.

The traditional transmitter mainly manipulates the amplitude and phase of the carrier signal through baseband IQ data, while each electromagnetic unit of the RIS-based transmitter can be independently and flexibly controlled based on a specific control signal, such as using a PIN/varactor diode. The biggest advantage of the RIS-based transmitter is that it can achieve low power consumption and flexible control of the signal.

4 RIS-Assisted V2X Communication System

4.1 Challenge of RIS-Assisted V2X Communication

V2X communication is an integral part of intelligent transportation systems, allowing vehicles to stay in touch with their surroundings and remote entities, and providing them with connectivity services anytime and anywhere. Due to the complex propagation environment, the quality of the propagation link established between the vehicle and the roadside unit (RSU) is easily degraded. In the weak coverage area of RSU due to obstacles, the use of RIS technology can provide an indirect LoS transmission link for moving vehicles, which is expected to provide support for energy-efficient V2X communication. In [39], the authors propose an architectural solution to improve the reliability of autonomous vehicular networks via placing real-time software-controlled RIS units along roadsides. The optimum locations of the RIS units are determined by solving the formulated optimization problem for the beamforming and focusing operating modes resulting in significant SNR gains. The authors in [40] investigate the joint RIS optimization and resource allocation for RIS-aided V2X networks with social trust, where multiple V2I links share the spectrum with V2V links. In [4], the authors focus on the prospective transmission design of RIS-aided V2X communications. In particular, two V2X sidelink modes are enhanced by exploiting RISs and their variants, followed by a customized transmission frame structure that partitions the transmission efforts into different phases. In [41], the authors derive closed-form expressions for the ergodic capacity, symbol error probability, and energy efficiency for a partial RIS selection design. Besides, they also compare its performances to the partial relaying design while considering: decode-and-forward and amplify-and-forward fixed gain. A sophisticated RIS-aided OFDM transmission is conceived for high-mobility scenario in the face of doubly selectivity Ricean fading in [42], where the RIS control feedback link exploited a priori knowledge about LoS statistics. Besides, a low-complexity RIS configuration algorithm was proposed for maximizing the received signal power at the high-mobility user. To enhance the ubiquitous coverage and QoS guarantee of V2X communications, authors in [43] demonstrate the performance benefits of deploying RISs in vehicular networks. Based on a spectrum-sharing model, a sum-V2I-capacity-maximization problem is formulated, subject to practical QoS requirements and RIS phases shifts. In [44], the authors propose to use practical deployment strategies namely hybrid vehicular-visible light communication (V-VLC)/vehicular-radio frequency (V-RF) with relaying and RIS aided V-RF solutions to improve the communication range for urban V2V communication. They present stochastic geometry-based analytical framework to analyze the performance of proposed solutions in terms of outage probability, throughput and delay outage rate.

Due to the highly dynamic change in the network environment, the use of RIS to assist V2X communications still faces many challenges. Such as: (I) The change in the distance between the RIS and the vehicle will affect the real-time channel state, and the vehicle's residence time is different, which makes the RSU need to consider the mobility of the vehicle while optimizing the communication quality. (II) In the case of obtaining the same service resources, the service quality of vehicles with a longer detention time will be degraded. Therefore, in order to ensure the fairness of the service as much as possible, the communication resources in this network scenario need to be reasonably allocated. The state information is usually unpredictable, and the RSU needs to continuously exchange various state information with the vehicle. Under such conditions, making online decisions based on real-time state information is necessary and practicable. Besides, methods based on deep reinforcement learning can make decisions considering the current state of the environment and require less prior information about the environment. In the following sections, we will present several application examples of RIS-assisted V2X communication systems.

4.2   RIS-Assisted Vehicular Communication Network

With the development of transportation systems and mobile communication, the data consumption of V2X is increasing rapidly. As a key research scenario in the 5G mobile communication technology, V2X has attracted the attention of all walks of life including researchers, telecom operators, and communication companies around the world. RIS has great potential for V2X communication systems and can be widely used in actual V2X communication scenarios in the future.

In the existing V2X communication system, serious signal fading, blockage by obstacles, and limited coverage of BSs are problems that need to be solved urgently. The research so far has not made a reasonable optimization and treatment of the above problems. Therefore, by introducing RIS into the V2X system is a promising way to improve the coverage performance of the system and the outage performance of signal transmission.

As illustrated in Fig. 3, considering a RIS-assisted downlink vehicle communication network system. The diagram shows a two-way highway with two lanes in each direction. Suppose a vehicle V_1 carrying a single-antenna target user is located in the outer lane of a certain direction, and the wireless communication link between it and another vehicle V_3 is blocked by vehicle V_2 in the inner lane. In order to enhance the communication transmission between vehicles, the RIS is deployed on the other side of the road to enhance the signal quality.

4.3   RIS-Assisted Vehicular Edge Computing

The computing and processing demand of vehicles, smart devices and IoT sensors are growing at an unprecedented rate with the proliferation of a large number of emerging computing-intensive applications. To liberate vehicles with limited resources from heavy computing workloads and provide them with high-performance, low-latency computing services, mobile edge computing (MEC) promotes the use of cloud

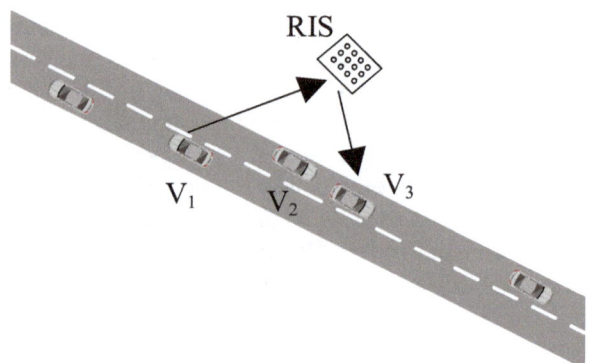

Fig. 3 RIS-assisted vehicular communication networks

computing capabilities at the edge of vehicular networks by integrating MEC servers on vehicles.

In order to further improve the uplink offloading performance of resource-limited vehicles, a low-cost and easy-to-deploy RIS can be deployed in the system, and the configuration of the reflection unit is adjusted to provide a more favorable wireless propagation environment for the system. Due to the discrete nature of reflection units, the optimal allocation of resources in RIS-assisted vehicular MEC systems is usually a non-convex problem with coupled variables. Using iterative algorithms to jointly optimize vehicular communication resources, energy resources, computing resources and RIS reflection coefficients can obtain the optimal solution to the optimization problem on the premise of ensuring convergence. However, such algorithms usually require high complexity and are difficult to achieve in the low-latency scenario to promote applications in the future network. Therefore, in order to solve the above problems, deep learning architecture is a promising method to providing an effective algorithm to achieve lightweight online optimization configuration through offline training.

As illustrated in Fig. 4, assuming that the vehicle $n \in \mathbb{N} = \{1, 2, \ldots, N\}$ in the system needs to perform task calculation within limited energy. Let s_n be the transmitted signal when the calculation is offloading, and assume that $|s_n| = 1$. Vehicles with computational offloading needs transmit signals in a multiplexed manner in a given time, so the received signal at the BS (or access point) with M antennas can be expressed as

$$y = \sum_{n=1}^{N} \sqrt{p_n} \left(h_{AP} \Phi h_{r,n} + h_{d,n} \right) s_n + n_0, \tag{2}$$

where $p_n = a_n P_n$ is the transmit power used to task offloading, $a_n \in [0, 1]$ is the power ratio used by the vehicle for offloading, where the remaining power is used for local computing, P_n is the total power consumption, $h_{d,n}$ is the direct channel gain between the vehicle and the BS, $h_{r,n}$ is the reflected channel between the vehicle

Fig. 4 RIS-assisted
vehicular edge computing

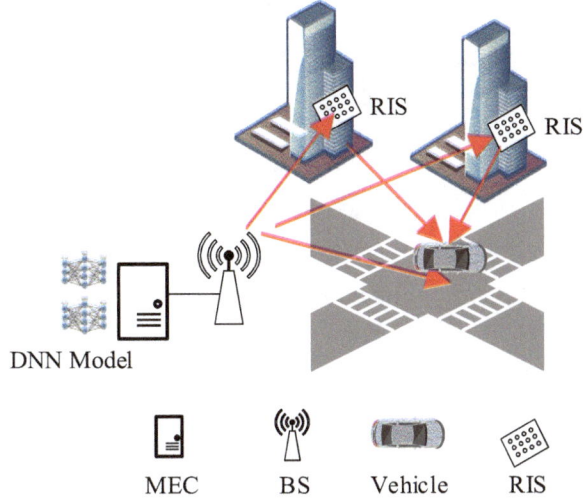

MEC BS Vehicle RIS

and the RIS, h_{AP} is the channel matrix between the RIS and the BS. Moreover, $\Phi = \text{diag}\{\phi\}$ is the RIS reflection coefficient matrix, where $\phi = [\phi_1, \phi_2, \ldots, \phi_K]^T$, $\phi_K = e^{j\theta_k}$, $k \in \{1, 2, \ldots, K\}$ is the phase shift of the k-th reflection unit. $n_0 \sim \mathcal{C}(0, \sigma^2 I_M)$ is the additive white Gaussian noise at the BS, and σ^2 is noise power. Assuming that linear beamforming is used at the BS to decode the vehicle's transmitted signal and defining ω_n is the vehicle's beamforming vector, then the estimated signal of the vehicle is

$$\overline{s_n} = \omega_n^H y = \omega_n^H \sum_{n=1}^{N} \sqrt{p_n}(h_{AP}\Phi h_{r,n} + h_{d,n})s_n + \omega_n^H n_0. \tag{3}$$

Then the uplink SINR of the offloading vehicle can be expressed as

$$\gamma_n(a, \omega_n, \phi) = \frac{a_n P_n |\omega_n^H (h_{AP}\Phi h_{r,n} + h_{d,n})|^2}{\sum_{i=1, i \neq n}^{N} a_i P_i |\omega_i^H (h_{AP}\Phi h_{r,i} + h_{d,i})|^2 + \sigma^2 |\omega_i^H|^2}, \tag{4}$$

where $a = [a_1, a_2, \ldots, a_N]$ represents the power distribution vector. Thus the capacity of the vehicle n for computing offloading can be expressed as $R_n^O = B_w T \log_2[1 + \gamma_n(a, \omega_n, \phi)]$. For the case where the vehicle performs local computing, it is assumed that dynamic voltage and frequency scaling techniques are used to improve computing energy efficiency by adaptively controlling the frequency of the central processor used for computing. A vehicle's capacity for local computing can be expressed as $R_n^L = \alpha_n \sqrt[3]{1 - a_n}$, where α_n represents a constant related to computing resources. In order to obtain the maximum capacity under the limited power P_n by jointly optimizing the power distribution coefficient a, the BS receiving

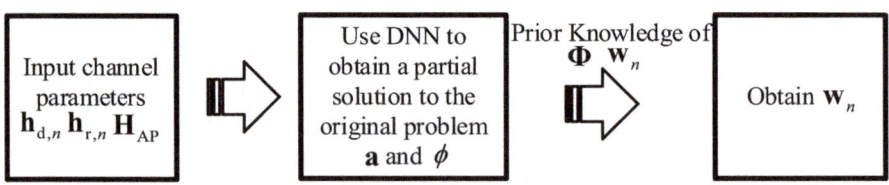

Fig. 5 DNN-based optimization algorithm architecture

beamforming vector ω_n and the RIS reflection coefficient ϕ, the optimization problem can be formulated as

$$\max_{a,\{\omega_n\},\phi} \sum_{n=1}^{N} R_n^O(a, \omega_n, \phi) + R_n^L(a_n), \tag{5a}$$

$$s.t. a_n \in [0, 1], \forall n = \{1, 2, \ldots, N\}, \tag{5b}$$

$$|\phi_n| \leq 1, \forall n = \{1, 2, \ldots, N\}, \tag{5c}$$

Due to the coupling relationship between a, ω_n and ϕ, the optimization problem is non-convex, and the common method is iterative optimization, but the time complexity is usually high. Deep neural networks (DNNs) are regarded as a general function approximator, and can effectively achieve online optimization, so the DNN framework can be used to solve the problem of excessively complex iterative optimization (Fig. 5).

By training and building a DNN offline using data samples generated from the iterative algorithm, the DNN is able to learn the algorithm's intrinsic mapping and output an efficient solution that mimics the algorithm. As a result, the desired optimization results can be predicted online using the trained neural network, greatly reducing computational complexity and running time. Specifically, a deep learning method is employed to obtain the original partial solutions, including RIS reflection coefficients ϕ and power distribution coefficients a. Then, using the relationship between the largest eigenvalue of beamforming vector ω_n and ϕ, we can directly obtain the receiving beamforming vector without the learning process. In this way, prior knowledge can be effectively combined with deep learning to obtain solutions of the proposed optimization problem, while reducing the cost of training, and testing DNNs.

5 Future Directions

One of the future research directions is to develop efficient baseband algorithms to support the widespread application of RIS technology in V2X systems. For different transmission scenarios, use flexible solutions to measure and model the RIS channel (including the modelling of the electromagnetic compatibility model of the RIS itself), such as modelling solutions based on a combination of statistical models and actual measurements, or based on electromagnetic calculations. Besides, explore interface solutions and key algorithms that support RIS-assisted V2X communication, such as AI-based channel information acquisition mechanism, beamforming solution design, distributed RIS cooperative transmission scheme, etc. Moreover, a system-level simulation platform and system verification platform can be developed. Based on this platform, comprehensive performance evaluation and actual measurement evaluation can be conducted in the new designed scenario. Last, the standardization process and industrialization process of RIS-assisted V2X communications should be promoted.

6 Conclusion

In this chapter, we introduce the literature review of RIS-assisted communications. Besides, modeling of the combination of V2X and RIS, including RIS-assisted vehicular communication network and RIS-assisted vehicular edge computing is presented. At last, we discuss the future development trend and challenge of RIS-assisted V2X communication systems.

References

1. J.B. Kenney, Dedicated short-range communications (DSRC) standards in the United States. Proc. IEEE **99**(7), 1162–1182 (2011)
2. X. Gu et al., Intelligent surface aided D2D-V2X system for low-latency and high-reliability communications. IEEE Trans. Veh. Technol. (2022)
3. K. Wu, P. Coquet, Q.J. Wang, P. Genevet, Modelling of free-form conformal metasurfaces. Nat. Commun. **9**, 3494 (2018)
4. Y. Chen, Y. Wang, J. Zhang, P. Zhang, L. Hanzo, Reconfigurable intelligent surface (RIS)-aided vehicular networks: their protocols, resource allocation, and performance. IEEE Veh. Technol. Mag. **17**(2), 26–36 (2022)
5. T.J. Cui, M.Q. Qi, X. Wan, J. Zhao, Q. Cheng, Coding metamaterials, digital metamaterials and programmable metamaterials. Light-Sci. Appl. **3**, 1–9 (2014)
6. L. Zhang, X. Chen, S. Liu, Q. Zhang, J. Zhao, J. Dai, G. Bai, X. Wan, Q. Cheng, G. Castaldi, V. Galdi, T.J. Cui, Space-time-coding digital metasurfaces. Nat. Commun. **9**, 1–11 (2018)
7. E. Basar, M. Di Renzo, J. de Rosny, M. Debbah, M.-S. Alouini, R. Zhang, Wireless communications through reconfigurable intelligent surfaces. arXiv:1906.09490 (2019)
8. M. Cui, G. Zhang, R. Zhang, Secure wireless communication via intelligent reflecting surface. IEEE Wirel. Commun. Lett. **8**(5), 1410–1414 (2019)

9. H. Guo, Y.-C. Liang, J. Chen, and E. G. Larsson, Weighted sum-rate optimization for intelligent reflecting surface enhanced wireless networks. arXiv:1905.07920 (2019)

10. X. Yu, D. Xu, R. Schober, MISO wireless communication systems via intelligent reflecting surfaces. arXiv:1904.12199 (2019)

11. H. Shen, W. Xu, S. Gong, Z. He, C. Zhao, Secrecy rate maximization for intelligent reflecting surface assisted multi-antenna communications. IEEE Commun. Lett. **23**(9), 1488–1492 (2019)

12. Q. Wu, R. Zhang, Towards smart and reconfigurable environment: intelligent reflecting surface aided wireless network. IEEE Commun. Mag. **58**(1), 106–112 (2020)

13. M. Di Renzo, M. Debbah, D.-T. Phan-Huy, A. Zappone, M.-S. Alouini, C. Yuen, V. Sciancalepore, G.C. Alexandropoulos, J. Hoydis, H. Gacanin, J. de Rosny, A. Bounceur, G. Lerosey, M. Fink, Smart radio environments empowered by reconfigurable AI meta-surfaces: an idea whose time has come. EURASIP J. Wirel. Commun. Netw. **2019** (2019)

14. M. Patzold, It's time to go big with 5G mobile radio. IEEE Veh. Technol. Mag. **13**(4), 4–10 (2018)

15. C. Liaskos, S. Nie, A. Tsioliaridou, A. Pitsillides, S. Ioannidis, I. Akyildiz, A new wireless communication paradigm through software-controlled metasurfaces. IEEE Commun. Mag. **56**(9), 162–169 (2018)

16. M. Di Renzo, J. Song, Reflection probability in wireless networks with metasurface-coated environmental objects: an approach based on random spatial processes. arXiv:1901.01046 (2019)

17. J. Zhao, X. Yang, J.Y. Dai, Q. Cheng, X. Li, N.H. Qi, J.C. Ke, G.D. Bai, S. Liu, S. Jin et al., Programmable time-domain digital-coding metasurface for non-linear harmonic manipulation and new wireless communication systems. Natl. Sci. Rev. **6**(2), 231–238 (2018)

18. E. Basar, Large intelligent surface-based index modulation: a new beyond MIMO paradigm for 6G. arXiv:1904.06704 (2019)

19. T. Jiang, Y. Shi, Over-the-air computation via intelligent reflecting surfaces. arXiv:1904.12475 (2019)

20. E. Bjornson, L. Sanguinetti, H. Wymeersch, J. Hoydis, T.L. Marzetta, Massive MIMO is a reality—What is next? Five promising research directions for antenna arrays. arXiv:1902.07678 (2019)

21. S. Hu, F. Rusek, O. Edfors, The potential of using large antenna arrays on intelligent surfaces, in *IEEE 85th Vehicular Technology Conference* (VTC Spring, 2017), pp. 1–6

22. S. Hu, F. Rusek, O. Edfors, Beyond massive MIMO: the potential of data transmission with large intelligent surfaces. IEEE Trans. Signal Process. **66**(10), 2746–2758 (2018)

23. S. Hu, F. Rusek, O. Edfors, Capacity degradation with modeling hardware impairment in large intelligent surface, in *IEEE Global Communications Conference (GLOBECOM)* (2018), pp. 1–6

24. M. Jung, W. Saad, Y. Jang, G. Kong, S. Choi, Performance analysis of large intelligence surfaces (LISs): asymptotic data rate and channel hardening effects. arXiv:1810.05667 (2018)

25. M. Jung, W. Saad, Y. Jang, G. Kong, S. Choi, Uplink data rate in large intelligent surfaces: asymptotic analysis under channel estimation errors, in *Uplink Data Rate in Large Intelligent Surfaces Asymptotic Analysis under Channel Estimation Errors* (2018). https://www.researchgate.net/publication/328827179

26. Q.-U.-A. Nadeem, A. Kammoun, A. Chaaban, M. Debbah, M.-S. Alouini, Asymptotic analysis of large intelligent surface assisted MIMO communication. arXiv:1903.08127 (2019)

27. C. Huang, A. Zappone, G.C. Alexandropoulos, M. Debbah, C. Yuen, Reconfigurable intelligent surfaces for energy efficiency in wireless communication. IEEE Trans. Wireless Commun. **18**(8), 4157–4170 (2019)

28. C. Huang, A. Zappone, G.C. Alexandropoulos, M. Debbah, C. Yuen, Large intelligent surfaces for energy efficiency in wireless communication. arXiv:1810.06934v1 (2018)

29. M. Fu, Y. Zhou, Y. Shi, Intelligent reflecting surface for downlink non-orthogonal multiple access networks. arXiv:1906.09434 (2019)

30. M. Jung, W. Saad, G. Kong, Performance analysis of large intelligent surfaces (LISs): uplink spectral efficiency and pilot training. arXiv:1904.00453 (2019)

31. B. Zheng, R. Zhang, Intelligent reflecting surface-enhanced OFDM: channel estimation and reflection optimization. IEEE Wirel. Commun. Lett. (2019)
32. Q.-U.-A. Nadeem, A. Kammoun, A. Chaaban, M. Debbah, M.-S. Alouini, Intelligent reflecting surface assisted multi-user MISO communication. arXiv:1906.02360 (2019)
33. A. Taha, M. Alrabeiah, A. Alkhateeb, Enabling large intelligent surfaces with compressive sensing and deep learning. arXiv:1904.10136 (2019)
34. Z. Gao, L. Dai, S. Han, C. I, Z. Wang, L. Hanzo, Compressive sensing techniques for next-generation wireless communications. IEEE Wirel. Commun. **25**(3), 144–153 (2018)
35. C. Zhang, P. Patras, H. Haddadi, Deep learning in mobile and wireless networking: a survey. IEEE Commun. Surv. Tutor. **21**(3), 2224–2287 (2019)
36. Z.-Q. He, X. Yuan, Cascaded channel estimation for large intelligent metasurface assisted massive MIMO. arXiv:1905.07948 (2019)
37. D. Mishra, H. Johansson, Channel estimation and low-complexity beamforming design for passive intelligent surface assisted MISO wireless energy transfer, in *IEEE International Conference on Acoustics, Speech and Signal Processing (ICASSP)* (2019), pp. 4659–4663
38. M. Munochiveyi, A.C. Pogaku, D.-T. Do, A.-T. Le, M. Voznak, N.D. Nguyen, Reconfigurable intelligent surface aided multi-user communications: state-of-the-art techniques and open issues. IEEE Access **9**, 118584–118605 (2021)
39. Y.U. Ozcan, O. Ozdemir, G.K. Kurt, Reconfigurable intelligent surfaces for the connectivity of autonomous vehicles. IEEE Trans. Veh. Technol. **70**(3), 2508–2513 (2021)
40. X. Gu, W. Duan, G. Zhang, Y. Ji, M. Wen, P.-H. Ho, Socially aware V2X networks with RIS: joint resource optimization. IEEE Trans. Veh. Technol. **71**(6), 6732–6737 (2022)
41. N. Mensi, D.B. Rawat, On the performance of partial RIS selection vs. partial relay selection for vehicular communications. IEEE Trans. Veh. Technol. **71**(9), 9475–9489 (2022)
42. C. Xu et al., Reconfigurable intelligent surface assisted multi-carrier wireless systems for doubly selective high-mobility Ricean channels. IEEE Trans. Veh. Technol. **71**(4), 4023–4041 (2022)
43. Y. Chen, Y. Wang, J. Zhang, M.D. Renzo, QoS-driven spectrum sharing for reconfigurable intelligent surfaces (RISs) aided vehicular networks. IEEE Trans. Wireless Commun. **20**(9), 5969–5985 (2021)
44. G. Singh, A. Srivastava, V.A. Bohara, Visible light and reconfigurable intelligent surfaces for beyond 5G V2X communication networks at road intersections. IEEE Trans. Veh. Technol. **71**(8), 8137–8151 (2022)

Chapter 6
Real-Time Object Detection for ITS Applications

Jianyong Song, Ziyi Hu, Yujie Song, Yu Wang, and Yue Cao

Abbreviations

5G	The 5th Generation Mobile Communication Technology
AP	Average Precision
BBox	Bounding Box
CAV	Connected-Automated Vehicle
CNN	Convolutional Neural Network
CSP	Cross Stage Partial
FLOPs	Floating Point Operations
FN	False Negative
FP	False Positive
FPN	Feature Pyramid Network
FPS	Frame Per Second
GIoU	Generalized Intersection over Union
ITS	Intelligent Transportation Systems
IoU	Intersection over Union
mAP	Mean of Average Precision

J. Song (✉) · Z. Hu · Y. Song · Y. Wang · Y. Cao
School of Cyber Science and Engineering, Wuhan University, Wuhan, China
e-mail: jianyongsong@whu.edu.cn

Z. Hu
e-mail: ZiyiHu@whu.edu.cn

Y. Song
e-mail: Y.Song@whu.edu.cn

Y. Wang
e-mail: wang.yu@whu.edu.cn

Y. Cao
e-mail: yue.cao@whu.edu.cn

NAS	Neural Architecture Search
NMS	Non-Maximum Suppression
PAN	Pyramid Attention Network
RCNN	Regions with Convolution Neural Networks
RNN	Recurrent Neural Networks
ROI	Region of Interest
TP	True Positive
TT100K	Tsinghua-Tencent 100 K
YOLO	You Only Look Once

1 Introduction

Intelligent Transportation Systems (ITS) can optimize vehicle flow and traffic efficiency, reduce fuel consumption and improve the safety of vehicle occupants. Through the deployment of sensors inside roadside infrastructure and vehicles, ITS is able to acquire road and vehicle information. With such help, ITS is able to provide hazardous environment detection, adaptive traffic management, and route planning capabilities.

The dramatic increase in the number of vehicles on the road over the last decade has created a number of problems, such as increased traffic accident and congestion etc. The development of new road construction may be a solution, but with high costs in maintenance and operation. Therefore, an efficient solution is to create a more efficient, safe, and reliable transportation system by using information, communication, and control technologies. ITS provides safety intelligence decisions for end users and drivers, by using data, images, and real-time video from roadside infrastructure and vehicle environments [1, 2], to achieve improved road conditions and enhance the safety of vehicle passengers.

The deep neural network based object detection algorithm is an advanced machine learning technique. It provides significant enhancements to road traffic safety and management, enabling ITS big data processing capability and intelligent decision-making ability. Object detection algorithms have a wide range of applications in ITS. It plays an instrumental role in scenarios such as traffic sign recognition, collision detection, pedestrian detection, and lane departure systems.

1.1 Motivation

With the development of network communication technology and the construction of network infrastructure, both the data transmission rates and transmission delays have been improved. The widespread use of Global Positioning System (GPS), roadside infrastructure, and Connected-Automated Vehicles (CAVs) generate massive amounts of traffic data. This makes it possible to train high-quality object detection models for ITS applications. Object detection models achieve high accuracy of prediction at the cost of high computation. However, embedded devices used in ITS are limited by cost and power consumption, making it difficult to achieve real-time object detection. Meanwhile, ITS requires real-time performance for traffic scenario task processing. Transferring data to servers in the cloud for processing and analysis has real-time and privacy issues. The high demand for computation and real-time performance with the low performance of embedded devices, is the major challenge in deploying existing object detection networks into ITS. Therefore, it is essential to significantly reduce the computational cost and model size of the object detection model without loss of detection accuracy, which makes it possible to accomplish real-time object detection tasks in ITS.

1.2   Main Contributions

Main contributions to this chapter are as follows:

(1) This chapter introduces common object detection algorithms, focusing on the YOLOv5 algorithm and network structure.
(2) This chapter describes the current challenges of object detection model deployment and summarizes common inference acceleration techniques for object detection models. The application of lightweight backbone networks and model compression is included in this chapter.
(3) This chapter details the knowledge distillation algorithm and describes how to train YOLOv5 models using the knowledge distillation algorithm to improve model accuracy.
(4) This chapter describes the training process of YOLOv5 model using the knowledge distillation algorithm.

1.3   Structural Organization

Section 2 introduces one-stage and two-stage object detection algorithms. Section 3 details common inference acceleration techniques for object detection models, including the application of lightweight backbone networks and model compression. Section 4 introduces the knowledge distillation algorithm, which is applied to train YOLOv5 models to improve their accuracy. Section 5 presents the experimental design and results of knowledge distillation for the YOLOv5 object detection model, and Sect. 6 summarizes the whole chapter.

2   Object Detection Algorithm

Object detection is a computer vision technique that can identify and localize targets in an image or video. Object detection can be divided into two parts: target localization and target classification. The former predicts the exact position of an object in an image and the latter defines which class it belongs to.

2.1   Overview of Object Detection Algorithm

Object detection algorithms based on deep neural network can be divided into two types: two-stage object detection algorithm and one-stage object detection algorithm. The two-stage object detection algorithm divides the task into two stages, stage one is to generate candidate regions using the backbone network, and stage two is to purify the candidate regions to regress the box [3]. R-CNN [4], improved versions Fast-RCNN [5] and Faster-RCNN [6] integrate the candidate region proposal, feature extraction, and bounding box regression modules of the object detection system, into a unified end-to-end learning framework, as the basic framework of the two-stage object detection algorithm. The two-stage detection algorithm has high detection accuracy but suffers from slow detection speed and high computational consumption.

In order to solve the problem of poor real-time performance of the two-stage object detection algorithm, the one-stage object detection algorithm applies the regression feature, to directly transform the bounding box localization problem into a regression problem, by forming a one-stage algorithm framework. In 2016, Joseph et al. proposed the first one-stage object detection network, named YOLO [7]. It divides the image into multiple regions, then predicts the bounding box and probability of each region. In the following YOLOv2 [8] and YOLOv3 [9], the detection accuracy and speed are further optimized, by improving the backbone network and introducing structures such as multi-scale residual blocks.

2.2   YOLOv5 Algorithm

YOLOv5 is the fifth generation of the YOLO family of inspection algorithms developed to date. Due to its lightweight feature, it has achieved excellent results in engineering applications.

YOLOv5 can be divided into five different versions according to the depth and width of the network, namely YOLOv5n, YOLOv5s, YOLOv5m, YOLOv5l, and YOLOv5x. The detection accuracy and model size of the five versions are improved in turn. Take YOLOv5s as an example. Its structure is shown in Fig. 1.

YOLOv5 is composed of three parts: Backbone, Neck and Head. The following paragraphs provide a more detailed description of each part.

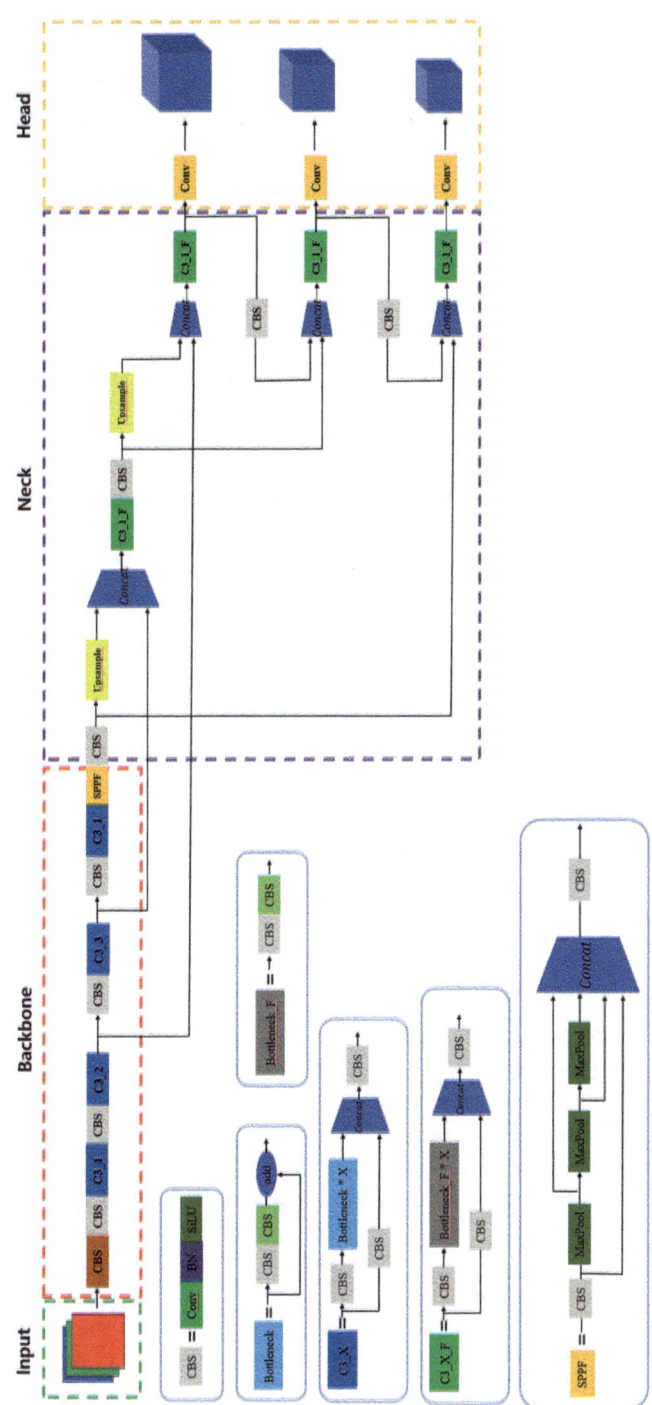

Fig. 1 YOLOv5 network structure

2.2.1   Backbone

The backbone consists of the Conv BatchNorm2d SiLu (CBS) module, the C3 module, and the Spatial Pyramid Pooling - Fast (SPPF) layer, which extracts the features of the input image.

CBS module: CBS module consists of Conv, BN and Silu activation function. The input image is first convolved to obtain different features of the input, then normalized by BN layer, and finally processed by Silu activation function to the next layer of convolution.

C3 module: YOLOv5 6.0 applies the C3 structure instead of the Bottleneck CSP structure in the backbone, whose structure is basically the same as the Cross Stage Partial (CSP) structure, except for a slight difference in the selection of correction units. The C3 structure makes the model more lightweight with low computational cost to improve the model learning capability.

SPPF: The SPPF structure can convert arbitrary size feature map into fixed size feature vector, to enhance feature map feature representation capability. In YOLOv5 6.0, the SPPF structure is applied instead of the SPP structure. SPPF only specifies one convolution kernel, and the output of each pooling becomes the input of the next pooling, which is faster than SPP.

2.2.2   Neck

The Neck network adopts a combination of FPN [10] and PAN [11], using a top-down FPN structure and a bottom-up PAN feature pyramid structure as the fusion part of the network. This network is mainly applied for stacking the extracted feature information and transmitting it to the output layer, and it enhances the network feature fusion capability.

2.2.3   Head

Head network is applied to complete the output of target detection results. YOLOv5 filters out the bounding box by Non-Maximum Suppression (NMS), and applies CIoU Loss [12] as the loss function to further improve the detection accuracy of the algorithm. The main part in Head is three detectors, when the input is 640×640, the feature maps on three scales are 80×80, 40×40, 20×20, and the grid-based anchor performs target detection on the feature maps at different scales.

3   Lightweight Method

The current trend in deep neural network applications is to deploy high performance models on edge devices, such as mobile and embedded devices, and to be able to run in real time. These devices are typically characterized by low memory resources, low processor performance, and limited power consumption. Because of the excessive memory and computational resource requirements, it is challenging to deploy the highest accuracy models on edge devices while satisfying real-time requirements. In order to accelerate model inference, two approaches are commonly employed. The first is to employ lightweight architectures. The second is to reduce the number of computations and parameters by various means, such as knowledge distillation, pruning, and quantification.

3.1   Lightweight Backbone

A well-designed feature extraction network can significantly improve the performance of object detection algorithms. In computer vision tasks, the network that performs feature extraction on images is referred as a backbone network. The research on lightweight backbone networks has focused on manually designed lightweight networks and automatic lightweight networks based on neural network structure search.

3.1.1   Manually Designed Lightweight Networks

Starting from Squeezenet [13], the design of Convolutional Neural Network (CNN) begin to focus on efficiency in resource-constrained scenarios. Researchers design lightweight and efficient convolutional neural network architectures, to effectively guarantee model accuracy while greatly reducing parameters. Some of the well-established lightweight networks include: MobileNet [14–16] series and EfficientNet [17] series from Google, ShuffleNet [18, 19] series from Megvii, GhostNet [20, 21] from Huawei, etc. SqueezeNet reduces parameters and computation by using low computational cost 1×1 convolutions and reducing the size of convolutional kernels. MobileNet extensively uses depthwise separable convolutions to distribute the computation of large-sized convolutional kernels to 1×1 convolutions, to significantly reduce computational consumption. MobileNet v2 further improves the model compression rate and running speed, by using resource efficient inverted residuals and linear bottlenecks. ShuffleNet applies the channel shuffle operation to compensate forinformation exchange between groups. This enables the network to apply pointwise group convolution, which not only reduces the main network computation but also increases the dimensionality of convolution. ShuffleNet v2 summarizes five design essentials for lightweight networks. Through channel splitting, the input features are split into two parts, resulting in a feature reuse effect similar to DenseNet [22].

3.1.2   NAS-Based Lightweight Networks

Although manually designed lightweight deep neural networks are widely available, manual methods require extensive experience in neural network design. They also require a large investment of manpower and time in designing the modules and hyperparameters of the overall network. With the development of reinforcement learning, lightweight methods based on neural architecture search [23] have emerged. Traditional lightweight networks such as MobileNet, MobileNet V2, ShuffleNet, and ShuffleNet V2 stack their respective basic units into a corresponding neural network structure. Hyperparameters applied in the stacked basic unit approach are in ordered series. Recurrent neural networks [24] are effective at learning ordered series. The main purpose of neural architecture search is to apply reinforcement learning methods to search for the most suitable hyperparameters in the basic units in the search space. Then it stacks the searched basic units to obtain a lightweight network. The one with more applications is MnasNet [25] proposed by Google, by considering model inference speed in the process of neural network structure search. This is able to search for a deep neural model with an optimal balance between model accuracy and model latency. While conventional lightweight networks apply FLOPs to evaluate model latency indirectly, MnasNet chooses to run the model directly on mobile devices to obtain realistic model latency parameters. It also proposes a decomposed hierarchical search space to obtain higher network performance in a smaller search space.

3.2   Model Compression

Model compression aims to reduce the number of neural network parameters. It can be achieved in a variety of ways. Common methods include pruning, quantization and knowledge distillation.

The main idea of network pruning is to remove the relatively insignificant weights from the weight matrix, and then fine-tune the network. Jaehong et al. [26] applied group sparsity to add sparse regularization to grouped features to prune off some columns of the weight matrix. Afterwards, exclusive sparsity was used to enhance the competitiveness of features among different weights in order to learn more effective filters. Both of them work together to achieve effective pruning results. Liu et al. [27] added a scaling factor to each channel, added sparsity regularization to these scaling factors, and then pruned the channels according to the size of the scaling factor. This achieves a slimming effect on the entire network.

The quantization adjusts the accuracy of the expression of the weights or activation values in the network structure, which can be classified as binary quantization, ternary quantization [28], and multi-valued quantization [29]. The idea of binary quantization is to represent the single-precision float in the weight matrix with two values, approximated by using symbolic functions or adding linearized symbolic functions. There is a challenge in distributed computing process, where each distributed server has a large amount of gradient information transmission process with the central server node, and this causes bandwidth limitation. Wen et al. [30] proposed TermGrad method, which takes the method of quantizing the gradient information to be transmitted into three values to accelerate the distributed computing effectively.

Knowledge distillation is the transfer of knowledge from one-or-more pre-trained teacher models to a single student model. Hinton [31] published "Distilling the Knowledge in a Neural Network" as the pioneering work in the field of knowledge distillation. He added a temperature constant T to control the degree of smoothing of the prediction probabilities, modified the softmax output of the original model as a soft target, and weighted it in combination with the true labels to compute the loss function for training small networks. Knowledge distillation will be introduced in detail in Sect. 4.

4   Knowledge Distillation Strategy

4.1   Knowledge Distillation

Knowledge Distilling is a method of model compression in which knowledge is transferred from one or more pre-trained teacher models to a single student model, with the aim that the lightweight student model will learn the "knowledge" of the teacher model and achieve the same performance as the teacher model. The narrow interpretation of "knowledge" here is that the output of the teacher model contains similarity that can be applied to transfer and assist the training of other models, which is called "dark knowledge" in the literature [31]. The broad interpretation is that the teacher model can be applied in all forms of knowledge, such as features, parameters, modules, etc. The "Distillation" refers to the process of amplifying the similarity of this knowledge and making it visible through certain methods, such as controlling parameters. The process of "distillation" refers to the process by which such knowledge can be amplified and made visible through certain methods, such as controlling parameters.

4.1.1   Distillation

In terms of knowledge distillation model evolution, "knowledge" is first realized at the output layer of the teacher model. Usually, the logits output by the model represents the category judgments. These judgments need to be passed through the softmax layer to obtain the predicted probability of the category, and then the loss of

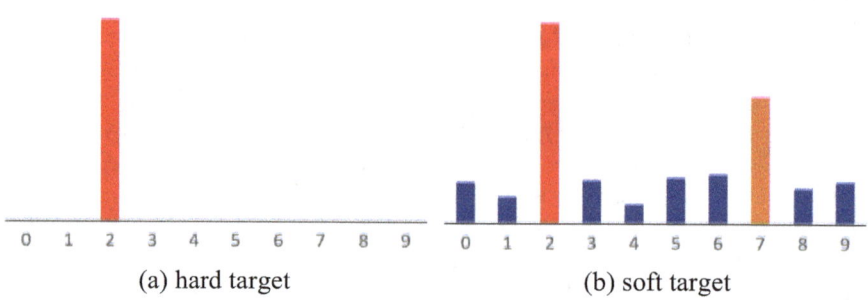

(a) hard target   (b) soft target

Fig. 2 Comparison between soft and hard

the model is calculated directly with Ground Truth. However, such category probabilities are relatively hard targets. The information about the similarity between
classes contained in the output probabilities is ignored, which largely affects the
generalization ability of the model. Hinton et al. [31] argue that similarity information in the output between classes is equally valuable. For example, in handwritten
character recognition, the number "2" and the number "7" are similar in appearance.
Therefore, if a hard target is applied to represent the number, we will obtain a true or
false representation as shown in Fig. 2a. However, the soft output after introducing
T gives the probability representation of each category as in Fig. 2b.

Therefore, the literature [31] introduces temperature T to soften softmax output
classification information. As shown in formula (1):

$$q_i = \frac{\exp\left(\frac{z_i}{T}\right)}{\sum_j \exp\left(\frac{z_j}{T}\right)} \tag{1}$$

where z_i, z_j are logits as input to the softmax layer, q_i corresponds to the output
probability of each category, and T represents the temperature to control the degree
of softness of the output probability. When $T = 1$, the formula degenerates into a
softmax function, and when the value of T is larger, a softer probability distribution
will be obtained. Experiments have proven that the softened category distribution is
easier to improve the learning effect of the student model as in Fig. 3.

4.1.2 Training Process

In the framework of knowledge distillation, the teacher model is generally a pre-
training model, and its model structure and design are usually relatively complex,
with good learning, representation and generalization capabilities. The output layer
of teacher model obtains soft labels in the manner of formula (1), and the soft labels
are used for student model training. The probability distribution of the teacher-student
network after softmax represents the distribution of information predicted by each
for the classification task. Therefore, the Cross Entropy loss is often used to measure

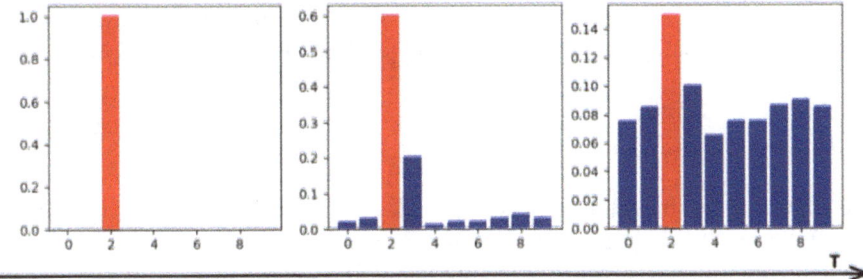

Fig. 3 As T increases, the output distribution of Softmax becomes more and more flat, and the information entropy will become larger and larger

the goal of knowledge transfer, to guide student learning and fit the probability distribution of the teacher's model. As shown in formulas (2) and (3):

$$p_i^T = \frac{\exp(\frac{z_i}{T})}{\sum_j^N \exp(\frac{z_j}{T})}, \quad q_i^T = \frac{\exp(\frac{z_i}{T})}{\sum_k^N \exp(\frac{z_k}{T})} \tag{2}$$

where p_i^T, q_i^T are the probability distributions of the teacher and student models after temperature T distillation.

$$L_{\text{soft}} = -\sum_j^N p_j^T \log(q_j^T), \quad L_{\text{hard}} = -\sum_j^N c_j \log(q_j^1) \tag{3}$$

where, c_j is the true label and q_j^1 denotes the case where $T = 1$. It is generally the case that the loss of the whole model consists of two elements: the loss L_{hard} between the predicted and real values of the student model and the loss L_{soft} between the teacher model and the student model, as described in the formula(4):

$$L = \alpha L_{\text{soft}} + (1 - a) L_{\text{hard}} \tag{4}$$

The balance factor α is applied to balance the contribution to the final loss. The student model needs to learn the generalization ability from the teacher model, meanwhile to correct the teacher model from the hard target.

4.2 Knowledge Distillation Based on YOLOv5

We perform knowledge distillation on the YOLOv5s model to improve accuracy. The design of the distillation algorithm is borrowed from the distillation algorithm used by Rakesh et al. [32] on the Tiny-Yolo model, which is an effective approach to introduce distillation loss into a single-stage object detection algorithm.

4.2.1 Objectness Scaled Distillation

The current distillation algorithm is mainly designed for the two-stage object detection algorithm, since introducing distillation loss directly in the one-stage object detection algorithm will cause some problems. In the case of the two-stage object detection algorithm, a very limited number of Regions Of Interest (ROIs) are sent to the detection network. Most of them are BBoxes that contain objects, therefore adding distillation losses will not be a significant issue. However, for one-stage algorithms like Yolo, a large number of BBoxes will be generated, and most of them are background. If a large number of background regions are passed to the student network, the network will predict the background region coordinates and categories. In this case, convergence of the model will be challenging. Thus, the authors limit the distillation loss by using the objectness of the YOLO network output. In other words, only the BBox with high objectness contributes to the loss of the student network.

4.2.2 Distillation Loss

The loss function of YOLO algorithm contains 3 parts:

- Objectness Loss: To determine whether there are objects in the predicted bounding box, this loss function helps the model to distinguish whether it is a background region or not.
- Classification Loss: It determines which category the object in the prediction box belongs to.
- Box Regression Loss: It is only applied when the prediction box contains objects.

$$L_{Yolo} = f_{obj}\left(o_i^{gt}, \hat{o}_i\right) + f_{cl}\left(p_i^{gt}, \hat{p}_i\right) + f_{bb}\left(b_i^{gt}, \hat{b}_i\right) \tag{5}$$

The loss function of the YOLO algorithm is shown in Formula (5), where $\hat{o}_i$, $\hat{p}_i$, $\hat{b}_i$ are the objectness, class probability and bounding box coordinates of the student network and o_i^{gt}, p_i^{gt}, b_i^{gt} are the values derived from the ground truth.

$$f_{obj}^{Comb}\left(o_i^{gt}, \hat{o}_i, o_i^T\right) = f_{obj}\left(o_i^{gt}, \hat{o}_i\right) + \lambda_D \cdot f_{obj}\left(o_i^T, \hat{o}_i\right) \tag{6}$$

The objectness loss does not apply objectness scaling. The formula mainly consists of two parts, the first part is detection loss, which is consistent with the objectness loss in the YOLO algorithm. The second part is distillation loss. A major difference between this part and the first part is that the input is not ground truth, but rather the objectness output of the teacher network. λ_D is used to balance the two loss.

$$f_{cl}^{Comb}\left(p_i^{gt}, \hat{p}_i, p_i^T, \hat{o}_i^T\right) = f_{cl}\left(p_i^{gt}, \hat{p}_i\right) + \hat{o}_i^T \cdot \lambda_D \cdot f_{cl}\left(p_i^T, \hat{p}_i\right) \qquad (7)$$

$$f_{bb}^{Comb}\left(b_i^{gt}, \hat{b}_i, b_i^T, \hat{o}_i^T\right) = f_{bb}\left(b_i^{gt}, \hat{b}_i\right) + \hat{o}_i^T \cdot \lambda_D \cdot f_{bb}\left(b_i^T, \hat{b}_i\right) \qquad (8)$$

Classification loss and regression loss apply objectness scaling. The distillation loss part of formulae includes $\hat{o}_i^T$, which is the objectness output of the teacher network, indicating the probability that each BBox contains an object. Whenever a BBox is in the background, its value will be negligible. Therefore, the distillation loss has a minimal effect on the loss calculation, which prevents the student network from incorrectly learning background information.

The YOLOv5 loss function using knowledge distillation is as follows:

$$L_{\text{final}} = f_{bb}^{Comb}\left(b_i^{gt}, \hat{b}_i, b_i^T, \hat{o}_i^T\right) + f_{cl}^{Comb}\left(p_i^{gt}, \hat{p}_i, p_i^T, \hat{o}_i^T\right) + f_{obj}^{Comb}\left(o_i^{gt}, \hat{o}_i, o_i^T\right)$$

$$(9)$$

5 Experiments and Discussions

5.1 Datasets

The dataset used in this chapter is Tsinghua–Tencent 100 K [33] (TT100K) jointly produced by Tsinghua University and Tencent (Fig. 4), which includes about 10,000 images and 30,000 traffic signs, all of which are 2048 × 2048 high-resolution images in real scenes.

 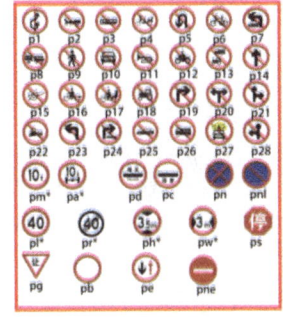 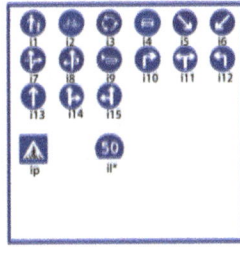

Fig. 4 Chinese traffic-sign classes

5.1.1 Dataset Processing

The number of training samples of different categories in TT100K varies greatly (Fig. 5), which will lead to the neural network being well trained for large sample categories but poorly trained for small sample categories. For this case, we run a script to filter the dataset and separate out the ones with more than 100 samples of sign information to make a traffic sign set containing 45 classes. To increase the number of small sample categories, we use the Copy-Paste algorithm [34] to enhance the training set offline and expand the number of traffic sign instances with less than 1000 samples to 1000. Then the training and validation sets are divided into 8:2 ratio.

5.2 Experiment Environment

Experimental environment of this chapter is based on AMD EPYC(TM) 7451, 128G RAM, GPU NVIDIA GeForce RTX 3090, 24G video memory, ubuntu 20.04 LTS, 64-bit operating system, PyTorch deep learning framework with Python programming language and GPU acceleration software CUDA11.1 and cuDNN8.0.5.

5.3 Evaluation Metrics

The evaluation metrics are mainly divided into two aspects: detection accuracy and detection speed.

The detection accuracy is used to evaluate the ability of the algorithm to accurately locate and classify the target, and the index to measure the accuracy is mean of Average Precision (mAP), and IoU is taken as 0.5, which is defined as follows:

$$mAP = \frac{1}{n} \sum_{j=1}^{n} AP(j) \tag{10}$$

where n is the total number of categories, the dataset we apply contains 45 categories of traffic signs, so $n = 45$. AP is the Average Precision of a category, defined as the average value of the precision rate under different recall rates, calculated as

$$AP = \int_{0}^{1} P(R)dR \tag{11}$$

$$P = \frac{TP}{TP+FP} \tag{12}$$

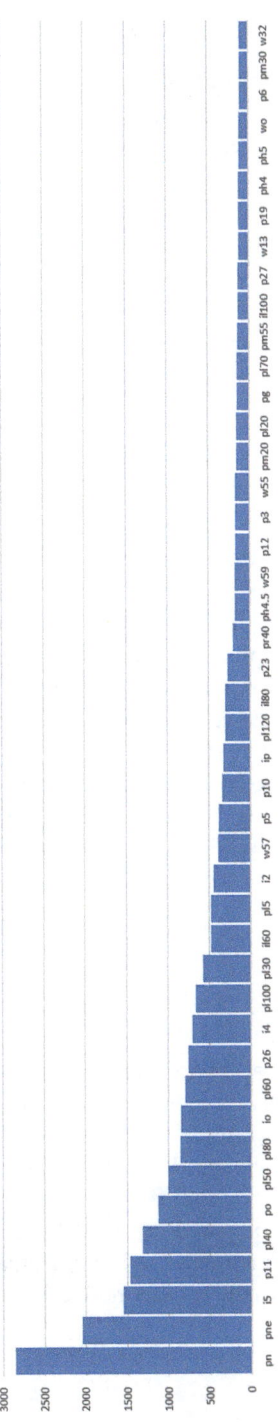

Fig. 5 Number of instances in each class, for classes with more than 100 instances

$$R = \frac{TP}{TP+FN} \tag{13}$$

where P is Precision, the probability that all positive samples are correctly predicted; R is Recall, the probability that all positive samples are detected, TP denotes True Positive, TN denotes True Negative, FP denotes False Positive, and FN denotes False Negative.

The detection speed is measured by FPS, which indicates the number of images that can be detected per second. The formula is as follows, where s is the time required to detect an image.

$$FPS = \frac{1}{s} \tag{14}$$

5.4 Model Training

As a means of verifying the effectiveness of the knowledge distillation strategy used in this chapter, YOLOv5m is used as the teacher network, and YOLOv5s is used as the student network. In this chapter, the network after YOLOv5s distillation was named YOLOv5s-distilled. In the process of model training, anchor is automatically calculated. The maximum iteration number is 100 epochs. An SGD optimizer is applied, and the initial learning rate is set at 0.01, the weight attenuation rate is set at 0.0005, and the batch size is set at 16.

As shown in Fig. 6, the average accuracy curve is calculated during algorithm training using the verification set. It can be seen from it that YOLOv5m algorithm converges to AP (IOU = 0.5) = 0.79 on the verification set, while the improved algorithm converges to AP (IOU = 0.5) = 0.82 after about 100 rounds of iteration.

5.5 Experimental Results

In order to verify the performance of the knowledge distillation strategy, comparative experiments with three YOLO series models are carried out on the test set. The experimental results are shown in Table 1.

After the introduction of knowledge distillation, the detection accuracy of the model has significantly increased in comparison to the previous one. YOLOv5s-distilled improved by 2.06% compared to mAP@0.5 before distillation, and mAP@0.5 decreased by 2.37%, parameter volume decreased by 66.03% and FPS improved by 39.68%, compared to the teacher network YOLOv5m.

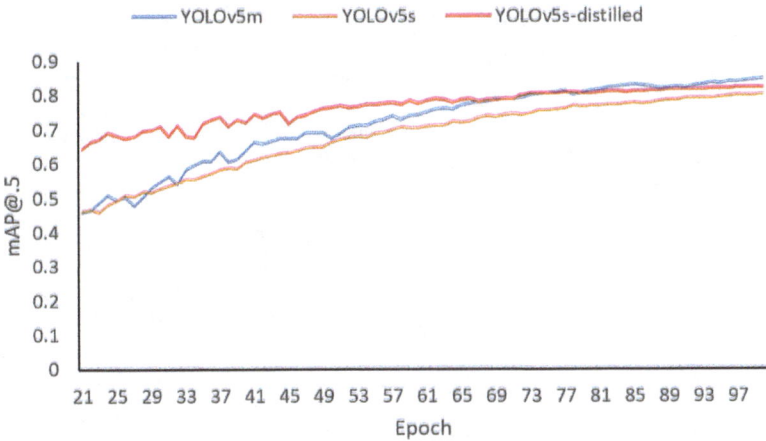

Fig. 6 Average precision accuracy variation curve

Table 1 Comparison of performance indicators of different detection algorithms

Model	Params (M)	FLOPs (B)	P%	R%	mAP@0.5%	FPS (RTX3090)
YOLOv5m	21.0	48.8	83.67	79.15	84.28	63
YOLOv5s	7.2	16.5	81.10	75.72	79.85	88
YOLOv5s-distilled	7.2	16.5	82.69	77.65	81.91	88

Figure 7 showed the comparison of partial detection effects of YOLOv5s before knowledge distillation and YOLOv5s after knowledge distillation. Through comparison, it can be found that the YOLOv5s-distilled proposed in this chapter has been improved in traffic sign detection, especially small target detection. It can correctly detect traffic signs, which are missed or incorrectly detected by the original YOLOv5s algorithm.

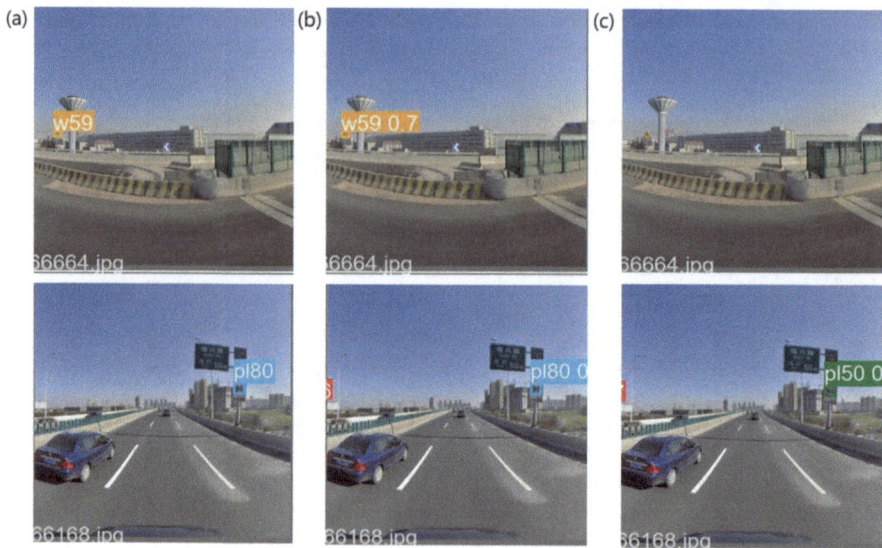

Fig. 7 Comparison of algorithm detection results. **a** ground truth; **b** YOLOv5s-distilled; **c** YOLOv5s

6 Conclusion

This chapter describes the application of object detection models in ITS, together with the requirements and challenges of object detection model deployment. Based on these requirements and challenges, this chapter introduces representative object detection and lightweight methods. Finally, this chapter trains the YOLOv5 model applying distillation loss for real-time traffic sign detection.

References

1. S. Wan, Z. Gu, Q. Ni, Cognitive computing and wireless communications on the edge for healthcare service robots, Comput. Commun. (2019)
2. M. Chen, V.C. Leung, S. Mao, Y. Yuan, Directional geographical routing for real time video communications in wireless sensor networks. Comput. Commun. **30**(17), 3368–3383 (2007)
3. W. Xu, L. Zou, Z. Fu et al., Two-stage 3D object detection guided by position encoding. Neurocomputing **501**, 811–821 (2022)
4. R. Girshick, J. Donahue, T. Darrell, et al., Rich feature hierarchies for accurate object detection and semantic segmentation, in *Proceedings of the IEEE Conference on Computer Vision and Pattern Recognition*, vol. 1 (2014). Pp. 580–587
5. R. Girshick, Fast r-cnn, in *Proceedings of The IEEE International Conference on Computer Vision*, vol. 1 (2015), pp. 1440–1448
6. S. Ren, K. He, R. Girshick et al., Faster R-CNN: towards real-time object detection with region proposal networks. IEEE Trans. Pattern Anal. Mach. Intell. **39**(6), 1137–1149 (2016)

7. J. Redmon S Divvala R Girshick et al 2016 You only look once: Unified, real-time object detection, in *Proceedings of the IEEE Conference on Computer Vision and Pattern Recognition*, vol. 1(2016), pp. 779–788
8. J. Redmon, A. Farhadi, YOLO9000: better, faster, stronger, in *Proceedings of the IEEE Conference on Computer Vision and Pattern Recognition*, vol. 1 (2017). pp. 7263–7271
9. J. Redmon, A. Farhadi, Yolov3: An incremental improvement. arXiv preprint arXiv:180402767 (2018)
10. T.Y. Lin, P. Dollár, R. Girshick, et al, Feature pyramid networks for object detection, in *Proceedings of the IEEE Conference on Computer Vision and Pattern Recognition* (2017). pp. 2117–2125
11. Li H, Xiong P, An J, et al. Pyramid attention network for semantic segmentation[J]. arXiv preprint arXiv:1805.10180, 2018.
12. Z. Zheng, P. Wang, W. Liu, et al., Distance-IoU loss: faster and better learning for bounding box regression, in Proceedings of the AAAI Conference on Artificial Intelligence, vol. 34, no. 7 (2020). pp. 12993–13000
13. F.N. Iandola, S. Han, M.W. Moskewicz, et al., SqueezeNet: AlexNet-level accuracy with 50x fewer parameters and < 0.5 MB model size. arXiv preprint arXiv:160207360 (2016)
14. A.G. Howard, M. Zhu, B. Chen, et al., Mobilenets: efficient convolutional neural networks for mobile vision applications. arXiv preprint arXiv:170404861 (2017)
15. M. Sandler, A. Howard, M. Zhu, et al., Mobilenetv2: inverted residuals and linear bottlenecks, in *Proceedings of the IEEE Conference on Computer Vision and Pattern Recognition* (2018). pp. 4510–4520
16. A. Howard, M. Sandler, G. Chu, et al., Searching for mobilenetv3, in *Proceedings of the IEEE/CVF International Conference on Computer Vision* (2019), pp. 1314–1324
17. M. Tan, Q. Le, Efficientnet: rethinking model scaling for convolutional neural networks, in *International Conference on Machine Learning* (PMLR, 2019). pp. 6105–6114
18. X. Zhang, X. Zhou, M. Lin, et al. Shufflenet: an extremely efficient convolutional neural network for mobile devices, in *Proceedings of the IEEE Conference on Computer Vision and Pattern Recognition* (2018). pp. 6848–6856
19. N. Ma, X. Zhang, H.-T. Zheng, et al. Shufflenet v2: Practical guidelines for efficient cnn architecture design, in *Proceedings of the European Conference on Computer Vision (ECCV)* (2018). pp. 116–131
20. K. Han, Y. Wang, Q. Tian, et al., Ghostnet: More features from cheap operations, in *Proceedings of the IEEE/CVF Conference on Computer Vision and Pattern Recognition* (2020). pp. 1580–1589
21. K. Han, Y. Wang, C. Xu et al., GhostNets on heterogeneous devices via cheap operations. Int. J. Comput. Vision **130**(4), 1050–1069 (2022)
22. G. Huang, Z. Liu, L. Van Der Maaten, et al., Densely connected convolutional networks, in *Proceedings of the IEEE Conference on Computer Vision and Pattern Recognition* (2017), pp. 4700–4708
23. B. Zoph, Q.V. Le, Neural architecture search with reinforcement learning. arXiv preprint arXiv: 1611.01578 (2016)
24. W. Zaremba, I. Sutskever, O. Vinyals, Recurrent neural network regularization. arXiv preprint arXiv:1409.2329 (2014)
25. M. Tan, B. Chen, R. Pang, et al., Mnasnet: platform-aware neural architecture search for mobile, in *Proceedings of the IEEE/CVF Conference on Computer Vision and Pattern Recognition* (2019), pp. 2820–2828
26. J. Yoon, S.J. Hwang, Combined group and exclusive sparsity for deep neural networks, in *International Conference on Machine Learning* (PMLR, 2017), pp. 3958–3966
27. Z. Liu, J. Li, Z. Shen, et al., Learning efficient convolutional networks through network slimming, in *Proceedings of the IEEE International Conference on Computer Vision* (2017), pp. 2736–2744
28. Y. Aratani, Y.Y. Jye, A. Suzuki, et al. Multi-valued quantization neural networks toward hardware implementation, in *IEEE International Conference on Artificial Life And Robotics (ICAROB)* (2017), p. 58

29. Y. Aratani, Y.Y. Jye, A. Suzuki et al., Multi-Valued Quantization Neural Networks toward Hardware Implementation. Proc Int Conf Artif Life Robot **22**, 132–135 (2017)
30. W. Wen, C. Xu, F. Yan, et al., Terngrad: Ternary gradients to reduce communication in distributed deep learning. Adv. Neural Inf. Process. Syst. **30** (2017)
31. G. Hinton, O. Vinyals, J. Dean, Distilling the knowledge in a neural network. arXiv preprint arXiv:1503.02531 (2015) **2**(7)
32. R. Mehta, C. Ozturk, Object detection at 200 frames per second, in *Proceedings of the European Conference on Computer Vision (ECCV) Workshops* (2018)
33. Z. Zhu, D. Liang, S. Zhang, et al., Traffic-sign detection and classification in the wild, in *Proceedings of The IEEE Conference on Computer Vision And Pattern Recognition* (2016), pp. 2110–2118
34. G. Ghiasi, Y. Cui, A. Srinivas, et al., Simple copy-paste is a strong data augmentation method for instance segmentation, in *Proceedings of the IEEE/CVF Conference on Computer Vision and Pattern Recognition* (2021), pp. 2918–2928

Chapter 7
3D Scene Perception for Autonomous Driving

Shuai Li, Huasong Zhou, Yanbo Gao, Xun Cai, Hui Yuan, and Wei Zhang

1 Introduction

With the booming of autonomous driving, robotics navigation, etc., 3D vision is attracting increasing interests. It provides a 3D realistic description of the scene with depth information in addition to the texture information. Therefore, it is very useful and important in depth required tasks such as 3D object detection in driving, subsequent 3D scene reconstruction and understanding [1]. As a basic and fundamental part of 3D vision, depth acquisition plays an important role in 3D vision. Depth acquisition in 3D vision is mainly performed in two ways: directly captured through active sensors (such as LiDAR/Radar probe and structured light) and indirectly estimated through color cameras [2]. Active sensors are usually of relatively high cost, and difficult to independently generate high-quality dense depth maps (usually produce sparse depth maps). Therefore, it is necessary to complete the depth map with the assistance of the corresponding color images.

S. Li · H. Zhou · H. Yuan · W. Zhang
School of Control Science and Engineering, Shandong University, Jinan, China
e-mail: shuaili@sdu.edu.cn

H. Zhou
e-mail: huasongzhou@mail.sdu.edu.cn

H. Yuan
e-mail: huiyuan@sdu.edu.cn

W. Zhang
e-mail: davidzhang@sdu.edu.cn

Y. Gao (✉) · X. Cai
School of Software, Shandong University, Jinan, China
e-mail: ybgao@sdu.edu.cn

X. Cai
e-mail: caixunzh@sdu.edu.cn

© The Author(s), under exclusive license to Springer Nature Singapore Pte Ltd. 2023   125
Y. Zhu et al. (eds.), *Communication, Computation and Perception Technologies for Internet of Vehicles*, https://doi.org/10.1007/978-981-99-5439-1_7

Compared to the active depth sensors that can directly measure the depth of the scene, vision based depth estimation requires further processing on the images to produce depth maps. Accuracy of the estimated depth maps depends on the input color images (single image, stereo/multi-view images, video) and the depth estimation algorithms. The cost of color cameras is much cheaper than the depth sensors and the quality of the estimated depth maps are acceptable and still improving, thus vision based depth estimation is very popular in both academia and industry communities. Therefore, this chapter focuses more on the vision based depth estimation approach. Vision based depth estimation approach can be roughly classified into three categories: single image based, stereo matching based and structure-from-motion based. Single image based method is to estimate the depth map using depth cues in the image such as parallel lines and edges. Stereo matching based method is using two images from the left and right views to imitate the human vision system and calculate the depth through 3D geometry. Its key lies in finding the corresponding matched pixels in two views. Structure from motion is to restore the three-dimensional structure from the motions of objects in a video or a collection of images taken at different times, where the moved cameras in different time can be regarded as different views.

With the great success of deep neural networks, deep learning based depth estimation has been actively studied. Similar as the conventional methods, based on the different types of inputs and different estimation methodology, deep learning-based depth estimation can also be roughly divided into monocular depth estimation and stereo depth estimation. Monocular depth estimation infers the dense depth map of the corresponding scene from the appearance features of the color image. Its key is to extract the depth-related information from the appearance and effectively decode it. Stereo depth estimation learns the scene structure information by matching the left-view and right-view patches to obtain the disparity value, which is then transformed to the depth values based on the camera intrinsic and extrinsic parameters. Its key is finding the pixel matching relationship between the stereo images and constructing the matching cost volume effectively. Since deep learning requires a large number of training data and ground-truth depth data is usually hard to obtain, each approach can be further classified with supervised and self-supervised methods according to the use of ground-truth depth for training. The supervised learning methods directly train the network by minimizing the error between the ground-truth from dataset and the predicted depth map. The self-supervised methods use view synthesis based on the estimated depth to reconstruct an image of an existing view or temporal frame, which can then be supervised by minimizing the reconstruction error. Since the color images can be obtained much easier than the depth maps, the self-supervised methods greatly reduced the training cost compared to the supervised methods. However, the reconstruction error of the synthesized view may deviate from the depth induced error due to the noise in different views or moving objects in different temporal frames, lowering the performance.

1.1   Motivation

Nowadays, 3D vision is becoming more and more popular and 3D vision related/ supported applications have significantly increased. 3D vision can not only provide the color information and geometric relationships on the two-dimensional plane, but also contains unique depth information, which is vital to certain applications. For example, autonomous driving [3] needs the depth information between the vehicle and the surrounding obstacles to plan the driving path and respond to various unexpected situations. For 3D scene reconstruction [4], the depth information can provide accurate description of the position relationship, and its acquisition is the footstone of the subsequent processing. Moreover, with the improvement of depth acquisition and its accuracy, depth information can also better assist other computer vision tasks by providing three-dimensional position relationships of scene objects [5, 6]. In summary, depth estimation in scene perception has large unexplored potential values in engineering applications and academic research, and is worth further investigation.

1.2   Main Contributions

Main contributions of this chapter can be summarized as follows:

(1) This chapter reviews and summarizes LiDAR sensor-based and vision-based scene perception approaches, especially focusing on vision based depth estimation using deep learning.
(2) This chapter presents the general vision-based scene perception framework and their use with different types of inputs, including single image, stereo pairs and monocular video sequence. Methodologies of different approaches are explained to unveil their working mechanism and make room for further improvement.
(3) This chapter discusses the future development on depth estimation in 3D vision, especially on temporal information exploration and fusion with cheaper Radar.

1.3   Structural Organization

Section 2 introduces the background and existing LiDAR sensor-based and vision-based scene perception methods. Section 3 elaborates vision-based scene perception in detail, including its framework and working mechanism. Section 4 presents the related datasets, especially large datasets used for the training of deep networks, and quality measure for depth estimation. Section 5 illustrates some future development directions of depth estimation. Section 6 summarizes the whole chapter.

2 Background

2.1 Existing LiDAR Sensor Based Scene Perception Methods

With the rapid development of autonomous driving, scene perception, as an insepa-
rable component of autonomous driving system, is attracting more and more attention. Scene sensing or perception aims to provide accurate environmental information
including the texture contents of the scene and the object (e.g., person and vehicle)
3D layout. It is important for tasks such as person, vehicle and traffic sign recognition, and 3D object localization. Currently, multiple sensors, including camera and
LiDAR (Light Detection and Ranging) sensor, are used for scene sensing in the
autonomous driving system.

LiDAR sensor projects laser signal to the target and measures the time for reflected
signal reflected to the receiver, determining the range of the target [7]. The data
collected by LiDAR sensor are called point cloud where each point contains the
three-dimensional coordinates of a 3D point on the object. The point clouds are
usually sparse due to the hardware capturing constraint [8]. Since LiDAR directly
provides the depth information of the scene, research on LiDAR for scene perception mainly focus on the depth completion, to complete the sparse point cloud to a
dense one [9], and signal enhancement such as filtering noises [10]. Under weather
conditions with poor visibility such as rain, snow and haze, the LiDAR data can also
be noisy. In, a deep learning algorithm was proposed for adverse weather denoising
on adjacent LiDAR. In [11], snow particles are removed in the signal while retaining
important environmental features. The following process such as 3D object detection
and recognition can also be performed on the LiDAR point cloud data as in [12, 13].

2.2 Existing Vision Based Scene Perception Methods

2.2.1 Conventional Methods

Conventional vision based scene perception methods mostly investigate the stereo
matching problem and the structure-from-motion problem. Stereo matching methods
focus on finding the corresponding relationship between two views, and can be
divided into global stereo matching, semi-global stereo matching and local stereo
matching. Generally, global matching methods achieve higher accuracy and local
matching methods perform more efficiently, while the semi-global matching balances
between speed and accuracy.

The global stereo matching algorithm [14] establishes a global energy function,
and performs optimization over the whole image. The energy, i.e., the matching cost
between two views, can be calculated by some handcrafted features such as SIFT
(scale-invariant feature transform) and SAD (sum of absolute difference) [15, 16].
Then the global optimization is performed with the dynamic programming method

[17], Markov random field [18], the graph cutting or belief propagation [19]. In contrast, the local stereo matching algorithm directly optimizes the block matching with some matching measures such as the SAD [20], or Rank transform and Census transform [21]. Another aspect of local stereo matching algorithm is to select an appropriate matching window size, which affects the matching accuracy. Veksler et al. [22] proposed to determine the size and shape of the matching region according to the feature points. Yoon et al. [23] proposed an earlier Adaptive Support Weight method to determine the shape and size of the matching window according to the difference of color and spatial location. Combining the above local and global algorithms, Hirschmuller et al. [24] proposed the classical semi-global optimization matching algorithm (SGM), which replaces the matching cost calculation in two-dimensional space with multiple one-dimensional calculations and aggregates one-dimensional cost from multiple directions.

Structure from Motion (SfM) is a problem of estimating camera parameters through multiple images or a video of a 3D scene and recovering the 3D structure information of the scene. The projection matrix of the camera is motion, also known as the pose, and the coordinates of the estimated 3D points are the structure, which contains the depth. Compared with stereo matching, SfM estimates both the depth and pose. Depending on how the pose is estimated, i.e., incrementally or globally optimized, SfM also contains two categories: incremental SfM and global SfM. In [25], an initial matching pair is first selected, then the camera parameters, i.e., internal and external parameters, are incrementally added together with depth information updated. Bundle Adjustment [26] is used for optimization to make the local structure and motion rigid. Snavely et al. [27] proposed to reconstruct a skeleton set of the scene first, and then incrementally complete the whole scene. On the other hand, global SfM focuses more on the estimation of camera motion, optimizing a global camera motion model that better fits the multi-view image set using linear fitting, belief propagation Bayesian inference, or Markov random field techniques [28–30].

2.2.2   Existing Deep Learning Based Depth Estimation Methods

Depth Estimation with a Single Image Methods

Deep learning based depth estimation with a single image directly establishes the hidden warping function between RGB images and depth maps. A hierarchical CNN combined with conditional random filed (CRF) method was proposed in [31] for depth prediction. The hierarchical CNN extracts multiscale features for predicting depth at super-pixel level, then CRF is used to refine the depth values within each super-pixel. Similarly, in [32], super-pixel level depth estimated is also used as a basic processing unit first, and the correlation among the super-pixels are explored with a CRF model to infer their geometric relationship. The above methods extract features at patch (super-pixel) level, focusing more on local information while the global information is concatenated or inferred with CRF models. These methods may not well explore the global information within the features. Thus, whole image-based methods have been proposed and become popular. A two-stage processing framework was proposed in [33], where a coarse processing network is first used to learn a coarse depth map, and then refined with a refinement network to obtain the final depth map. In [34], the depth estimation task is decomposed into two tasks, first predicting the mid-level representation with depth cue such as depth derivatives of different orders, orientations and scales, then infer the depth based on these mid-level representations. In the depth generation process, usually multi-scale processing is used where the resolution of a depth map is progressively increased. To enhance this upsampling process, Song et al. [35] used Laplacian pyramid to process the input image, where the different scales of input information are used to assist the upsampling of the corresponding depth map. To reduce the effect of style information which is not useful for depth estimation, in [36], Chen et al. proposed a S2R network (Synthetic to Real DepthNet) with an image translation encoder exploring a synthetic dataset of various styles. In [37, 38], the Transformer architecture is used to better explore the spatial global information.

In addition to directly predict the depth map as a sole task, there are also multi-task based methods, which not only predict depth information, but also learn other tasks to improve the depth estimation. The rationale behind this is that the multiple tasks are related to each other and the learning of similar features can reinforce each other. For example, a network to simultaneously learn three tasks including depth prediction, surface normal estimation, and semantic segmentation was proposed in [39]. In [40], surface curves are first learned from RGB and then the surface curvatures are used to make more accurate estimates of depth and surface normals. In [41, 42], depth estimation and semantic segmentation are learned together from a single RGB image using a deep CNN with a fully connected CRF. The contextual relationships and interactions between semantics and depth are explore by the network to improve the accuracy of the final result. The semantics are also explored in [43], where the features for the depth estimation and semantic segmentation are combined using a cross-task multi-embedding attention module.

Depth Estimation with a Stereo Pair Methods

Deep learning based depth estimation with a stereo pair usually simulates the traditional stereo matching methods by calculating the correlation between the stereo pair and estimating the disparity that is finally transformed to the depth value. In [44], a correlation estimation layer, which measures the feature similarity from the stereo pair, was proposed to build a 3D cost volume and then an encoder-decoder network is used for disparity regression. In [45], the cross-view consistency error based on the initial depth is further introduced to the network to refine the depth. Considering the characteristics of the depth map with sharp edges, in [46], the depth estimation and edge derivation are learned together and the feature information of the edge derivation branch and the output depth map is also explored for depth estimation. While the above methods use correlation to obtain a 3D cost volume to represent the relationship of the stereo pair features, there are also methods that concatenate them to avoid information loss. In [47], a Geometry and Context Stereo Matching Network (GCNet) was proposed, which constructs a 4D cost volume by concatenating the features from the stereo pair over different disparities. Then 3D convolution is used to aggregate the features. A spatial pyramid pooling is further used in [48] to obtain a cost volume of multiple scale features. In order to take advantage of both the feature similarity information in the 3D cost volume and the feature information in the 4D cost volume, Zhang et al. [49] fuses the two cost volumes together for final estimation.

In addition to directly exploring the cross-view correlation between the stereo pair for depth estimation, there are also methods exploring the stereo pair for supervision instead of using them both as input [50, 51]. One view image is used to estimate the depth map, which is then used to synthesize the other view image, and thus supervised by the original other view image. This provides a self-supervised approach without the ground truth depth map. Essentially, this is also single image based depth estimation with a new supervision approach, but certain correlation between the views can also be explored to enhance the performance. In [51], the depth result using the conventional stereo matching methods are also used for supervision. Left–right view consistency was explored in [52, 53], where the estimated left-view and right-view depth maps need to be consistent when projected to the other view. In [54], an adversarial learning framework was adopted, where the synthesized images using the estimated depth map are used. Considering that the disparity only exists in the horizontal direction in the depth estimation of a stereo pair, in [55], a data augmentation strategy was proposed where the data grafting is used to insert perturbation in the vertical direction to reduce the overfitting.

Depth Estimation with a Monocular Video Methods

Depth estimation with a monocular video is similar as that with a stereo pair, where the two images now come from a video sequence instead of stereo views. It can also be classified into two categories: a video is used together as input to estimate the depth, and a video is only used for training while only one image is used to estimate the depth.

In the first category where a video is used together as input, the temporal dependency is explored for depth estimation. In [56, 57], convolutional LSTM is used to extract both the spatial coherence and temporal coherence. Yang et al. [58] used Bayesian inference to fuse the depth maps of multiple frames based on their error variances. Watson et al. [59] designed a multi-frame feature fusion network by borrowing stereo matching, where a cost volume between the current frame and the warped previous frame using pose is constructed for depth estimation. Long et al. [60] proposed an Epipolar Spatio-Temporal (EST) transformer to specify the geometric linkage and temporal correlation of multi-view depth estimates.

In the second category where a video is only used for training, the relative pose of adjacent frames also needs to be estimated together with the depth map, in order to assist the synthesis of the following frame for supervision [61]. Accordingly, the encoder and decoder for single image based depth estimation can also be used for the monocular video based one. On the other hand, different from the depth estimation with a stereo pair where the images are captured with a stereo camera at the same time, the different frames in a video are of different time and objects may move. Therefore, such object pixels cannot be aligned with the other frame under depth projection. To solve this problem, a motion mask, as a latent variable, is generated in [62, 63] to remove the supervision of such pixels. In [64], the object motion is predicted separately and the pose among frames are also updated iteratively. In [65], the 3D geometry, in the form of point clouds drawn from the estimated depth, is explicitly considered. The consistency between the point clouds of adjacent frames is enforced with the ego-motion between frames to construct a 3D loss.

2.3 Existing LiDAR Sensor and Vision Fusion Methods

LiDAR sensors can collect generally very accurate scene point clouds with depth information but they are sparse with limited amount of data and high cost. On the other hand, vision based depth estimation is cheap with lower quality. Moreover, vision images have abundant textures and color information for object detection and classification which LiDAR sensors lack. Therefore, LiDAR sensor and vision fusion based methods are getting popular.

Most fusion algorithms generally learn to complete the sparse LiDAR depth maps with the color information. Considering the point cloud data from the LiDAR sensor is sparse, a sparse convolution is used in [66], where only available pixels are used for convolution and this visibility state is propagated to the subsequent layers. In [67], normalized convolution was proposed, which introduces the confidence of pixels and propagation to the consecutive layers. Different ways of extracting texture features to assist depth completion have also been investigated including using the global information to assist the local processing [68] or processing under multiple scales [69].

On the other hands, there are also some works on using the spare depth data to assist the vision based depth estimation. In [9], the sparse point clouds first are projected onto the image plane with 3D coordinates and fused with the image features for processing. A LiDAR and image alignment method was proposed in [70, 71] before fusing the features, where a cross-attention module is proposed to dynamically align them (Fig. 1).

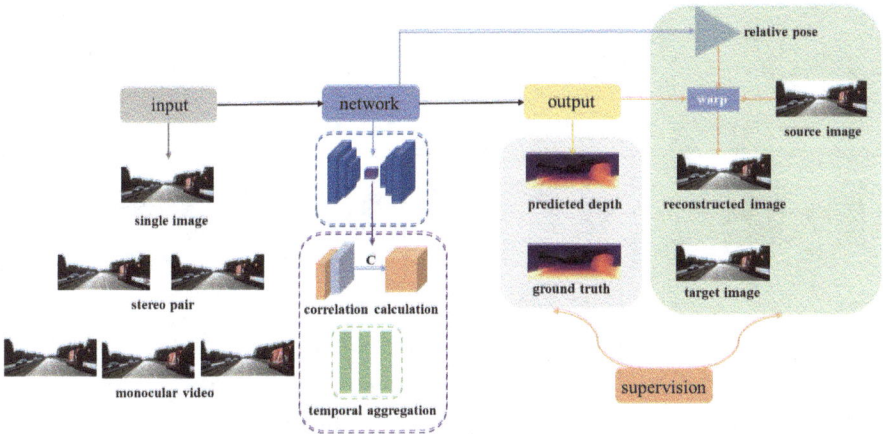

Fig. 1 The typical learning framework for depth estimation

3   Vision Based 3D Scene Perception

3.1   *Overview*

Let $I = \{I_k, k = 1, \ldots n\}$ be a set of scene RGB images captured by color cameras with optional camera extrinsic parameters T. Depth estimation aims to recover a depth map D from I at certain view perspectives. Considering that deep learning based depth estimation methods are widely investigated and achieve the state-of-the-art results, hereafter this chapter only describes the deep learning based methods. Learning-based depth estimation methods can be summarized as learning a predictor f_θ that inferences the scene depth map $\widehat{D}$ based on color images I, by minimizing the loss function $L = d(f_\theta(I), D)$ to make the predicted depth map $\widehat{D}$ approximate the real depth map D as closely as possible. Here, $d(., .)$ is a measure of difference between the predicted depth map $\widehat{D}$ and the scene real depth map D. It can be a direct measure such as MSE, or it can be an indirect measure based on the error of the synthesized images using the corresponding depth maps.

The predictor f_θ is in the form of deep neural networks for the current deep learning based depth estimation methods. It can directly produce the depth maps learning from the texture features, or first estimate disparity maps using the binocular disparity principle and then convert them to depth maps. A typical learning framework generally consists of input, network of a certain architecture, output and supervision, as shown in Fig. 2, which can then be trained with the existing optimization methods such as the stochastic gradient descent (SGD). In the aspect of input, a single image, a monocular video or a stereo image pair can all be used, where for the stereo image pair, the disparity calculation is generally used. In the aspect of network structure, an encoder-decoder structure is generally used as the backbone, where the encoder consists of sequential convolutional blocks to extract multi-level features from local

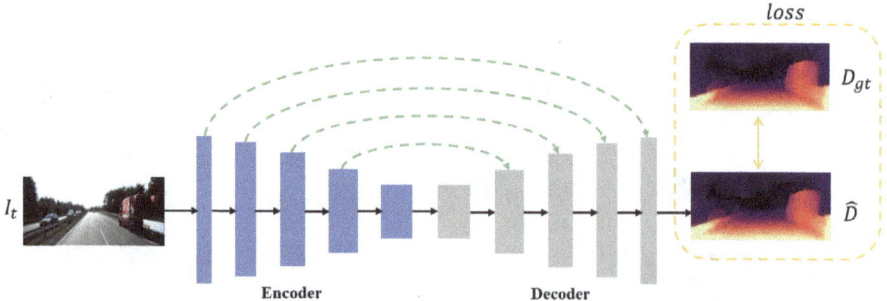

Fig. 2 The general network architecture of supervised depth estimation with a single image

to global and the decoder contains several up-sampling to scale up the features and corresponding generated depth map gradually. When processing a stereo image pair or a monocular video, a cost volume construction module, containing the correlation between the two views (directly related to the disparity) or containing the temporal information aggregated in a video (directly related to the temporal depth consistency), is also used. In the aspect of output and supervision, depth is usually regressed from the texture features or the disparity enhanced features. When the groundtruth depth map is available, it can be directly used for supervision. Otherwise, when only one image of a stereo image pair or a monocular video is used as input, the estimated depth can be used to synthesize the other view image and using the reconstruction error for supervision. In the synthesis process, depth is incorporated and thus can be indirectly supervised. Also pose information is also required which can be obtained for a stereo image pair or estimated for a monocular video. In the following, depth estimation with a single image, a stereo image pair and a monocular video with different architectures are sequentially described.

3.2 Depth Estimation with a Single Image

The learning based depth estimation with a single image approach follows the framework shown in Fig. 2. It takes one image as input and adopts an encoder-decoder architecture for processing. The encoder consists of multiple blocks, e.g., convolutional blocks, residual block or dense blocks, to extract the features. Usually, a multiscale processing is used in the shape of UNet or using a spatial pyramid, in order to extract both local and global features. The decoder contains several up-sampling and convolutional blocks alternately and gradually recovers the depth maps to the same resolution of the input. The decoder layer is enhanced with the features of the corresponding encoder layer with the same resolution via a skip connection to enrich the local information that may be lost in the final encoder global information. In addition to the basic architecture, other depth-related information can also be explored to assist depth estimation such as using the plane normal as extra input, or using semantic segmentation as an auxiliary task.

For supervision with groundtruth depth map, the objective function is designed based on the difference between the predicted depth values and the ground-truth depth, generally in terms of MSE:

$$L = \frac{1}{N} \sum_{i=0}^{N} (\widehat{D} - D_i)^2 \tag{1}$$

For self-supervised depth estimation methods, the reconstruction error of the synthesized view to the other frame in a stereo pair or a video is used for supervision. When processing a stereo pair, the camera parameters are usually known and can be directly used to synthesize the other view with the estimated depth map. However, for a monocular video, the camera parameters between the frames, also known as the pose, is unavailable and needs to be predicted in order to perform the view synthesis. A typical framework of using view synthesis for supervision in a video-based method is shown in Fig. 3. Given three consecutive frames $\{t-1, t, t+1\}$ in a video, denote the neighboring frames by the source frames I_s, $s = \{t-1, t+1\}$, the target frame by I_t and its corresponding to-be-predicted depth map by $\widehat{D}_t$. The whole network consists of the above encoder-decoder based depth estimation network and a pose prediction network. The pose prediction network produces the relative pose of the target frame to the adjacent frame $T_{t \to s}$, based on the features of the source and target frames. The adjacent frame then synthesizes a new frame at the position of the target frame $I_{s \to t}$ (via backward 3D warping) using the estimated depth map and the pose. This process can be represented by

$$I_{s \to t} = I_s \langle proj(\widehat{D}_t, T_{t \to s}, K) \rangle \tag{2}$$

where $proj(.)$ is the two-dimensional coordinate of the predicted depth $\widehat{D}_t$ in I_s and $\langle . \rangle$ is the sampling operator. K is the camera intrinsic parameters.

The photometric reconstruction error L_p is used as the loss function with both L1 and SSIM measures as follows

$$L_p = \sum_s pe(I_t, I_{s \to t})$$

$$pe(I_a, I_b) = \frac{\alpha}{2}(1 - SSIM(I_a, I_b)) + (1 - \alpha)\|I_a - I_b\|_1 \tag{3}$$

Since this error based on the reconstructed frame is related to the estimated depth map, it can be used to supervise the depth estimation process. Moreover, the errors are supposed to come mainly from the depth to avoid optimizing the differences between the frames themselves. In this case, the scene in two frames is assumed to be static with no moving object, no occlusion considered and the object surface being Lambertian (i.e., pixel value of the same object in different frames are the

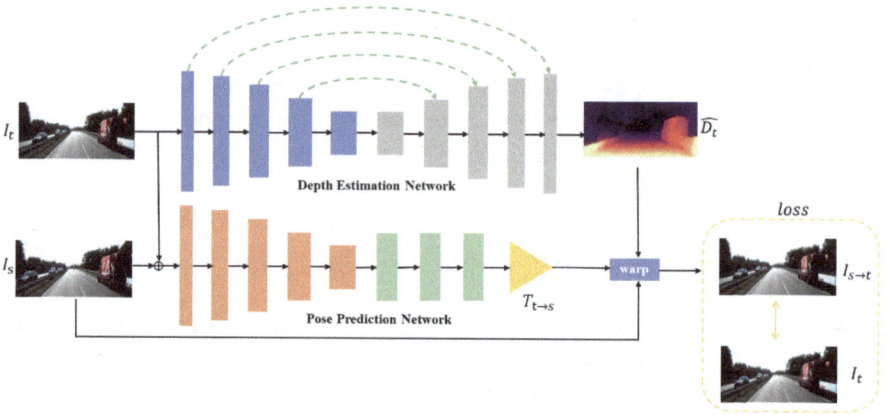

Fig. 3 The general network architecture of depth estimation using reconstruction error for supervision

same). However, this cannot be met in practical cases, reducing the supervision effect, especially the moving objects. In order to remove the effect of such moving objects, an auto-masking method is used to filter out the moving objects. It outputs a per-pixel mask $\widehat{E_s}$, indicating whether the error of target frame pixels can effectively supervise the corresponding depth value, to guide the supervision as $\widehat{E_s}L_P$. To avoid the trivial solution where $\widehat{E_s}$ predicted to be all zero, a regularization term $L_{reg}(\widehat{E_s})$ by the cross-entropy loss with constant label 1 is used to encourage it to produce nonzero mask values.

In addition to the reconstruction loss function, an edge smoothing function is used to encourage local smoothness, where $\partial_x d_t$ denotes the depth gradient and $\partial_y d_t$ denotes the image gradient, with the following equation:

$$L_s = |\partial_x d_t| e^{-|\partial_x I_t|} + |\partial_y d_t| e^{-|\partial_y I_t|} \tag{4}$$

The final loss function L for network training can be represented as follows:

$$L = \widehat{E_s}L_P + L_{reg}(\widehat{E_s}) + L_s \tag{5}$$

3.3   Depth Estimation with a Stereo Pair

For depth estimation with a stereo pair, the key is to construct a cost volume to formulate the correlation between the views, which leads to the disparity. The general network architecture is shown in Fig. 4. It mimics the traditional stereo matching and

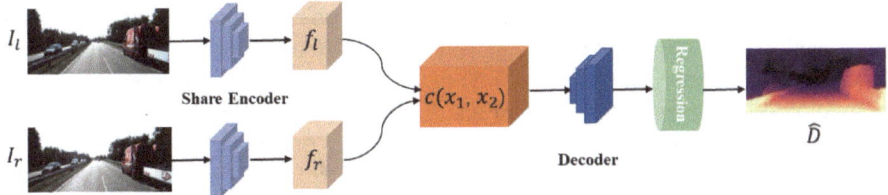

Fig. 4 The general depth estimation architecture using stereo matching

consists of four building blocks: feature extraction, cost volume construction, cost volume regularization and disparity estimation.

Similar to the above single-image based networks, it also adopts an encoder-decoder architecture for processing, where the feature extraction module plays the role of the encoder to obtain high-level features from a stereo pair, respectively. Since stereo depth estimation learns the scene structure information by comparing the similarity of local left-view and right-view patches, a correlation volume is built with the cost volume construction module. Specifically, it measures the feature similarity between the stereo pair feature patches within the pre-defined disparity range using the correlation distance, which forms a 3D cost volume (in the shape of height, width, disparity) as follows:

$$c(x_1, x_2) = \sum_{o \in [-k,k] \times [-k,k]} \langle f_1(x_1 + o), f_2(x_2 + o) \rangle \qquad (6)$$

where $\langle . \rangle$ is the convolution symbol used to calculate the correlation, f_1, f_2 are the two feature maps, x_1, x_2 are the center coordinates of the two patches with o being the patch size. The distance between x_1 and x_2 are within a pre-defined disparity range. In this way, the output correlation $c(x_1, x_2)$ is a three dimensional feature where the third dimension is of the disparity range and each element contains the correlation between two features with its disparity shift (indicated by its channel index). Moreover, considering the feature itself contains structure information of the scene, the feature of a view, generally the left-view feature, is then concatenated to the correlation volume, which is then used as the final 3D cost volume.

On the other hand, it is argued that the correlation, a handcrafted calculation, may not be able to fully capture the relation between two views and preserve full contextual information of a stereo pair. Therefore, a 4D cost volume (height, width, disparity, feature) is built simply by concatenating the left-view and right-view encoded features with a disparity shift. Then convolutional layers can be used to automatically learn to extract the relation between two views under a disparity shift, instead of using a fixed correlation calculation. Generally, the 3D cost volume can run faster, while the 4D cost volume can acquire higher accuracy.

With the 3D or 4D cost volume, a cost volume regularization module is then used, which acts as the decoder to process the cost volume to disparity or depth. Generally, 2D convolutional or 3D convolutional layers can be used. Specifically, for a 4D cost volume, a pooling or a squeezing operation over the channel dimension is used after the convolution to reduce the 4D cost volume back to a three-dimensional feature. Therefore, regardless of the cost volume dimension, the cost volume regularization module produces a three-dimensional feature for disparity estimation.

Finally, the regularized features are converted to the probability volume c_d in each disparity d, which is then processed with the softmax operation $\sigma(.)$ across the disparity dimension to achieve the final probability of each disparity. In order to achieve the subpixel level disparity, instead of using the disparity with the largest probability, the final predicted disparity d' is obtained as the weighted sum of the disparities, which can be represented by:

$$d' = \sum_{d=0}^{D_{max}} d \times \sigma(c_d) \tag{7}$$

When the probability volume refers to the feature similarity as in the 3D cost volume, the higher the similarity, the higher the possibility of the corresponding disparity value. Therefore, the softmax can directly normalize the probability volume to the probability of final disparity. On the contrary, when the probability volume refers to the matching cost as in the 4D cost volume, the higher the matching cost, the lower the probability of the corresponding disparity. Therefore, the final probability is obtained by normalizing the negative volumes, and the final depth is then obtained in the same way using a weight summation as $d' = \sum_{d=0}^{D_{max}} d \times \sigma(-c_d)$.

3.4 Depth Estimation with a Monocular Video

A Monocular video can also be used for depth estimation. The strong temporal correlation and the structure change among frames can also be explored to facilitate the depth estimation. Therefore, fusing the multi-frame information can effectively improve the quality of the estimated depth map. On the other hand, inconsistent depth values are prone to occur in single image based depth estimation, especially in the case of self-supervised learning using a monocular video. To be specific, the predicted depth map and the relative pose between adjacent frames are warped together, producing independent and relative depth values for each frame, leading to temporal inconsistency. Therefore, a temporal domain processing module, which can also be considered as a cost volume construction module over time, is used to extract the temporal coherence from video and alleviate the problem of inconsistent depth values in a video.

A natural selection of the temporal cost volume construction is using the recurrent neural network (RNN) [72] to aggregate the information among features. Similar as the depth estimation with a stereo pair, an encoder to extract the features at the beginning and a decoder to process the cost volume to depth at the end is also used together with the temporal cost volume. The general framework of depth estimation

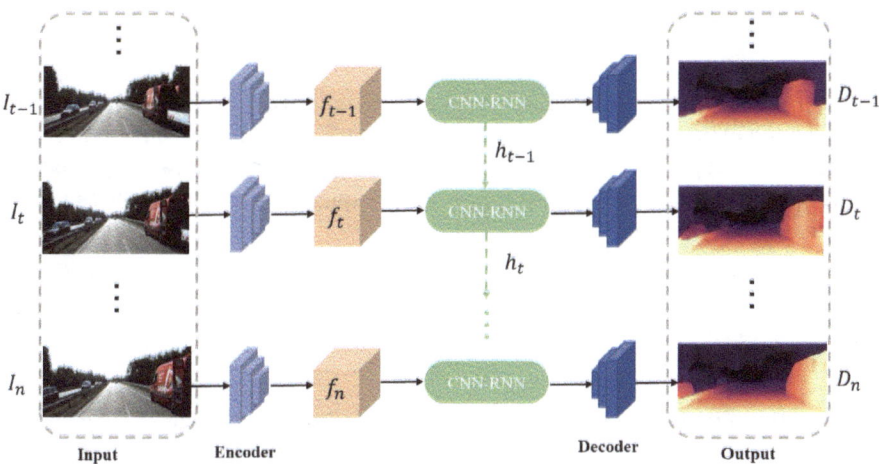

Fig. 5 The general depth estimation architecture with temporal cost volume construction

is shown in Fig. 5. The encoder first extracts the spatial features from multiple frames in a video, then a temporal cost volume construction module with RNN aggregates the information temporally. Considering the task is to predict a per-pixel depth map, the location information of the features matters to the result. Therefore, convolutional RNN is generally used, where instead of processing the features with a linear operation, convolution is used to both input processing and recurrent processing to keep the location information. Then a decoder processes the aggregated temporal features to produce the final depth estimation.

On the other hand, the temporal information can also be explicitly explored in a form of structure/feature change among frames. Similar as the stereo pair, the relationship between adjacent frames also forms a 3D warping process considering the two frames are capturing the same scene with a moving camera (moving objects ignored). Therefore, with an estimated pose, the correlation between the adjacent frames can also be explored for the depth estimation. For a frame whose depth is to be estimated, the features of its adjacent frames are first warped to its position with an estimated pose and multiple candidate depth values from a predefined depth range. Then the feature difference between the warped feature and the current frame feature is used to construct the temporal cost volume. A small feature difference indicates an accurate 3D warping among the frames and thus an accurate depth value. Together with the feature of the current frame containing the spatial structure, the temporal cost volume can be used to infer the final depth map.

4 Datasets for Evaluation

For deep learning based methods, a large dataset is required for training. The conventional multiview image/video plus depth dataset such as the Middlebury is not large enough for training a deep network. In the following, we introduce two large datasets that have been widely used in the literature and the quality measures for the estimated depth maps.

4.1 KITTI

KITTI [73] is a large dataset widely used for evaluating autonomous driving and 3D vision research in vehicle environment, such as depth estimation, optical flow, 3D object detection and tracking. It was captured using a variety of sensors, including two high-resolution color and grayscale stereo cameras, a Velodyne 3D laser scanner for sparse depth maps, and a high-precision GPS/IMU inertial navigation system. It contains serval subsets for evaluating different tasks. The stereo matching and optical flow estimation benchmark consists of 194 training and 195 test image pairs, totaling 389 stereo image pairs, with a resolution of 1240×376 for each image. The 3D visual odometry/SLAM dataset contains 22 stereo sequences covering road with a total length of 39.2 km. The 3D object detection benchmark contains over 200 K 3D objects annotated (using 3D bounding boxes) with labels of cars, vans, trucks, pedestrians, pedestrians (sitting), bicycles, trams and miscellaneous.

For depth estimation, Moritz Menze et al. [66] annotated 400 dynamic scenes and provided 400 street scene images in two consecutive frames of original KITTI dataset with light flow, disparity and depth ground truth. Jonas Uhrig et al. [74] further created a large-scale dataset based on the original KITTI dataset, which consists of 93 K frames with semi-dense depth maps. The KITTI raw dataset provides depth information in the form of Velodyne 3D laser scanner with some noise. Therefore, the depth in the new dataset is processed by comparing the scan depth to the results of stereo estimation methods using semi-global matching to remove outliers from laser scans. Nowadays, this large dataset is widely used for evaluating the depth estimation methods.

4.2 NuScenes

The NuScenes dataset [75] is a large-scale dataset captured with the AV sensor suite, including 6 cameras, 1 lidar, 5 mm Wave radar, GPS and IMU data. Compared to KITTI, NuScenes contains $7 \times$ more object annotations, includes complicated roads, weather conditions, etc. NuScenes is divided into 1000 scenes. The data, collected from Boston and Singapore, are the most complex driving scenarios in urban environments. It contains 1.4 million stereoscopic images, and rich depth information derived from 390,000 LiDAR scans for intensive depth estimation tasks. The 3D object detection and tracking benchmark contains a combination of RGB cameras, Radar and LiDAR, and 1.4 million 3D object bounding boxes annotated with 23 object classes over the entire dataset. Object-level properties such as visibility, activity and posture are also annotated. In the Lidar semantic segmentation benchmark, each LiDAR point in a keyframe is annotated with one of the 32 possible semantic labels, resulting in a total of 1.4 billion annotation points, spanning 40,000 point clouds and 1000 scenarios (850 scenarios for training and validation and 150 for testing).

4.3 Quality Measure

For evaluating the quality of the generated depth map, a diverse set of quality metrics have been used in the literature, including the Absolute Relative Error (AbsRel), Root Mean Square Error (RMSE), logRMSE, Square Relative Error (SqRel) and accuracy (%correct) indicator measure, as described below.

$$Abs\,Rel = \frac{1}{N} \sum \frac{|d_i - d_i^*|}{d_i}$$

$$RMSE = \sqrt{\frac{1}{N} \sum |d_i - d_i^*|^2}$$

$$RMSE\,(log) = \sqrt{\frac{1}{N} \sum |log\,d_i - log\,d_i^*|^2} \qquad (8)$$

$$Sq\,Rel = \frac{1}{N} \sum \frac{|d_i - d_i^*|^2}{d_i}$$

$$\%correct : \max\left(\frac{d_i}{d_i^*}, \frac{d_i^*}{d_i}\right) = \delta < T$$

d_i and d_i^* are the ground truth, where i represents the pixel index and N is the sum of all pixels. T represents the threshold, which is usually set to be 1.25, 1.25^2, 1.25^3.

5 Future Directions

5.1 Temporal Information Exploration

Although the existing depth estimation methods have achieved relatively good results, the depth maps flicker frequently in consecutive frames of the video due to the independent depth estimation of each frame. Moreover, there is a strong correlation among adjacent frames, which can be explored for consistent depth estimation. Although there have been some research on extracting temporal information and fusing multi-frame information, the investigation is still premature. There still exists a large improvement space for video based depth estimation with temporal information exploration.

5.2 Fusion with Cheap Radar

Considering the high cost of LiDAR, there is a tendency towards using cheaper Radar to assist depth estimation. Radar works by radiating electromagnetic energy and detecting echoes returned from reflecting objects. It works similarly as LiDAR but not as accurate as LiDAR. Rader have been used in most vehicles as the cheap substitutes with the continuous advances in accuracy and resolution. Moreover, there is also millimeter wave Radar, which operates in the millimeter wave band. The ranging principle is the same as the general radar, but the wavelength of millimeter wave is between centimeter wave and light wave. So, to some extent, millimeter wave Radar has the advantages of both Radar and LiDAR. It is of small size, light weight, high spatial resolution, and has strong ability to penetrate fog, smoke, and dust. However, currently the detection range is small. With the population of the general Radar and millimeter wave Radar and their future developments, it is desirable to investigate the cheap Radar and video fusion based depth estimation.

6 Conclusion

In this chapter, 3D scene perception for autonomous driving is discussed, especially the depth estimation which is the keystone of 3D vision. A comprehensive summarization and review on depth estimation with the recent developments and their general working mechanism is provided. This chapter also introduces three vision based depth estimation approaches, analyzes their frameworks and working processes. Finally, this chapter proposes two future directions of depth estimation development from temporal information exploration and multimodal information fusion.

Acknowledgements This chapter was supported in part by the National Natural Science Foundation of China under Grant 62271290 and Grant 61901083; and in part by SDU QILU Young Scholars Program. The authors would also like to acknowledge the contributions from Hongwei Xu, Xianye Wu and Jianguo Wang for providing relative materials to this chapter.

References

1. A. Smolic, 3D video and free viewpoint video—from capture to display. Pattern Recogn. **44**, 1958–1968 (2011)
2. V. Guizilini, R. Ambruş, W. Burgard, A. Gaidon, Sparse auxiliary networks for unified monocular depth prediction and completion, in *IEEE/CVF Conference on Computer Vision and Pattern Recognition (CVPR)* (2021), pp. 11073–11083
3. W. Yan, W. Chao, Pseudo-lidar from visual depth estimation: bridging the gap in 3d object detection for autonomous driving, in *IEEE/CVF Conference on Computer Vision and Pattern Recognition (CVPR)* (2019)
4. S. Izadi, D. Kim, Kinectfusion: real-time 3d reconstruction and interaction using a moving depth camera, in *Proceedings of the 24th Annual ACM Symposium on User Interface Software and Technology* (2011), pp. 559–568
5. S. Song, J. Sun, RGB-D: a RGB-D scene understanding benchmark suite, in *Proceedings of the IEEE Computer Society Conference on Computer Vision and Pattern Recognition (CVPR)* (2015), pp. 567–576
6. P.L. Lin, T. Zhou, R. Tucker et al., Depth prediction without the sensors: leveraging structure for unsupervised learning from monocular videos. IEEE Robot. Autom. Lett. 315–326 (2018)
7. S. Royo, M. Ballesta-Garcia, An overview of lidar imaging systems for autonomous vehicles. Appl. Sci. **9**(19), 4093 (2019)
8. M. Himmelsbach, A. Mueller, T. Lüttel, H.J. Wünsche, LIDAR-based 3D object perception, in *Proceedings of 1st International Workshop on Cognition for Technical Systems* (2008)
9. L. Caltagirone, M. Bellone, L. Svensson, M. Wahde, LIDAR–camera fusion for road detection using fully convolutional neural networks. Robot. Auton. Syst. (2019)
10. A. Seppänen, R. Ojala, K. Tammi, *4DenoiseNet: Adverse Weather Denoising from Adjacent Point Clouds* (2022). arXiv preprint arXiv:2209.07121
11. J.I. Park, K.S. Kim, Fast and accurate desnowing algorithm for LiDAR point clouds. IEEE Access 160202–160212 (2020)
12. L. Caltagirone, M. Bellone, L. Svensson, M. Wahde, R. Sell, Lidar-camera semi-supervised learning for semantic segmentation. Sensors **21**(14), 4813 (2021)
13. G. Yan, J. Pi, C. Wang, X. Cai, Y. Li, *An Extrinsic Calibration Method of a 3D-LiDAR and a Pose Sensor for Autonomous Driving* (2022). arXiv preprint arXiv:2209.07694
14. Z. Cui, P. Tan, Global structure-from-motion by similarity averaging, in *IEEE International Conference on Computer Vision (ICCV)* (2015), pp. 864–872
15. Y. Zhai, L. Zeng, A SIFT matching algorithm based on adaptive contrast threshold, in *Conference on Consumer Electronics, Communications and Networks (CECNet)* (2011), pp. 1934–1937
16. T.T. San, N. War, Stereo matching algorithm by hill-climbing segmentation, in *Global Conference on Consumer Electronics (GCCE)* (2017), pp. 1–2
17. J. Cai, Integration of optical flow and dynamic programming for stereo matching. Image Process. **6**(3), 205–212 (2012)
18. J. Sun, N.N. Zheng, H.Y. Shum, Stereo matching using belief propagation. IEEE Trans. Pattern Anal. Mach. Intell. **25**(7), 787–800 (2003)
19. P.F. Felzenszwalb, D.P. Huttenlocher, Efficient belief propagation for early vision. Int. J. Comput. Vision **70**(1), 41–54 (2006)

20. Y. Chang, Y. Ho, Modified SAD using adaptive window sizes for efficient stereo matching, in *International Conference on Embedded Systems and Intelligent Technology* (2014), pp. 9–11
21. R. Zabih, J. Woodfill, Non-parametric local transforms for computing visual correspondence, in *European Conference on Computer Vision (ECCV)* (1994), pp. 151–158
22. O. Eksler, Fast variable window for stereo correspondence using integral images, in *IEEE Computer Society Conference on Computer Vision and Pattern Recognition* (2003)
23. K.J. Yoon, I.S. Kweon, Adaptive support-weight approach for correspondence search. IEEE Trans. Pattern Anal. Mach. Intell. **28**(4), 650–656 (2006)
24. H.H. Stereo, Processing by semiglobal matching and mutual information. IEEE Trans. Pattern Anal. Mach. Intell. **30**(2), 328–341 (2007)
25. N. Snavely, S.M. Seitz, R. Szeliski, Modeling the world from internet photo collections. Int. J. Comput. Vision **80**(2), 189–210 (2008)
26. C. Wu, S. Agarwal, B. Curless, S.M. Seitz, Multicore bundle adjustment, in *IEEE Conference on Computer Vision and Pattern Recognition (CVPR)* (2011), pp. 3057–3064
27. N. Snavely, S.M. Seitz, R. Szeliski, Skeletal graphs for efficient structure from motion, in *IEEE Conference on Computer Vision and Pattern Recognition (CVPR)* (2008), pp. 1–8
28. V.M. Govindu, Combining two-view constraints for motion estimation, in *IEEE Computer Society Conference on Computer Vision and Pattern Recognition* (2001)
29. D. Devarajan, R.J. Radke, Calibrating distributed camera networks using belief propagation. EURASIP J. Adv. Signal Process. 1–10 (2006)
30. P. Moulon, P. Monasse, R. Marlet, Global fusion of relative motions for robust, accurate and scalable structure from motion, in *IEEE International Conference on Computer Vision (ICCV)* (2013), pp. 3248–3255
31. B. Li, C. Shen, Depth and surface normal estimation from monocular images using regression on deep features and hierarchical CRFS, in *IEEE Conference on Computer Vision and Pattern Recognition (CVPR)* (2015), pp. 1119–1127
32. F. Liu, C. Shen, G. Lin, Deep convolutional neural fields for depth estimation from a single image. Comput. Vision Pattern Recogn. (CVPR) (2015)
33. D. Eigen, C. Puhrsch, R. Fergus, Depth map prediction from a single image using a multi-scale deep network. Adv. Neural Inform. Process. Syst. 2366–2374 (2014)
34. A. Chakrabarti, J. Shao, G. Shakhnarovich, Depth from a single image by harmonizing overcomplete local network predictions. Adv. Neural Inform. Process. Syst. 2658–2666 (2016)
35. M. Song, S. Lim, W. Kim, Monocular depth estimation using Laplacian pyramid-based depth residuals. IEEE Trans. Circ. Syst. Video Technol. **31**, 4381–4393 (2021)
36. X. Chen, Y. Wang, X. Chen, W. Zeng, S2R-DepthNet: learning a generalizable depth-specific structural representation, in *Proceedings of the IEEE/CVF Conference on Computer Vision and Pattern Recognition (CVPR)* (2021), pp. 3034–3043
37. R. Ranftl, A. Bochkovskiy, V. Koltun, Vision transformers for dense prediction, in *IEEE/CVF International Conference on Computer Vision (ICCV)* (2021), pp. 12179–12188
38. A. Agarwal, C. Arora, *Attention Everywhere: Monocular Depth Prediction with Skip Attention* (2022). arXiv preprint arXiv:2210.09071
39. D. Eigen, R. Fergus, Predicting depth, surface normals and semantic labels with a common multi-scale convolutional architecture, in *International Conference on Computer Vision (ICCV)* (2015), pp. 2650–2658
40. T. Dharmasiri, A. Spek, T. Drummond, Joint prediction of depths, normals and surface curvature from RGB images using CNNS, in *IEEE/RSJ International Conference on Intelligent Robots and Systems (IROS)* (2017), pp. 1505–1512
41. P. Wang, X. Shen, Z. Lin, S. Cohen, Towards unified depth and semantic prediction from a single image, in *IEEE Conference on Computer Vision and Pattern Recognition (CVPR)* (2015), pp. 2800–2809
42. A. Mousavian, Pirsiavash, Joint semantic segmentation and depth estimation with deep convolutional networks, in *Fourth International Conference on 3D Vision (3DV)* (2016), pp. 611–619

43. H. Jung, E. Park, Fine-grained semantics-aware representation enhancement for self-supervised monocular depth estimation, in *IEEE/CVF International Conference on Computer Vision (ICCV)* (2021), pp. 12642–12652
44. N. Mayer, E. Ilg, P. Hausser, A large dataset to train convolutional networks for disparity, optical flow, and scene flow estimation. Comput. Vision Pattern Recogn. (CVPR) 4040–4048 (2016)
45. J.H. Pang, W.X. Sun, J.S.J. Ren, Cascade residual learning: a two-stage convolutional neural network for stereo matching, in *IEEE International Conference on Computer Vision Workshops* (2017), pp. 878–886
46. X. Song, X. Zhao, H.W. Hu, L.J. Fang, EdgeStereo: a context integrated residual pyramid network for stereo matching, in *Asian Conference on Computer Vision* (2018)
47. A. Kendall, H. Martirosyan, End-to-end learning of geometry and context for deep stereo regression, in *IEEE International Conference on Computer Vision (ICCV)* (2017)
48. J.R. Chang, Y.S. Chen, Pyramid stereo matching network, in *IEEE/CVF Conference on Computer Vision and Pattern Recognition (CVPR)* (2018), pp. 5410–5418
49. S. Zhang, Z. Wang, Q. Wang, et al., EDNet: efficient disparity estimation with cost volume combination and attention-based spatial residual, in *IEEE/CVF Conference on Computer Vision and Pattern Recognition (CVPR)* (2021), pp. 5433–5442
50. J. Xie, R. Girshick, A. Farhadi, Deep3d: Fully automatic 2d-to-3d video conversion with deep convolutional neural networks, in *European Conference on Computer Vision (ECCV)* (2016), pp. 842–857
51. R. Garg, G. Carneiro, I.D. Reid, Unsupervised CNN for single view depth estimation: Geometry to the rescue, in *European Conference on Computer Vision (ECCV)* (2016), pp. 740–756
52. C. Godard, O.M. Aodha, G.J. Brostow G. J. (2017). Unsupervised monocular depth estimation with left-right consistency, in *2017 IEEE Conference on Computer Vision and Pattern Recognition (CVPR)* (2017), pp. 6602–6611
53. A. Wong, S. Soatto, Bilateral cyclic constraint and adaptive regularization for unsupervised monocular depth prediction, in *IEEE/CVF Conference on Computer Vision and Pattern Recognition (CVPR)* (2019), pp. 5637–5646
54. A. Pilzer, D. Xu, M. Puscas, Un-supervised adversarial depth estimation using cycled generative networks, in *International Conference on 3D Vision (3DV)* (2018), pp. 587–595
55. R. Peng, R. Wang, Y. Lai, et al., Excavating the potential capacity of self-supervised monocular depth estimation, in *IEEE/CVF International Conference on Computer Vision (CVPR)* (2021), pp. 15560–15569.
56. H. Zhang, C. Shen, Y. Li, Y. Cao, Y. Liu, Y. Yan, Exploiting temporal consistency for real-time video depth estimation, in *IEEE/CVF International Conference on Computer Vision (ICCV)* (2019), pp. 1725–1734
57. R. Wang, S.M. Pizer, J. Frahm, Recurrent neural network for (Un-)supervised learning of monocular video visual odometry and depth, in *IEEE/CVF Conference on Computer Vision and Pattern Recognition (CVPR)* (2019), pp. 5550–5559
58. X. Yang, Y. Gao, H. Luo, C. Liao, K. Cheng, Bayesian DeNet: monocular depth prediction and frame-wise fusion with synchronized uncertainty. IEEE Trans. Multimedia **21**, 2701–2713 (2019)
59. J. Watson, O. Mac Aodha, V. Prisacariu, et al., The temporal opportunist: Self-supervised multi-frame monocular depth, in *IEEE/CVF Conference on Computer Vision and Pattern Recognition (CVPR)* (2021), pp. 1164–1174
60. X. Long, L. Liu, W. Li, et al., Multi-view depth estimation using epipolar spatio-temporal networks, in *IEEE/CVF Conference on Computer Vision and Pattern Recognition (CVPR)* (2021), pp.8258–8267
61. T. Zhou, M. Brown, N. Snavely, Unsupervised learning of depth and ego-motion from video, in *IEEE Conference on Computer Vision and Pattern Recognition (CVPR)* (2017), pp. 6612–6619
62. Z. Yin, J. Shi, Geonet: Unsupervised learning of dense depth, optical flow and camera pose, in *IEEE/CVF Conference on Computer Vision and Pattern Recognition (CVPR)* (2018), pp. 1983–1992

63. C. Godard, O. Mac Aodha, M. Firman, et al., Digging into self-supervised monocular depth estimation, in *The IEEE/CVF International Conference on Computer Vision (ICCV)* (2019), pp. 3828–3838
64. T.-W. Hui, RMDepth: unsupervised learning of recurrent monocular depth in dynamic scenes, in *Proceedings of the IEEE/CVF Conference on Computer Vision and Pattern Recognition (CVPR)* (2022)
65. R. Mahjourian, M. Wicke, A. Angelova, Unsupervised learning of depth and ego-motion from monocular video using 3d geometric constraints, in *IEEE Conference on Computer Vision and Pattern Recognition (CVPR)*, pp. 5667–5675 (2018)
66. J. Uhrig, N. Schneider, L. Schneider, U. Franke, T. Brox, A. Geiger, Sparsity invariant CNNS, in *International conference on 3D Vision (3DV)* (2017), pp. 11–20
67. A. Eldesokey, M. Felsberg, F.S. Khan, *Propagating Confidences Through CNNS for Sparse Data Regression* (2018). arXiv preprint arXiv:1805.11913
68. W. Van Gansbeke, D. Neven, B. De Brabandere, L. Van Gool, Sparse and noisy lidar completion with RGB guidance and uncertainty, in *International Conference on Machine Vision Applications (MVA)* (2019), pp. 1–6
69. S. Shivakumar, T. Nguyen, I.D. Miller, S.W. Chen, V. Kumar, C.J. Taylor, Dfusenet: deep fusion of RGB and sparse depth information for image guided dense depth completion, in *Intelligent Transportation Systems Conference (ITSC)* (2019), pp. 13–20
70. X. Bai, Z. Hu, X. Zhu, Q. Huang, Y. Chen, H. Fu, C.L. Tai, Transfusion: robust lidar-camera fusion for 3d object detection with transformers, in *IEEE/CVF Conference on Computer Vision and Pattern Recognition (CVPR)* (2022), pp. 1090–1099
71. Y. Li, A. Yu, Deepfusion: Lidar-camera deep fusion for multi-modal 3d object detection, in *IEEE Conference on Computer Vision and Pattern Recognition (CVPR)* (2022), pp. 17182–17191
72. S. Li, W. Li, C. Cook, et al., Independently recurrent neural network (INDRNN): building a longer and deeper RNN, in *IEEE conference on computer vision and pattern recognition (CVPR)* (2018), pp. 5457–5466
73. A. Geiger, P. Lenz, R. Urtasun, Are we ready for autonomous driving? the Kitti vision benchmark suite, in *IEEE conference on computer vision and pattern recognition (CVPR)* (2012), pp. 3354–3361
74. M. Menze, A. Geiger, Object scene flow for autonomous vehicles, in *IEEE Conference on Computer Vision and Pattern Recognition (CVPR)*, (2015), pp. 3061–3070
75. H. Caesar, V. Bankiti, A.H. Lang, S. Vora, V.E. Liong, Q. Xu, O. Beijbom, Nuscenes: a multimodal dataset for autonomous driving, in *IEEE/CVF conference on computer vision and pattern recognition (CVPR)* (2020), pp. 11621–11631

Chapter 8
Multi-sensor Fusion for Perception in Complex Traffic Environments

Qian Huang, Kainan Zhu, Kan Wu, Wei Hua, and Yongdong Zhu

Abbreviations

AVs	Autonomous Vehicles
BEV	Bird's-eye View
IoU	Intersection over Union
Lidar	Light Detection and Ranging
RGB-D	RGB-depth
ROIs	Region of Interests

1 Introduction

To gain reliable and accurate perception for autonomous driving in complex traffic environments is a big challenge, which will not completely come true without the cooperative perceptions based on on-board and roadside sensors. The perception of

Q. Huang · K. Zhu · K. Wu · W. Hua · Y. Zhu (✉)
Zhejiang Lab, Interdisciplinary Innovation Research Institute, Hangzhou, China
e-mail: zhuyd@zhejianglab.com

Q. Huang
e-mail: huangq@zhejianglab.com

K. Zhu
e-mail: zhukainan@zhejianglab.com

K. Wu
e-mail: kanwu@zhejianglab.com

W. Hua
e-mail: huawei@zhejianglab.com

© The Author(s), under exclusive license to Springer Nature Singapore Pte Ltd. 2023   147
Y. Zhu et al. (eds.), *Communication, Computation and Perception Technologies for Internet of Vehicles*, https://doi.org/10.1007/978-981-99-5439-1_8

on-board system is mainly based on sensors mounted on the vehicle, while perception of roadside unit is based on sensors mounted on the roadside. Generally, sensors mounted on the vehicle include but are not limited to mechanical LiDAR (Light Detection and Ranging), camera, millimeter wave radar, ultrasonic radar. However, due to the blind spots of on-board sensors it is difficult to fully access information of dynamic obstacles and stationary obstacles, especially in complex traffic environments, which easily leads to driving safety problems. On the other hand, sensors mounted on the roadside generally include solid-state LiDAR, camera, millimeter wave radar, whose purposes are to obtain comprehensive perception of the roadside view and can serve as a powerful aid to the on-board perception [1].

Multi-sensor fusion is a fundamental perception technology for automatic driving in complex traffic environments. Detailed perception provides an accurate understanding of the environments which is essential for autonomous vehicles, helping to make accurate and safe driving decisions [2]. In traffic environments, sensing devices detect and respond to some types of inputs from the physical environments, to obtain descriptive information about the appearance and position of obstacles in the environments. A single sensor, such as camera, deeply impacted by the weather and lighting conditions. And can easily get confused by unexpected input, such as a car on a billboard, reflection in the water, odd light and shadow behavior, etc. In addition, cameras are unable to obtain accurate depth information. Contrary to a camera, Lidar generates 3D point cloud but not an image, which can delivers reliable information about the distance and position of every entity the laser beam reaches. However, the spatial density of the point cloud is lower than that of the image, and the point cloud lack color and texture information about the entity.

But, sensors can complement with each other to gain a more overall comprehensive description about environments, including color, texture, distance and the position of entity in traffic. At present, multi-sensor fusion conducts the perception tasks on multiple sensors [3], and is widely used to achieve high-precision perception [4–6] in complex traffic environments.

The perception tasks include but are not limited to object detection, semantic segmentation, depth completion and behavior prediction. Due to the limited computing power of on-board computing units, the feature-level fusion method with real-time performance is more favored by on-board perception. Since roadside computing power is sufficient, more complex tasks, such as results ensemble based on the decision-level fusion method, can also implemented. In addition, roadside perception usually takes multi-source sensing data to achieve redundant confirmation, by exchanging accurate information with more computation.

The revolutionary progress in environmental perception would not be possible without deep learning methods [7, 8]. Recently, many multi-sensor fusion methods [9, 10] with the highest perceptual accuracy are based on the deep learning models. The fusion methods mentioned in this chapter are mostly based on deep learning. It is necessary to summarize these fusion methods and give a more reasonable division rule, to distinguish the characteristics of various methods and give a clearer promotion path for future research directions.

1.1 Motivation

As we all know, on-board perception is a necessary module of autonomous driving system, but easily leads to driving safety problems. At present, **most of studies are concerned about the on-board fusion perception, ignoring roadside perception. It is necessary to combine on-board perception and roadside perception for autonomous driving system in complex traffic environments to alleviate security problems**.

At present, the single-mode perception method cannot meet the demand for a complete highly accurate description of the complex traffic environments. Multi-sensor fusion is a technology to combine information from several sensor data sources in order to form a unified description [11], and has been proven to have great potential in improving perception accuracy especially in complex traffic environments. However, most literature divided fusion methods into three major classes, such as data-level fusion, feature-level fusion, proposal-level fusion. **This division rule is difficult to distinguish fusion methods and cannot completely cover all deep learning based fusion methods. It is necessary to propose a more reasonable division rule to separate them.**

In addition, due to the heterogeneous dimensions and resolution of multi-mode data resources, lots of information has been lost after fusion. Alignment deviation also occurs when the projection matrix is inaccurate or noisy. **In this chapter, we propose some feature research directions to alleviate these problems.**

1.2 Main Contributions

In this chapter, we will give a review of the recent classical papers focusing on multi-sensor fusion perception tasks. By distinguishing and summarizing the characteristics of existing multi-sensor fusion methodologies, we propose a more reasonable division rule to separate them. The main contributions of this chapter can be summarized as the following:

- This chapter surveys deep learning based multi-sensor fusion methodologies, and proposes a more reasonable division rule to divide fusion methods into decision-level fusion, decision-feature-level fusion, and feature-level fusion three categories.
- Except for on-board fusion perception of autonomous driving system, we make a detailed introduction and analysis about another key applied scenario–roadside.
- This chapter presents some open challenges and possible research directions in the future to alleviate problems encountered in the multi-sensor fusion domain.

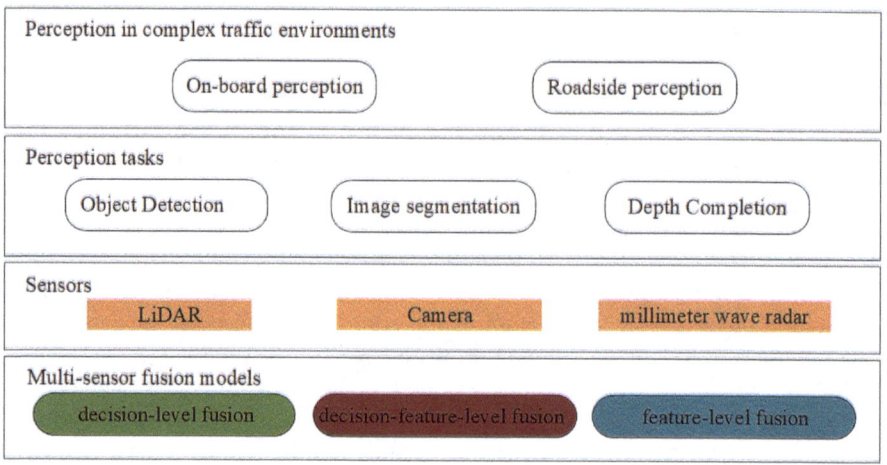

Fig. 1 The framework of this chapter

1.3 Structural Organization

The framework of this chapter as illustrated in Fig. 1. Section 2 introduces the background of perception in complex traffic environments, which includes two key scenarios: on-board perception and roadside perception. And some classical perception tasks and data sources also presented in Sect. 2. In Sect. 3, we present many classical fusion models in detail, and divide them into three categories: decision-level fusion, decision-feature-level fusion, and feature-level fusion. Section 4 elaborates on the open challenges encountered by the multi-sensor fusion method and possible research directions in the future. Section 5 summarizes the whole chapter.

2 Background

2.1 Cooperative Perceptions

At present, most of studies are concerned about the on-board fusion perception for autonomous driving system. However, the blind spots of on-board sensors always cause driving safety problems, especially in complex traffic environments. The roadside perception have a long-range global perspective beyond on-board perception and can give a global trajectory prediction, which can cooperate with on-board perception and help to reduce safety problems [12].

For example, a pedestrian walking behind a parked vehicle might suddenly crash into a moving vehicle since on-board sensors fail to detect abrupt changes in the environments owing to the limited perceptual range or heavy occlusion. At the same time, the roadside view provides global perception which is capable of behavior prediction promptly, sending pedestrian trajectory information on time to vehicles through the high bandwidth network, such as 5G communications [13, 14]. Besides, roadside perception is cost-efficient since information from roadside cameras can broadcast to all surrounding AVs (Auto vehicles). Moreover, the roadside perception can also facilitate smart traffic control and flow management with global real-time perception information. The critical contributions of roadside perceptual system has been acknowledged in many works [15, 16].

2.2 Tasks

The purpose of the perception task is to provide complete information about the dynamic and stationary obstacles. Perception tasks include Object Detection, Image segmentation, Depth Completion and Trajectory Prediction, and so on. Here, we mainly focus on the first two tasks, which helps to detect obstacles, traffic lights, traffic signs, and to segment lanes or free space. We also briefly introduce some other remaining perception tasks.

Object Detection

All the deep learning detectors follow a similar idea: extract the feature from the input data with the backbone network to generate proposals and then classify and locate the objects with a 2D/3D bounding box with the detection head. Depending on whether region proposals are generated or not, the object detectors can be classified into two-stage [17] and one-stage [18, 19] detectors. Two stage detectors generate the last results from the ROIs (region of interests) proposed from the feature map, while one-stage detectors generate the last results from the high-level feature map directly. The function of object detection is to generate instances' bounding box and categories. Generally, 2D object detection result can be expressed as (x, y, h, w, c), while 3D detection can be expressed as $(x, y, z, h, w, l, c, \theta)$.

Image Segmentation

Image segmentation is typically used to locate objects and boundaries of objects within image. Image Segmentation methods can be classified into semantic segmentation [20, 21] and instance segmentation [22, 23], which are both crucial for scene understanding of autonomous driving. Semantic segmentation focuses on per-point semantic label prediction so as to partition a scene into several parts with certain classes (i.e., per-point one class label), while instance segmentation aims at finding the edge of instances of interest (i.e., per-object one mask and class label). Image Segmentation based on deep learning often using feature pyramid network [24] to merge multi low-and-high level feature together, and make prediction mask for each pixel or object. Semantic segmentation is often applied to segment moving objects and unmoving objects [25], which help to improve vigilance against moving objects and ignore unmoving objects at same time.

Free-space detection [26, 27] is a typical semantic segmentation, which is used to classify the ground pixel into drivable or non-drivable parts. Lane-detection methods [28] are used to distinguish different lanes on the road. 3D semantic segmentation [29, 30] based on multi-sensor is used to obtain sematic information and position information of targets at same time. Feature-level fusion method is a centralized research domain of 3D segmentation, and is mainly based on 3D feature encoder network branch with the input of point cloud and fuse sematic feature from other image encoder network branch to fusion both source data information.

Some other Perception Tasks

Other common perception tasks in autonomous driving include object classification, depth evaluation, and trajectory prediction. Object classification mainly determines the category of objects within given point clouds and images. The depth evaluation tasks focus on predicting the distance of every pixel in the image based on LiDAR' point cloud and camera' image data. The trajectory prediction focuses on predicting the trajectories of moving targets, by generating its position or speed in the feature.

In addition, there are many other perception tasks not described in this chapter, but most of them are the variants of object detection and image segmentation, which can be easily extended.

2.3 Data Representations for LiDAR and Image

At present, multi-sensor fusion methods are usually based on deep learning methods, and the deep learning model process the raw sensor source data by neural network feature extractor. Usually, the data representation needs to be selected according to the characteristics of the model. Since LiDAR and Image are most common raw sensor data, we will introduce the representations of both.

2.3.1 Image Representation

The most common camera types for perception are monocular camera, binocular camera, and RBG-D (RGB-depth) camera. The monocular camera usually captures RGB image decomposing in the red, blue, green channel. The RGB image is rich in texture information [31] with high resolution and compose pixels in order, which make it good at identifying targets accurately. Binocular camera captures depth of each pixel as value with high computation. The RGB-D camera captures RGB color image and depth image at the same time.

Image is widely used for object detection and semantic segmentation tasks and has been achieved remarkable results in the field of visual perception. Due to the abundant description about shape and category of objects, image has been widely used in perception task of both on-board view and roadside view. However, image can be easily affected by weather and lighting conditions, other supplementary sensors need to be combined under harsh situations.

2.3.2   LiDAR Representation

Mechanical LiDAR and solid-state LiDAR are commonly mounted for either on-board view or roadside view perception. Mechanical laser radar has rotating parts to control laser emission angle, while solid-state laser radar does not own mechanical rotating parts, but some electronic parts, including Optical Phased Array, Photonic IC, Far-Field Radiation Pattern, to control laser emission angle. Mechanical laser radar is larger and more expensive in general, with relatively high measurement accuracy. Solid-state LiDAR is smaller in size and lower in cost, but its measurement accuracy is relatively low. A more obvious advantage of mechanical laser radar is its 360° field of view, while the solid-state laser's field of view angle is within 120° in general.

The existing mainstream LiDAR emits laser to hit the surface of an object, and the reflected laser carries information of the target object such as location and distance. Moreover, LiDAR is robust to diverse weather conditions. The LiDAR point cloud is an unordered sparse points set representing the spatial distribution of targets and the characteristics of the target surface related to the reflectivity of the point. There are three major LiDAR data representations basically implemented in deep learning based methods as input of neural network: (1) point-based representation; (2) voxel-based representation; and (3) multi-view-based representation. Point-based representation conducts deep learning methods directly on the point cloud [32]. Voxel-based representation discretizes the 3D point cloud space into many of voxels with fixed volume size [33]. The local 3D information can be preserved, since points adjacent in space has been organized together. However, the 3D information will be lost if the resolution of the voxels is small due to the limited computing power and memory of computation device. The multi-view-based representation includes bird-eye view and front view, and is commonly used to match the corresponding view of image original data source or intermediate feature.

3   Fusion Methodology

In this section, we propose a more reasonable multi-sensor fusion division rule to divide multi-sensor fusion methodologies into three categories, as illustrated in Fig. 2, taking camera and LiDAR for example. The decision-level fusion, merges decision level results from different sensor. While decision-feature-level fusion method fuses decision level information from one branch while data level or feature level information from the other branch. Feature-level fusion method directly fuses feature from the LiDAR branch but data-level or feature-level information from the other branch.

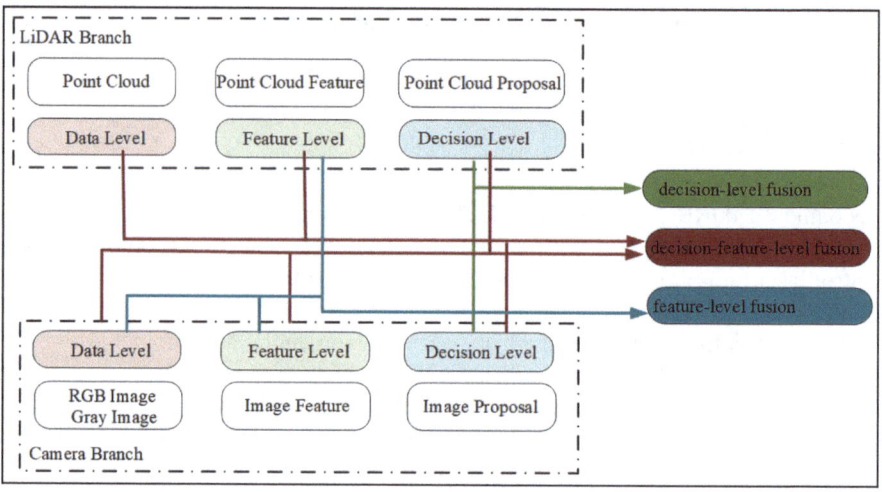

Fig. 2 Fusion methodology overview

3.1 Decision-Level Fusion

The decision-level fusion, also known as object-level fusion, get multi-model sensor perception results firstly, and then turn them into unifies coordinate space to merge results. This method is always used for redundancy confirmation and multi-source information integration. Redundancy confirmation, for example, by projecting one LiDAR 3D detection bounding box to image, if the project location exactly lies in a 2D detection bounding box with the same class, which means there's a huge probability of the existence of object. Multi-source information integration is to synthesize different source perception results, for example, integrating spatial position information from LiDAR with appearance information from the image [34]. Besides, traditional methods like Kalman Filter [35], IoU (Intersection over Union) overlap computation is usually used for redundancy confirmation and information integration process.

CLOCs [36] leverages output from both the point cloud branch and image branch and makes the final prediction based on the results of two modalities. It refines every 3D proposal's score by combining with 2D proposal's implementation from image branch. Reference [37] focuses on 2D objects detection, it combines the proposals from two branches along with confidence score, and synthesizes the final IoU score from both branches.

3.2 Decision-Feature-Level Fusion

This method fuses decision-level information from one branch while data-level or feature-level information from other branches. A classical strategy is that the decision-level information from one branch is used as the initial detection proposals, later fusion work will based on these proposals. The critical point of such fusion method is to unify the data or features under different coordinate systems to the same coordinate system, either 3D point cloud coordinate system or 2D image coordinate system.

F-PointNet [38] first generates 2D object region proposals basis of the RGB image through a CNN feature extractor. Each 2D proposal is extruded to a 3D viewing frustum and then predicts a 3D bounding box for the object based on the points in frustum by PointNet [39] module. However, one 3D frustum can only predict one bounding box, which is not suitable for a crowded scene, especially for far away small objects. IPOD [40] separates point cloud into foreground and background by utilizing sematic segmentation results from image branch. And then generates massive candidate 3D bounding boxes proposals based on foreground point cloud. IPOD [40] works better within a scene in which a large number of objects exists with occlusion than F-PointNet [38] by using rich image semantic information.

MV3D [41] is a typical work basis of 3D proposals which is generated from bird's-eye view point cloud. These 3D proposals are then projected to front-view and fused with features from raw-level point cloud and image, to ensure more original features can be gained. AVOD [42] generated 3D proposals by uniform sampling under BEV (bird's-eye view) through prior knowledge, and then fuse with image feature and point cloud feature to produce the final detection results.

Generally, this method which fuse different features based on the initial object proposals inevitable lead to the loss of spatial details.

3.3 Feature-Level Fusion

Feature-level fusion methods directly fuse cross-modal data at the feature level for the LiDAR branch but data-level or feature-level for the image branch. The common method is to use feature extractor to acquire the embedding feature representation of LiDAR point cloud and semantic features of camera image respectively and then fuse features from these two modalities by a series of downstream modules [43, 44]. This method leverages both raw and high-level LiDAR feature which represent different sematic level information in a cascading way, which varies with fusion architecture designs. A typical feature-level fusion model is shown in Fig. 3.

Because feature-level fusion is often used for stereoscopic perception purpose, and mainly relies on the LiDAR branch, which can reserves more spatial information. PointPainting [45] projects point cloud to image's semantic segmentation mask, the points fall in the instance area will be added semantic information. Point cloud with

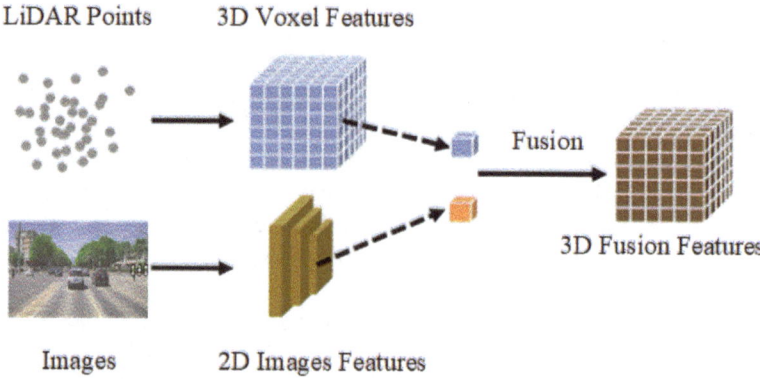

LiDAR Points   3D Voxel Features

Fusion

3D Fusion Features

Images   2D Images Features

Fig. 3 A typical feature-level fusion model

more rich semantic information can be dealt with any 3D detection network, likes PointRCNN [46], VoxelNet [17], PointPillar [47] and so on.

However, the 2D image sematic segmentation results attached to point cloud are high-level information, leading to massive raw-level information lost. It is usually considered that raw-level feature can keep more information. MVX-Net [48] utilizes the mapping relationship between point cloud and image to attach the raw-level images feature to point cloud. The point cloud is fused with raw-level image feature and then used to generate 3D detection results. The 3D detection results obtained by these two methods show that point cloud level fusion is better than voxel level fusion, which further illustrates that lower level fusion is more beneficial.

MVP [49] generates dense 3D virtual points to augment original sparse 3D point cloud. This approach makes points up-sample which is called virtual points falling within 2D detections pixel area, and adds the 2D result-level semantic information to the point by the mapping relationship between point cloud and image. These points after up-sample can naturally be integrated into any standard Lidar-based 3D detectors. This method makes far-away or small objects which turn out to be driving hazards more easily visible. However, high-level sematic images information is still not fully used.

TransFusion [7] is a LiDAR-Camera feature-level fusion for 3D Object Detection with Transformer [50] module. The attention mechanism of the transformer enables fusion model adaptively determining where and what information should be taken from both branches, leading to a robust and effective fusion strategy. This method turns a hard association of LiDAR points and image pixels which established by calibration matrices to a soft-association, helping to tackle inferior image conditions, e.g., bad illumination and sensor misalignment. The first layer of the decoder predicts initial bounding boxes from a LiDAR point cloud by fusing a sparse set of image object queries, and the second decoder layer adaptively fuses with low-level image features, leveraging both structural and contextual relationships.

Despite the attention mechanism of the transformer make feature-level fusion more effective, the different resolution between LiDAR and camera still make semantic loss and geometric distortion. BEVFusion [51] unifies multi-model feature in the shared BEV representation space, which nicely preserves both geometric and semantic information. A convolution-based BEV encoder is used to alleviate the local misalignment between different modal features. This method is task-agnostic and seamlessly support different 3D perception tasks with almost no architectural changes.

4 Future Directions

4.1 Information Loss Due to Heterogeneous Dimensions

Due to heterogeneous dimensions of data source or extracted feature, intrinsic and extrinsic parameters between cameras and LiDARs are used to transform one mode to another. However, the transformation between different dimension leads to information loss. Generally, it is common to utilize a projection calibration matrix to project LiDAR point cloud to the corresponding pixels or vice versa, e.g., mapping the 3D LiDAR point cloud into 2D image. The dimension reduction operation will inevitably lead to 3D geometric information loss and geometric distortion. In addition, this pixel to point alignment also results in semantic density loss of camera features. For the feature-level fusion method, the transformation of input data and feature space results in projection misalignment and semantic confusion.

The information loss due to heterogeneous dimensions mainly result from unreasonable transformation, a solution is to unify multi-modal features in the shared representation space, which nicely preserves both geometric and semantic information. BEVFusion [51] is a novel work that unifies multi-modal features in the shared BEV representation space and achieves new state-of-the-art performance on the nuScenes [52] benchmark. This indicates unified high-dimensional geometric space is a good solution for multi-sensor fusion. Therefore, feature research can explore a new data representation space to keep data natural information, for example, 3D stereoscopic space. This method keeping the three-dimensional advantage of point cloud and embedding 3D image information is more suitable for current 3D perception tasks. It is crucial to estimate the 3D information of pixels and fusion dense pixels' information adaptively in future work.

4.2 Misalignment Due to Noisy Projection Matrix

Multi-sensor fusion methods utilize an intrinsic and extrinsic calibration matrix to project all the LiDAR points directly to the corresponding pixels or vice versa.

It's possible to get a mapping result with unsatisfying accuracy due to the noisy projection matrix. Although the calibration has been carried out in advance, the calibration matrices are prone to change due to equipment vibration or slight damage. Existing fusion methods are easily affected by such conditions, due to the fact that the association of LiDAR points and image pixels is highly dependent on accurate calibration matrices which is hard to be corrected.

Up to now, calibration matrices are produced manually at most, or semi-automatic. Besides, equipment vibration can lead to calibration matrices offset, and it is not convenient to do calibration frequently. It is necessary to propose a soft-association mechanism to make up the shortage of hard association mechanism which wholly dependent on the normal calibration. Transformer-based fusion model performs fine grained fusion in an attentive manner, which allows the model to adaptively determine where and what information should be taken from image [7], and can tolerant projection noise within a certain extent. It is crucial to find a mechanism to solve severe projection noise or to reduce impact of misalignment in future work.

4.3 Conflicts Due to Data Resolution

Sensors from different modalities often have different resolutions. Generally, the spatial density of LiDAR is notably lower than that of the image, e.g., the 64-line mechanical LiDAR produces ten thousand points in maximum while image with 1080p owns one million pixels. No matter what projection method is adopted, some information must be eliminated because the one-to-one relationship cannot be found. This may lead the model to be dominated by the data of one specific modality, whether it is due to the different resolution between feature vector or the imbalance of original information. Furthermore, the conflicts of data resolution lead to the fusion process always under one branch, which makes the other branch not fully utilized. It is also the bottleneck of perception accuracy.

The low resolution of LiDAR data can be solved slightly by points augmentation, and has been proven to be able to improve far-away or small objects' detection rates. Furthermore, points augmentation has better performance of rigid objects rather than non-rigid objects. Besides, the effect of non-semantic points augmentation is poor, while semantic points augmentation takes too much time. A deeper understanding of source data and a new data representation compatible with sensors in different spatial resolutions is required, to fuse source data in a way that can balance the data of different resolution.

5   Conclusion

In this chapter, we review a large number of related work about multi-sensor fusion for perception in complex traffic environments. Firstly, we talk about the roadside perception along with on-board perception, which is cooperative to autonomous driving and provides traffic administrator with abundant perception information. Then we summarize these fusion methods and give a more reasonable division rule to divide them into three categories, namely, decision-level fusion, decision-feature-level fusion, and feature-level fusion. At last, this chapter gives some recommendations about the research directions in the future to alleviate problems that have been encountered in the field of multi-sensor fusion.

References

1. X. Ye et al., Rope3D: the roadside perception dataset for autonomous driving and monocular 3D object detection task, in *Proceedings of the IEEE Conference on Computer Vision and Pattern Recognition (CVPR)* (2022), pp. 21341–21350
2. J. Zhang, Sensor data validation and driving safety in autonomous driving systems (2022). arXiv preprint arXiv:2203.16130
3. K. Huang, B. Shi, X. Li, X. Li, S. Huang, Y. Li, Multi-modal sensor fusion for auto driving perception: a survey. arXiv preprint arXiv:2202.02703
4. A.V. Malawade, T. Mortlock, M.A. Al Faruque, HydraFusion: context-aware selective sensor fusion for robust and efficient autonomous vehicle perception, in *2022 ACM/IEEE 13th International Conference on Cyber-Physical Systems (ICCPS)* (2022), pp. 68–79
5. M. Liang, B. Yang, Y. Chen et al., Multi-task multi-sensor fusion for 3D object detection, in *Proceedings of the IEEE/CVF Conference on Computer Vision and Pattern Recognition (CVPR)* (2019), pp. 7345–7353
6. Z. Wang, X. Zeng, S. L. Song, Y. Hu, Towards efficient architecture and algorithms for sensor fusion (2022). arXiv preprint arXiv:2209.06272
7. M. Sommer, M. Stang, M. Ferdinand, E. Sax, TalkyCars: a distributed software platform for cooperative perception among connected autonomous vehicles based on cellular-V2X communication, in *IEEE Intelligent Vehicles Symposium (IV)* (2020), pp. 701–707
8. J. Cui, H. Qiu, D. Chen, P. Stone, Y. Zhu, COOPERNAUT: end-to-end driving with cooperative perception for networked vehicles, in *Proceedings of the IEEE Conference on Computer Vision and Pattern Recognition (CVPR)* (2022), pp. 17252–17262
9. Z. Xie, Y. Song, J. Wu, Z. Li, C. Song, Z. Xu, MDS-Net: a multi-scale depth stratification based monocular 3D object detection algorithm (2022). arXiv preprint arXiv:2201.04341
10. A.V. Malawade, T. Mortlock, M.A. Al Faruque, EcoFusion: energy-aware adaptive sensor fusion for efficient autonomous vehicle perception, in *Proceedings of the 59th ACM/IEEE Design Automation Conference (DAC)* (2022), pp. 481–486
11. X. Bai et al., TransFusion: robust LiDAR-camera fusion for 3D object detection with transformers, in *Proceedings of the IEEE/CVF Conference on Computer Vision and Pattern Recognition (CVPR)* (2022), pp. 1090–1099
12. Z. Bai, G. Wu, X. Qi, Y. Liu, K. Oguchi, M.J. Barth, Infrastructure-based object detection and tracking for cooperative driving automation: a survey, in *2022 IEEE Intelligent Vehicles Symposium (IV)* (2022), pp. 1366–1373
13. C. Chang, J. Zhang, K. Zhang et al., BEV-V2X: cooperative birds-eye-view fusion and grid occupancy prediction via V2X-based data sharing. IEEE Trans. Intell. Veh., 1–18 (2023)

14. S. Zheng, C. Xie, S. Yu et al., A robust strategy for roadside cooperative perception based on multi-sensor fusion, in *2022 International Conference on Sensing, Measurement & Data Analytics in the era of Artificial Intelligence (ICSMD)* (2022), pp. 1–6
15. Z. Bai, G. Wu, M. J. Barth, Y. Liu, E. A. Sisbot, K. Oguchi, PillarGrid: deep learning-based cooperative perception for 3D object detection from onboard-roadside LiDAR, in *2022 IEEE 25th International Conference on Intelligent Transportation Systems (ITSC)* (2022), pp. 1743–1749
16. X. An, Research on multi-sensor fusion perception method of vehicle-infrastructure collaboration for smart automobiles, in *Proceedings of the 2021 1st International Conference on Control and Intelligent Robotics (ICCIR)* (2021), pp. 164–175
17. J. Deng, S. Shi, P. Li, W. Zhou, Y. Zhang, H. Li, Voxel R-CNN: towards high performance voxel-based 3D object detection, in *National Conference on Artificial Intelligence (AI)* (2021), pp. 1201–1209
18. S. Lang, F. Ventola, K. Kersting, DAFNe: a one-stage anchor-free deep model for oriented object detection (2021). arXiv preprint arXiv:2109.06148
19. T. Yin, X. Zhou, P. Krahenbuhl, Center-based 3D object detection and tracking, in *Proceedings of the IEEE/CVF Conference on Computer Vision and Pattern Recognition (CVPR)* (2021), pp. 11784–11793
20. A. Garcia-Garcia, S. Orts-Escolano, S. Oprea, V. Villena-Martinez, J. Garcia-Rodriguez, A review on deep learning techniques applied to semantic segmentation (2017). arXiv preprint arXiv:1704.06857
21. J. Long, E. Shelhamer, T. Darrell, Fully convolutional networks for semantic segmentation, in *Proceedings of the IEEE Conference on Computer Vision and Pattern Recognition (CVPR)* (2015), pp. 3431–3440
22. D. Bolya, C. Zhou, F. Xiao, Y.J. Lee, YOLACT: real-time instance segmentation, in *Proceedings of the IEEE/CVF International Conference on Computer Vision (ICCV)* (2019), pp. 9157–9166
23. A.M. Hafiz, G.M. Bhat, A survey on instance segmentation: state of the art. Int. J. Multimed. Inf. Retr. **9**(3), 171–189 (2020)
24. S. Seferbekov, V. Iglovikov, A. Buslaev, A. Shvets, Feature pyramid network for multi-class land segmentation, in *Proceedings of the IEEE Conference on Computer Vision and Pattern Recognition Workshops (CVPRW)* (2018), pp. 272–275
25. D. Riehle, D. Reiser, H.W. Griepentrog, Robust index-based semantic plant/background segmentation for RGB-images. Comput. Electron. Agric. **169**, 105201 (2020)
26. Q. Xie, R. Liu, Z. Sun, S. Pei, F. Cui, A flexible free-space detection system based on stereo vision. Neurocomputing **485**, 252–262 (2022)
27. P. Cerri, P. Grisleri, Free space detection on highways using time correlation between stabilized sub-pixel precision IPM images, in *Proceedings of the 2005 IEEE International Conference on Robotics and Automation (CRA)* (2005), pp. 2223–2228
28. C. Lee, J.-H. Moon, Robust lane detection and tracking for real-time applications. IEEE Trans. Intell. Transp. Syst. **19**(12), 4043–4048 (2018)
29. Y. He et al., Deep learning based 3D segmentation: a survey (2021). arXiv preprint arXiv:2103.05423
30. Q. Huang, W. Wang, U. Neumann, Recurrent slice networks for 3D segmentation of point clouds, in *Proceedings of the IEEE Conference on Computer Vision and Pattern recognition (CVPR)* (2018), pp. 2626–2635
31. E. Ataer-Cansizoglu, Y. Taguchi, S. Ramalingam, T. Garaas, Tracking an RGB-D camera using points and planes, in *Proceedings of the IEEE International Conference on Computer Vision Workshops (ICCVW)* (2013), pp. 51–58
32. Y. Li et al., Deep learning for LiDAR point clouds in autonomous driving: a review. IEEE Trans. Neural Netw. Learn. Syst. **32**(8), 3412–3432 (2021)
33. J. Yang, Z. Kang, Voxel-based extraction of transmission lines from airborne LiDAR point cloud data. IEEE J. Sel. Top. Appl. Earth Obs. Remote Sens. **11**(10), 3892–3904 (2018)
34. Y. Cui et al., Deep learning for image and point cloud fusion in autonomous driving: a review. IEEE Trans. Intell. Transp. Syst. **23**(2), 722–739 (2021)

35. G. Welch, G. Bishop, An introduction to the Kalman filter (1995)
36. S. Pang, D. Morris, H. Radha, CLOCs: camera-LiDAR object candidates fusion for 3D object detection, in *2020 IEEE/RSJ International Conference on Intelligent Robots and Systems (IROS)* (2020), pp. 10386–10393
37. A. Asvadi, L. Garrote, C. Premebida, P. Peixoto, and U. J. Nunes, Multimodal vehicle detection: fusing 3D-LIDAR and color camera data. Pattern Recognit. Lett. **115**, 20–29 (2018)
38. P. Cao, H. Chen, Y. Zhang, G. Wang, Multi-view frustum pointnet for object detection in autonomous driving, in *2019 IEEE International Conference on Image Processing (ICIP)* (2019), pp. 3896–3899
39. C. R. Qi, H. Su, K. Mo, L.J. Guibas, PointNet: deep learning on point sets for 3D classification and segmentation, in *Proceedings of the IEEE Conference on Computer Vision and Pattern Recognition (CVPR)* (2017), pp. 652–660
40. Z. Yang, Y. Sun, S. Liu, X. Shen, J. Jia, IPOD: intensive point-based object detector for point cloud (2018). arXiv preprint arXiv:1812.05276
41. X. Chen, H. Ma, J. Wan, B. Li, T. Xia, Multi-view 3D object detection network for autonomous driving, in *Proceedings of the IEEE Conference on Computer Vision and Pattern Recognition (CVPR)* (2017), pp. 1907–1915
42. J. Ku, M. Mozifian, J. Lee, A. Harakeh, S.L. Waslander, Joint 3D proposal generation and object detection from view aggregation, in *2018 IEEE/RSJ International Conference on Intelligent Robots and Systems (IROS)* (2018), pp. 1–8
43. A. Mahmoud, S.L. Waslander, Sequential fusion via bounding box and motion pointpainting for 3D objection detection, in *2021 18th Conference on Robots and Vision (CRV)* (2021), pp. 9–16
44. J. Deng, K. Czarnecki, MLOD: a multi-view 3D object detection based on robust feature fusion method, in *2019 IEEE intelligent transportation systems conference (ITSC)* (2019), pp. 279–284
45. S. Vora, A.H. Lang, B. Helou, O. Beijbom, PointPainting: sequential fusion for 3D object detection, in *Proceedings of the IEEE/CVF Conference on Computer Vision and Pattern Recognition (CVPR)* (2020), pp. 4604–4612
46. S. Shi, X. Wang, H. Li, PointRCNN: 3D object proposal generation and detection from point cloud, in *Proceedings of the IEEE/CVF Conference on Computer Vision and Pattern Recognition (CVPR)* (2019), pp. 770–779
47. A. H. Lang, S. Vora, H. Caesar, L. Zhou, J. Yang, O. Beijbom, PointPillars: fast encoders for object detection from point clouds, in *Proceedings of the IEEE/CVF Conference on Computer Vision and Pattern Recognition (CVPR)* (2019), pp. 12697–12705
48. V. A. Sindagi, Y. Zhou, O. Tuzel, MVX-Net: multimodal VoxelNet for 3D object detection, In *2019 International Conference on Robotics and Automation (ICRA)* (2019), pp. 7276–7282
49. T. Yin, X. Zhou, P. Krähenbühl, Multimodal virtual point 3D detection. Adv. Neural Inf. Process. Syst. **34**, 16494–16507 (2021)
50. N. Parmar et al., Image transformer, in *International Conference on Machine Learning (ICML)* (2018), pp. 4055–4064
51. Z. Liu et al., BEVFusion: multi-task multi-sensor fusion with unified bird's-eye view representation, in *2023 IEEE International Conference on Robotics and Automation (ICRA)* (2022), pp. 2774–2781
52. H. Caesar et al., nuScenes: a multimodal dataset for autonomous driving, in *Proceedings of the IEEE/CVF Conference on Computer Vision and Pattern Recognition (CVPR)* (2020), pp. 11621–11631

Chapter 9
A Cooperative Positioning Enhancement for Blockchain-Enabled Vehicular Networks

Xuting Duan, Ao Zhang, and Jianshan Zhou

Abbreviations

CV	Connected Vehicles
CVIS	Cooperative Vehicle-Infrastructure Systems
DNN	Deep Neural Network
GNSS	Global Navigation Satellite Systems
GPS	Global Positioning System
IoV	Internet of Vehicles
V2V	Vehicle-to-Vehicle Communication
V2I	Vehicle-to- Infrastructure Communication
VANET	Vehicular Ad-hoc Network
NN	Neural Network
ANN	Artificial Neural Network
BP	Back-Propagation
CNN	Convolutional Neural Network
HPAVs	High Positioning Accuracy Vehicles
LPAVs	Low Positioning Accuracy Vehicles
DGPS	Differential Global Positioning System
VCC	Vehicular Cloud Computing
RSUs	Road Side Units

X. Duan (✉) · A. Zhang · J. Zhou
School of Transportation Science and Engineering, Beihang University, Beijing, China
e-mail: duanxuting@buaa.edu.cn

A. Zhang
e-mail: aozhang@aliyun.com

J. Zhou
e-mail: jianshanzhou@foxmail.com

© The Author(s), under exclusive license to Springer Nature Singapore Pte Ltd. 2023   163
Y. Zhu et al. (eds.), *Communication, Computation and Perception Technologies for Internet of Vehicles*, https://doi.org/10.1007/978-981-99-5439-1_9

1 Introduction

Accurate vehicle positioning technology is key to ensuring safety and improving travel efficiency in traffic application scenarios. Positioning and navigation signals from Global Navigation Satellite Systems (GNSS), such as the Global Positioning System (GPS), GLONASS, the BeiDou navigation satellite system, and Galileo, can be used in most cases to enhance navigation and are widely used. The combination of high-precision positioning technology with vehicle status data information can serve location-based services for Internet of Vehicles (IoV) applications, such as autonomous driving and vehicle emergency rescue. Nevertheless, due to inaccurate GPS positioning of vehicles in urban canyons and under heavy traffic environments and because auxiliary positioning sensors have limited energy storage in high mobility scenarios, supporting application scenarios of automatic driving and other high-precision positioning requirements remains challenging.

Under the influence of building occlusion and the multipath effect in urban areas [1], GNSS cannot meet the positioning accuracy requirements of intelligent vehicle infrastructure cooperative systems and automatic driving applications. Thus, it is necessary to study other methods to improve positioning accuracy and support more application scenarios through auxiliary positioning technology. Researchers are currently working to enhance the overall positioning performance in vehicle application scenarios by using advanced sensors [2].

1.1 Motivation

With the emergence of the IoV, vehicles can obtain more extensive information and further support more applications [3]. Vehicles use off-board information to form cooperative positioning systems in which they benefit from information regarding nearby vehicles (V2V) and roadside infrastructure (V2I). We define vehicle positioning technology in which the information of adjacent vehicles is combined with a vehicle's own sensing information as V2V positioning technology, where a vehicular ad-hoc network (VANET) is formed by V2V communication to obtain each other's position and trajectory. In a sensor-based positioning system, vehicle positioning depends on a sensor or even multi-sensor fusion technology. Nevertheless, it is presently very difficult to incorporate sensors in all vehicles, and human-driven vehicles will still account for a large proportion of vehicles on the road in the coming years. Therefore, it is significant to enhance the positioning precision of vehicles by reducing GNSS error and to study cooperative positioning in a traffic scene where vehicles with different positioning abilities coexist.

On the other hand, collaborative vehicles and mobile edge computing nodes may provide false information to other vehicles, causing malicious attacks. Blockchain technology, as an emerging technology with features such as unauthorized, decentralized and immutable, could become favorable for vehicle location data sharing and safe storage and can also encourage more vehicles to participate in providing positioning error references to further enhance the positioning accuracy of vehicles.

1.2   Main Contributions

Main contributions to this chapter are as follows:

(1) Starting from the actual problem, the traffic positioning scenario where vehicles with different positioning capabilities coexist is considered, and the system positioning system architecture is formed by making full use of roadside facilities;
(2) The vehicle positioning error prediction model based on the neural network is trained by making full use of the vehicular historical data, and the positioning error of the vehicle positioning correction signal is broadcast to the requesting vehicle to enhance the vehicle positioning accuracy;
(3) In the process of information transmission, blockchain technology is combined to realize the cooperative positioning between vehicles with different positioning capabilities under the support of trusted mechanism.

1.3   Structural Organization

Relevant vehicle positioning technologies are introduced in Sect. 2. Section 3 shows the system model and discusses the assumptions that our work is based on. Section 4 presents the positioning error prediction algorithm and positioning accuracy enhancement method. In Sect. 5, a vehicular blockchain system and smart contracts are designed. Section 6 provides and discusses the numerical results of the simulation and presents the conclusions.

2   Related Works

2.1   Machine Learning Applied in Vehicle Positioning

Neural network (NN) technology imitating biological NNs uses the connection between neurons for transmission and feedback to realize the learning and modeling of historical data. Machine learning based on a NN is widely used in the field of transportation. El-Sheimy et al. proposed a method based on an artificial NN (ANN) for two different architectures to assist navigation and localization systems [4]. A feedback NN is more suitable for the position prediction of moving objects. By adjusting the weights and thresholds in the network through back propagation, the learning model can obtain a better prediction effect. Amy et al. proposed a method that applies a back-propagation (BP) NN to cooperative vehicle positioning, and it could estimate the position of a vehicle from the acquired image points [5]. Gurghian et al. presented a method to estimate positions from an image obtained from a camera using a DNN [6]. Baek et al. used the machine learning method based on a Kalman filter to correct vehicle positioning errors and trained vehicle GPS trajectories based on a model-free NN [7]. Peng et al. introduced deep reinforcement learning algorithms to vehicle positioning scenarios and achieved improved vehicle positioning accuracy [8]. Wang et al. studied vehicle positioning based on a convolutional NN (CNN) combined with deep learning to enhance positioning efficiency [9].

The studies above have enhanced vehicle positioning accuracy to variable degrees, proving that machine learning technology can be used in high-precision vehicle positioning research.

2.2   Blockchain Technology for Internet of Vehicles

Blockchain is a decentralized technology that uses cryptography to secure the upload and download of data, enabling consistent storage and increased security of data. Blockchain has the advantages of decentralization, irrevocability, traceability, fault tolerance, etc. [10]. The development of blockchain mainly consisted of public blockchain with public access, consortium blockchain with a consensus of designated nodes, and private blockchain with organizational nodes [11]. With the increase of communicable facilities in the IoV, large-scale data must be managed, which makes the trustworthiness of data transmission in the network a concern [12]. Many studies have explored the use of a blockchain network to build a reliable and transparent system to achieve goals in Intelligent Transportation Systems [13]. Some blockchain technologies have been applied in the IoV to integrate IoV applications with blockchain technology. Yuan et al. built blockchain systems for transportation and reliable architectures for the transportation management system to provide a foundation for subsequent research [14]. Jiang et al. investigated the extension of blockchain technology to the field of IoV and constructed blockchain integration programs for IoV applications [15]. The work in [16] proposes blockchain-centric architecture in which entities, such as smart cars, develop overlay networks that communicate with each other to support safety and reliability. Firdaus et al. proposed consortium blockchains and smart contracts that allow vehicles to check the trust level of neighbor information to achieve decentralized management for data using an incentive mechanism [17]. Additionally, some research studies have applied blockchain systems in the field of vehicle localization. Li et al. introduced a blockchain-based process for localization of information to improve vehicle localization accuracy [18]. Song et al. used a blockchain system for cooperative positioning in a vehicle network to reduce positioning error [19].

Although these efforts have contributed a lot to IoV, especially in the field of vehicle positioning, there is still room for improvement in practical applications. For application scenarios, in particular, vehicles with different positioning modes on the road at present constitute mixed traffic scenarios wherein vehicles with different positioning accuracy abilities coexist. It is meaningful to use high-precision positioning vehicles to provide positioning error information for low-precision positioning vehicles. Additionally, data security in the process of vehicle positioning information sharing is a concern in cooperative positioning. Hence, we construct a DNN with an optimized initial value to predict vehicle positioning error to improve vehicle positioning accuracy and combine it with blockchain technology to realize cooperative positioning between vehicles with different positioning capabilities supported by a trust mechanism.

3 System Model

3.1 Positioning Scenarios

3.1.1 Vehicle Description

There are two kinds of vehicles with different positioning abilities in the vehicle positioning scenario: high positioning accuracy vehicles (HPAVs) and low positioning accuracy vehicles (LPAVs). The GPS receiver is the same for all vehicles. Besides using GPS to obtain positioning information, the HPAVs are equipped with other on-board sensors or a differential GPS (DGPS) to assist in precise positioning. Therefore, the HPAVs can obtain their coordinates through existing high-precision positioning technology. Nevertheless, there is a large error in the positioning information obtained by GPS for LPAVs.

3.1.2 Communication and Computing Capability

The vehicle can realize V2V and V2I communication through the DSRC communication module. HPAVs have strong computing power and large data storage space, and all vehicles have access to road ride units (RSUs) within the communication range. Communication technology enables data and resources to be shared between vehicles.

3.1.3 Vehicular Cloud Computing (VCC)

VCC allows authorized users to dynamically access a set of resources that coordinate vehicles [20]. In VCC, vehicle resources such as computing, storage and sensing are shared to assist in vehicle cooperative positioning. On-device data collection and processing starts with device-level data collection, such as on-board sensors. The

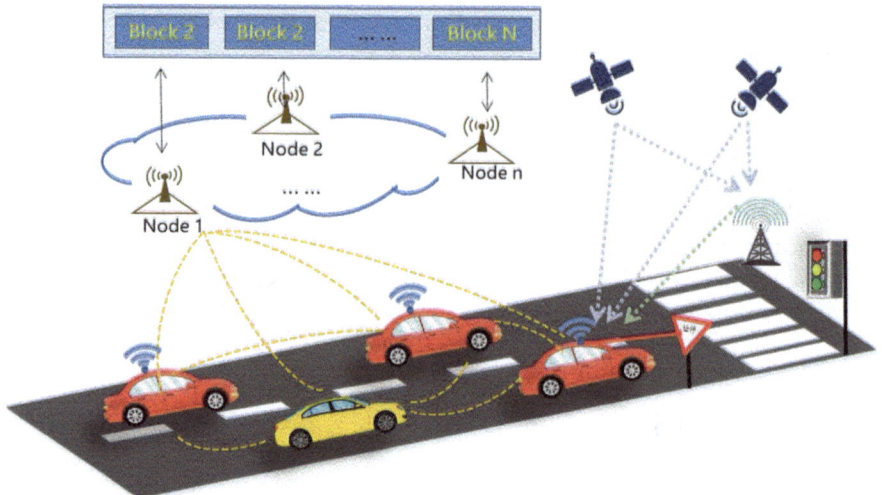

Fig. 1 System architecture

data is sent to the vehicle's local repository for low-level data processing. The second level is V2V communication for resource and information sharing. The VANET infrastructure in the vehicle positioning scenario is abstracted as a vehicular cloud. VCC uses a self-organizing model of the local environment to solve queries. That is, vehicles effectively form a cloud in which services are generated, maintained, and consumed.

Overall system architecture is shown in Fig. 1. In this vehicle positioning scenario, we solve the positioning error problem for two types of vehicles with different positioning abilities:

(1) For the HPAV, the positioning error prediction model is trained by providing positioning error information as the data provider.
(2) For the LPAV, the data requester requests to share the positioning error prediction model to correct the positioning error.
(3) For the positioning scenario where vehicles with different positioning abilities coexist, the positioning accuracy is enhanced through the cooperative positioning method of high-precision positioning vehicles and vehicles with positioning errors.

3.2 Feasibility Analysis of Sharing Positioning Error

The error sources of GPS usually come from GPS satellites, the satellite signal propagation process, and the ground receiving equipment. Some researchers divide these kinds of error sources into systematic and random natures [21]. The GPS positioning error can be expressed as:

$$E = E_{sys} + E_{ran} \tag{1}$$

where E_{sys} is the systematic error and E_{ran} is the random error. Random errors are negligible for larger systematic errors.

In the positioning scenario of this paper, the satellite clock difference and ionospheric delay, which are the sources of systematic error, are basically the same under the condition that the satellite combination observed by different vehicles is basically the same in a similar time period and when the GPS receiver is the same type. The position difference between two vehicles driving on the same road section is far less than the distance between vehicles and the satellite.

4   Positioning Error Prediction Algorithm

4.1   Design of the DNN Algorithm

Deep learning models can not only simulate multilayer nonlinear mapping but also have high fault tolerance and adaptive ability [22]. In the vehicular positioning problem, many factors affect the positioning accuracy, and the correlation of each factor is complex. Therefore, a DNN with nonlinear problem solving ability can be used to predict vehicular positioning error. The HPAVs using positioning information and movement status act as DNN training data. The proposed DNN prediction algorithm runs on RSUs with powerful computing power and data storage space.

(1) DNN structure: The designed DNN as shown in Fig. 2. A fully connected NN is used to predict position errors. The number of nodes in each hidden layer is the same if the hidden layer is greater than 1.

(2) Input and output nodes: We mainly consider the motion state of the driving vehicle as the input node. There are seven input nodes in total, including the vehicle GPS position $p_{Gi}^x(t)$, $p_{Gi}^y(t)$, the vehicle speed $v_i(t)$, heading $h_i(t)$, and acceleration $a_i(t)$ at the current moment. The vehicle GPS position is given by $p_{Gi}^x(t-1)$ and $p_{Gi}^y(t-1)$ at the last moment.

The output nodes are the positioning errors in the X and Y directions we need to predict.

(3) Hidden nodes and layers: The number of hidden layers can influence the effect of network training, while too many layers will also lead to time-consuming problems. Thus, we need to balance network accuracy and time cost with the number of the hidden layers between the number of input and output nodes [23].

The number of nodes in each layer is a critical parameter affecting network performance. An appropriate number of neurons will enable the network to have excellent information processing ability without reducing training efficiency or falling into a local optimal solution. In this paper, the number of hidden layers nodes of the network that does not lead to overfitting [24] is defined as:

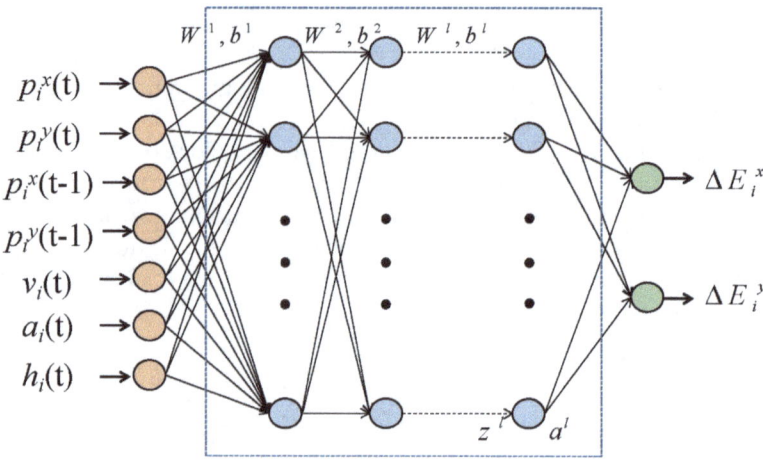

Fig. 2 The structure of DNN

$$N = \frac{S}{\lambda * (I + O)} \tag{2}$$

where S is data set size. The number of input and output nodes is represented by I and O. The constant λ is a scaling factor within [2, 10].

As shown in Fig. 2, z^l indicates the input vector, a^l indicates the output vector, and b^l indicates the bias vector of the l'th layer. Therefore, the relationship between these parameters is as follows:

$$z^l = W^l * a^{l-1} + b^l \tag{3}$$

where W is the weight of output and input between two hidden layers. By defining the activation function $\sigma(\cdot)$, we can get

$$a^l = \sigma(z^l) = \sigma(W^l * a^{l-1} + b^l) \tag{4}$$

To reduce the dependency between parameters and prevent overfitting, rectified linear units (ReLU) is used as the activation function

$$\sigma(x) = \begin{cases} z, \ z \geq 0 \\ 0, \ z < 0 \end{cases} \tag{5}$$

In the process of training, the cost function is constantly minimized. To make it applicable to numerical prediction problems, we choose the mean square error (MSE) of the output value and target value as the cost function; that is

$$J(W, b, x, y) = \frac{1}{2n} \sum_{i=1}^{n} (y_x - a_x^L)^2 \tag{6}$$

where y_x and a_x^L are the target value and output value of each sample, respectively. There are a total of L hidden layers, and W and b are the weight matrix and sample training bias, respectively. We can get the gradient

$$\frac{\partial J}{\partial W^L} = \frac{\partial J}{\partial a^L} \frac{\partial a^L}{\partial z^L} \frac{\partial z^L}{\partial W^L} = (a^L - y) \circ \sigma'(z^L)(a^{L-1})^T \tag{7}$$

and

$$\frac{\partial J}{\partial b^L} = \frac{\partial J}{\partial a^L} \frac{\partial a^L}{\partial z^L} \frac{\partial z^L}{\partial b^L} = (a^L - y) \circ \sigma'(z^L) \tag{8}$$

Finally, we can obtain

$$\frac{\partial J}{\partial b^l} = (W^{l+1})^T \delta^{l+1} \circ \sigma'(z^l) \tag{9}$$

and

$$\frac{\partial J}{\partial W^l} = (W^{l+1})^T \delta^{l+1} \circ \sigma'(z^l)(a^{l-1})^T \tag{10}$$

Therefore, the adjustment rules of W^l and b^l are

$$W^l = W^l - \alpha * \delta^l (a^{l-1})^T \tag{11}$$

and

$$b^l = b^l - \alpha * \delta^l \tag{12}$$

4.2 EPSO Optimized DNN Algorithm

At the same time, the initialization methods of weights and thresholds in the DNN algorithm will affect algorithm convergence, whereas the random initialization methods commonly used in DNN may affect the convergence efficiency.

The particle swarm optimization algorithm (PSO) can achieve ideal convergence at the beginning of optimization, but premature convergence afterward leads to low efficiency [25]. In this section, an evolutionary PSO (EPSO) algorithm based on the selection of crossover factors is used to solve problems encountered in the PSO convergence process.

The EPSO algorithm is based on the idea of combined crossover and mutation in the genetic algorithm and increases the tracking range of particle swarms by using a crossover factor to generate new populations representing new solution sets. This process allows populations to show a better adaptive capacity to search for the global optimal solution. In the crossover factor method, the corresponding fitness values are sorted after updating the position and velocity of the particle swarm each time. The top half of the particles can be directly used as the next-generation particles. At the same time, the position and velocity of the same half particles with better fitness are used to replace the other half particles to keep the individual extremum of the latter unchanged. In the crossover mechanism, the last half of the particles carry out the crossover operation in the genetic algorithm to generate offspring. Then, half of the particles with the top-ranking fitness value enter the next generation. By crossing, the diversity of particles can be increased to escape from the local optimum, and the convergence speed can be accelerated. In the crossover operation, the real-number crossover method is adopted, and two arbitrary particles in the population can be matched.

The network structure parameters of the DNN are determined by EPSO (EPSO-DNN) such that the optimized DNN model is more suitable for data prediction. In this paper, the weights and thresholds of the network are taken as the particle positions of the particle swarm. Combining the problem of data prediction with the meaning of the loss function, the loss function in DNN as the fitness function of the EPSO algorithm can affect the model prediction effect. In our algorithm, the individual local optimum and the group global optimum are selected. The crossover factor is introduced to enhance the excellent characteristics of swarm particles, get out of the local optimum, and accelerate the convergence speed. Experimental results show that the proposed algorithm greatly improves the ability of jumping out of local convergence and has strong global search ability, and can effectively avoid the premature convergence problem of conventional algorithms. It can maintain a balance of exploration and development to facilitate the algorithm to perform at maximum capacity.

4.3 Positioning Accuracy Enhancement Method

LPAVs use the positioning information of HPAVs in the identification range of traffic signs as training sets to predict the current positioning error of vehicles. The positioning accuracy of LPAVs is so low that it needs to be enhanced by positioning error information provided by HPAVs. The following three methods are proposed to enhance the positioning accuracy of vehicles using the precise positioning information provided by HPAVs. The workflow of the proposed positioning accuracy enhancement method is as follows:

(1) Data sharing:

The positioning information data of HPAVs are uploaded to edge nodes as a training set to train the DNN model to predict positioning errors. The LPAVs send requests to the RSUs within the communication range to obtain the trained DNN model and predict vehicle positioning errors by combining their own GPS positioning and vehicle motion state information.

(2) Model sharing:

The model parameters and positioning errors of J HPAVs were averaged on the RSU, and a hidden layer was added to the DNN model to continue training with the transfer learning strategy. This allows us to obtain a new error evolution to predict the positioning errors of other vehicles. LPAVs can send a request to the nearby RSU and can then use the trained DNN prediction model to predict the vehicle positioning error combined with their own GPS positioning and vehicle motion state information.

(3) EPSO-DNN sharing:

Since HPAVs have computing and storage capabilities, they can use their own historical data to train EPSO-DNN models and upload the trained model parameters and positioning errors to mobile edge computing nodes. Then, the steps are repeated in model sharing for error evolution to predict the positioning errors of other vehicles.

5 Blockchain and Smart Contracts

5.1 Vehicular Blockchain

HPAVs can obtain their positioning information through their own high-precision positioning technology. From the derivation Sect. 4, the HPAV information can be provided to LPAVs in the immediate area at similar moments. In other words, the positioning error of LPAVs can be predicted by obtaining the error prediction model of HPAVs to make a positioning error correction. RSU is used as the information storage center to ensure the timeliness and validity of positioning error sharing. Additionally, the vehicle may provide incorrect data to attack the RSU, making the error prediction model provided by the RSU for other vehicles unreliable. Therefore, blockchain is used to make the storage and sharing of positioning error information safe and reliable.

Since vehicle positioning information and movement status are only shared on the nearby roadway, we use consortium blockchain to reduce the cost of traditional blockchain and speed up the propagation, as shown in Fig. 3. In this scenario, HPAVs upload the encrypted data to the nearby RSU for reward as data providers, while LPAVs send the positioning error correction request to the RSU to obtain the positioning error prediction model as the data requesters. Noteworthily, the RSU acts as both a storage center and an execution controller during data sharing.

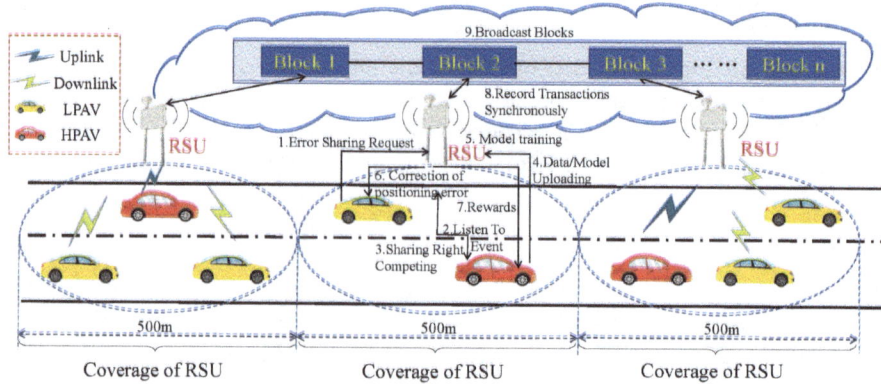

Fig. 3 Smart contract process for blockchain

5.2 *Secure Data Sharing Scheme for Enhanced Positioning Accuracy*

Smart contracts can avoid the impact of malicious behavior on the data sharing process without a third party, so such that data storage and sharing are efficient and make transactions are traceable and safe. According to the sharing method in Sect. 4, the shared data include positioning error data, positioning error prediction model parameters, or optimized positioning error prediction model parameters. We use blockchain technology to avoid malicious attacks when sharing data, which can ensure that the data requester selects the appropriate data provider to provide itself with positioning error correction information. The proposed sharing scheme of enhanced positioning accuracy as follows:

(1) Positioning Error Correction Requesting

LPAV nodes can send positioning error correction requests to nearby RSU nodes to obtain positioning error data or positioning error prediction model parameters. RSU nodes will receive the encrypted message sent by the vehicle with the corresponding reward and process the encrypted message.

(2) Sharing Right Competing

All HPAV nodes in the communication range will provide their own prediction model training accuracy to share data and get rewarded. When the model training accuracy is higher than a certain threshold, the HPAV node can obtain the right to data sharing and become a data provider.

(3) Data Sharing

The data provider encrypts the positioning error correction information using the private key and sends it to the RSU. The RSU collects the positioning error correction information sent by the data providers and retrains the model. Then, the RSU uses the public key of the data requester to encrypt the trained model parameters.

(4) Receiving and Rewarding

The LPAV requester receives the data or model parameters from the RSU and uses the private key to decrypt the received encrypted data and corrects the positioning error based on its own vehicle positioning information. All data providers are rewarded with vehicle coins when the LPAV requester receives the encrypted data. In addition, similar to [26], each vehicle has a wallet account for storing and managing personal vehicle coins. LPAV transfers vehicle coins from its wallet to a wallet address given by the HPAV.

(5) Transactions Recording

The RSU node records and periodically packages the complete positioning error correction process into blocks and broadcasts to the system. Practical Byzantine fault tolerance (PBFT) algorithms can reach consensus in practical scenarios where a small number of nodes do evil (such as forging messages), which is suitable for consortium blockchain [27]. At least $3f + 1$ RSUs are randomly selected as consensus nodes, where $f \geq 1$. The above methods improve security by ensuring that the randomly selected nodes are different. The above methods improve security by ensuring that the randomly selected nodes are different. The execution flow is shown in Fig. 4.

Deep learning is a representation learning method of data in machine learning. Reinforcement learning obtains an optimal strategy by exploring the unknown environment, establishing the environment model and learning. Deep reinforcement learning combines the perception ability of deep learning with the decision-making ability of reinforcement learning, which can be controlled directly according to the

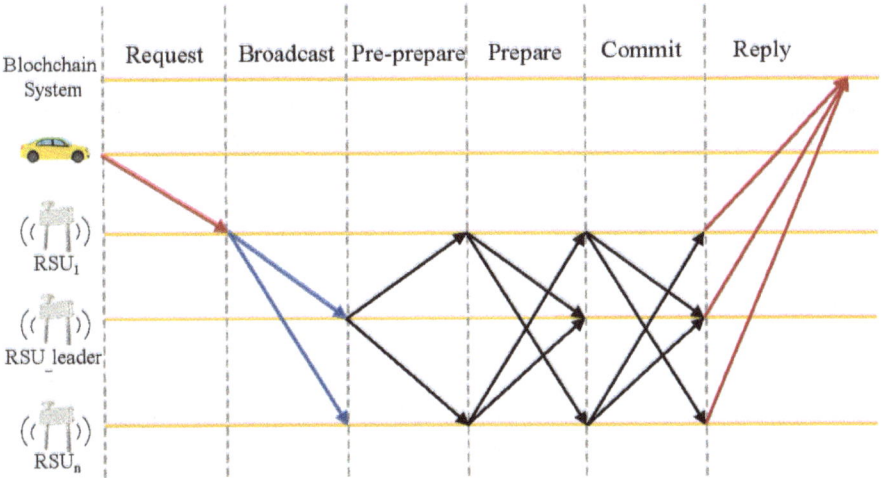

Fig. 4 Consensus mechanism for enhanced positioning accuracy

input information. In work [28], deep reinforcement learning method is effectively used to decide how to better allocate taxi drivers to users in taxi-hailing apps. Most of the current parking resource management strategy can only ensure the optimal profit in the current or a short future time slot, while deep reinforcement learning is able to provide long-term solutions. However, it faces the same problem as blockchain technology, that is to say, it is difficult to implement in actual scenarios.

- Broadcast: The RSUs receive the positioning error correction request from the LPAVs. The RSU node will broadcast the request to the consensus RSUs in the blockchain network, which is composed of all RSU nodes. Each node collects and stores events sent by other nodes.
- re-Prepare: Each node determines whether it is the RSU_leader (primary node). The RSU_leader will package and validate the transaction and then broadcast the block validation result along with the ordered transaction information as a Pre-Prepared message to the whole blockchain network.
- Prepare: After receiving the Pre-Prepared message sent by the RSU_leader, other RSU nodes will verify the authenticity and validity of the content in the block. After other RSU nodes confirm whether the message comes from the RSU leader and whether they accept it for the first time, the verification results will be broadcast in the blockchain network as a Prepare message.
- Commit: The RSU node verifies and compares the Prepare message received from other RSU nodes with the message received from the RSU_leader in the second phase and gives feedback on whether the block is successfully generated. The Prepare message enters the commit state when it has more than $2f$ feedback agreed. The RSU node broadcasts a Commit message to the other RSU nodes.
- Reply (write block): Each RSU node receives the Commit message. When the node judges that the message has obtained $2f + 1$ (self-included) certifications, it is recognized that the blockchain system has reached a consensus on the transaction and generates a new block.

Each RSU node will receive the positioning error correction request of LPAVs within the coverage area in a period of time and generate records for collection to the RSU_leader node. The transaction is certified and sent to other RSUs in the network for validation to determine whether the new block is successfully generated.

6   Simulation and Conclusion

6.1   *Positioning Error Prediction and Correction*

We conducted experiments on the positioning error prediction model for the following three scenarios:

(1) Open Road: Receive satellite signal well

First, the positioning error correction effect of the open road section is discussed. In this scenario, the satellite signal can maintain a continuous receiving state. The number of visible satellites is commonly higher than the required four [29]. The data in Fig. 5 is the average positioning error of the whole road section after correction. In the case of the same number of vehicles, the method of sharing data and the sharing model can correct the positioning error to a certain extent, whereas the modified average error of the EPSO-DNN method can achieve a submeter positioning accuracy of 0.37 m for the whole road section when three vehicles share positioning accuracy enhancement information.

(2) Boulevard: Semi-occlusion affects signal reception

The positioning accuracy enhancement effect of a boulevard is unstable because of intermittent tree occlusion. In Fig. 6, when the two positioning accuracy enhancement methods of shared data and the shared model are used. However, error correction under the EPSO-DNN method can still reach 0.46 m average positioning error and achieve submeter positioning accuracy. Although the intermittent occlusion of trees can affect the correction effect of the method, the three methods are still feasible according to the accuracy enhancement results of the positioning error.

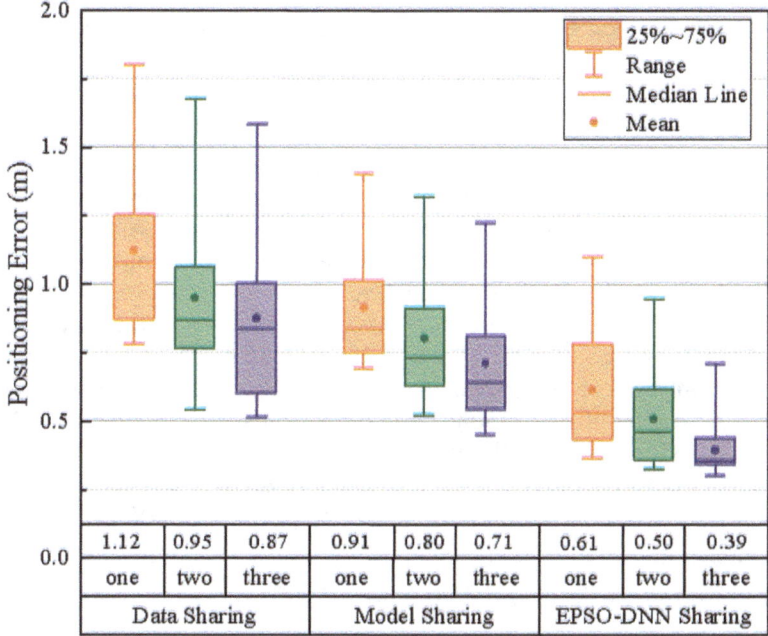

Fig. 5 Comparison of positioning errors between different positioning methods on open road

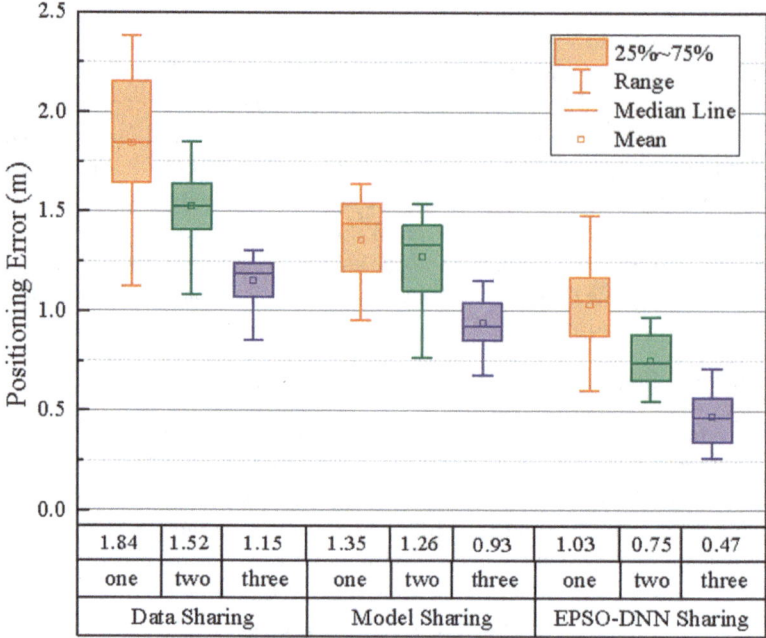

1.84	1.52	1.15	1.35	1.26	0.93	1.03	0.75	0.47
one	two	three	one	two	three	one	two	three
Data Sharing			Model Sharing			EPSO-DNN Sharing		

Positioning Method

Fig. 6 Comparison of positioning errors between different positioning methods on boulevard

(3) Obscured Road: Tall buildings block satellite signals

Under the influence of a block of high buildings, the satellite signal is unstable. The positioning errors of the 30 sampling points in Fig. 7 are corrected by the three methods. In this scenario, the maximum error of the sampling point is close to 11 m, and each of the positioning accuracy enhancement methods can achieve a reduction in positioning error. The least effective of these methods is data sharing, which has a mean error of 2.70 m. LPAVs use the positioning error data provided by the shared positioning information vehicles to train the model for direct prediction, whereas the weight and threshold of the DNN model are not well selected. There are still large errors in the method of data sharing, but both model sharing and the EPSO-DNN sharing method can achieve the ideal effect of positioning error correction by sharing positioning error information among multiple vehicles.

6.2 Comparison Between Different Methods

The positioning error correction based on three positioning accuracy enhancement methods was carried out for three scenarios. From the perspective of the number of

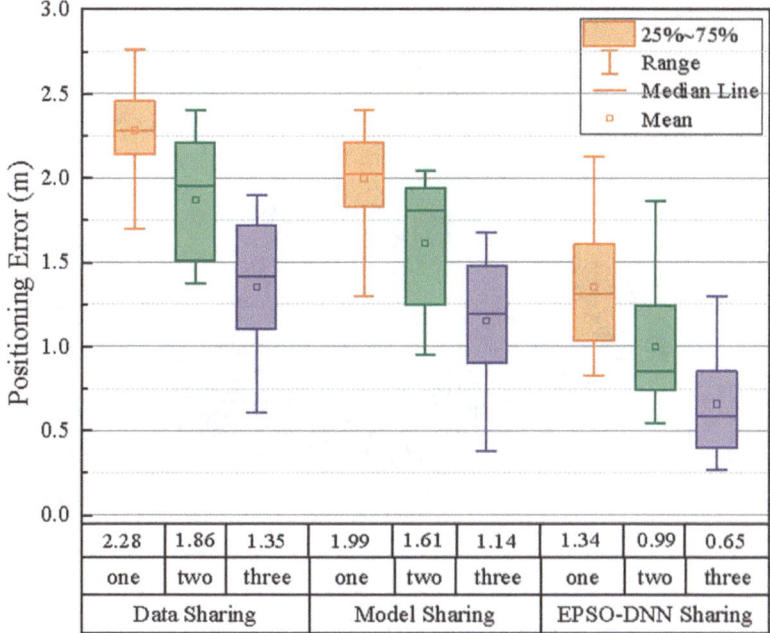

2.28	1.86	1.35	1.99	1.61	1.14	1.34	0.99	0.65
one	two	three	one	two	three	one	two	three
Data Sharing			Model Sharing			EPSO-DNN Sharing		

Positioning Method

Fig. 7 Comparison of positioning errors between different positioning methods on obscured road

shared vehicles, the three methods are based on the shared information of shared vehicles to correct the positioning error. Therefore, the positioning accuracy of vehicles using the positioning accuracy enhancement method increases with the increase of the number of shared vehicles. The weights and thresholds of the DNN model in the model sharing method and the EPSO-DNN sharing method are optimized before training; the positioning accuracy enhancement performance results in the three scenarios are better than that for data sharing. Figure 8 shows the MSE of the model prediction results and the actual error of the three positioning accuracy enhancement methods in the road sections of the three scenarios when the number of vehicles sharing positioning error information is three. By contrast, the maximum MSE of data sharing is 1.58 in the scenario of an obscured road. This is because the occlusion of highrise buildings affects satellite reception signals and directly using the positioning error information of shared vehicles for training is not conducive to the prediction of positioning error. Under the condition of positioning information of multivehicle cooperative sharing, the EPSO-DNN sharing method has the best performance, and the minimum MSE on the open road is only 0.09. This shows that the prediction model based on the EPSO-DNN sharing method can more accurately describe the experimental data and that it is reliable for enhancing the positioning accuracy of vehicles.

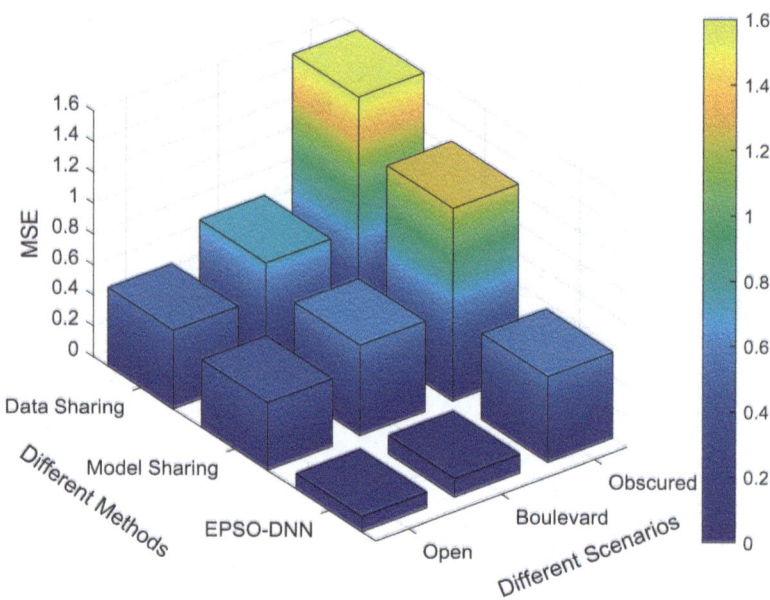

Fig. 8 The MSE of positioning error prediction results and true error

6.3 Conclusion

This chapter considers the traffic positioning scenario where vehicles with different positioning abilities coexist in accordance with an actual situation. Cooperative positioning between vehicles with different positioning abilities is realized by combining existing high-precision positioning technology and IoV technology to enhance vehicle positioning accuracy and the safety and privacy of the system. Based on the characteristics of different vehicle positioning capabilities, three DNN-based vehicle positioning accuracy enhancement methods are proposed; that is, positioning error information of HPAVs is used to enhance the positioning accuracy of LPAVs. Moreover, to make the system more reliable and efficient, we designed a blockchain system based on the PBFT consensus algorithm. Additionally, we conducted simulation experiments for three scenarios. Experiments proved that the EPSO-DNN method can effectively correct the positioning error and further enhance the positioning accuracy of vehicles. The method is feasible, accurate, and safe. In addition, due to the limited experimental conditions and equipment, we adopted a positioning system in 2D. If the hybrid positioning system can be used when conditions permit in the future, the DNN model will have more input nodes and the training effect will be more accurate. In future work, improvement of the efficiency of data sharing for location information stored on the blockchain requires further study.

References

1. M. Rohani, D. Gingras, D. Gruyer, A novel approach for improved vehicular positioning using cooperative map matching and dynamic base station DGPS concept. IEEE Trans. Intell. Transp. Syst. **17**(1), 230–239 (2016)
2. S. Kuutti, S. Fallah, K. Katsaros, M. Dianati, F. Mccullough, A. Mouzakitis, A survey of the state-of-the-art localization techniques and their potentials for autonomous vehicle applications. IEEE Internet Things J. **5**(2), 829–846 (2018)
3. F. Yang, S. Wang, J. Li, Z. Liu, Q. Sun, An overview of internet of vehicles. China Commun. **11**(10), 1–15 (2014)
4. N. El-Sheimy, K.W. Chiang, A. Noureldin, The utilization of artificial neural networks for multisensor system integration in navigation and positioning instruments. IEEE Trans. Instrum. Meas. **55**(5), 1606–1615 (2006)
5. M.S. Ifthekhar, N. Saha, Y.M. Jang, Stereo-vision based cooperative-vehicle positioning using OCC and neural networks. Opt. Commun. **352**, 166–180
6. A. Gurghian, T. Koduri, S.V. Bailur, K.J. Carey, V.N. Murali, DeepLanes: end-to-end lane position estimation using deep neural networks, in *2016 IEEE Conference on Computer Vision and Pattern Recognition Workshops (CVPRW), Las Vegas, NV* (2016), pp. 38–45
7. S. Baek, C. Liu, P. Watta, Y.L. Murphey, Accurate vehicle position estimation using a Kalman filter and neural network-based approach, in *2017 IEEE Symposium Series on Computational Intelligence (SSCI), Honolulu, HI* (2017), pp. 3058–3065
8. B. Peng, G. Seco-Granados, E. Steinmetz, M. Frohle, H. Wymeersch, Decentralized scheduling for cooperative localization with deep reinforcement learning. IEEE Trans. Veh. Technol. **68**(5), 4295–4305 (2019)
9. Y. Wang, Y. Feng, H. Sun, Research on vehicle intelligent wireless location algorithm based on convolutional neural network. Neural Comput. Appl. **33**(14), 8131–8141 (2020)
10. C. Peng, C. Wu, L. Gao, J. Zhang, Y. Ji, Blockchain for vehicular internet of things: recent advances and open issues. Sensors **33**(14), 8131–8141 (2020)
11. J. Xie, H. Tang, T. Huang, F.R. Yu, R. Xie, J. Liu et al., A survey of blockchain technology applied to smart cities: research issues and challenges. IEEE Commun. Surv. Tutor. **21**(3), 2794–2830 (2019)
12. L.M. Ang, K.P. Seng, G.K. Ijemaru, A.M. Zungeru, Deployment of IoV for smart cities: applications, architecture, and challenges. IEEE Access **7**, 6473–6492 (2019)
13. M.B. Mollah, J. Zhao, D. Niyato, Y.L. Guan, L.H. Koh, Blockchain for the internet of vehicles towards intelligent transportation systems: a survey. IEEE Internet Things J. **8**(6), 4157–4185 (2021)
14. Y. Yuan, F.Y. Wang, Towards blockchain-based intelligent transportation systems, in *IEEE International Conference on Intelligent Transportation Systems* (IEEE, Rio de Janeiro, BRAZIL, 2016), pp. 2663–2668
15. T. Jiang, H. Fang, H. Wang, Blockchain-based internet of vehicles: distributed network architecture and performance analysis. IEEE Internet Things J. **6**(3), 4640–4649 (2019)
16. A. Dorri, M. Steger, S.S. Kanhere, R. Jurdak, Blockchain: a distributed solution to automotive security and privacy. IEEE Commun. Mag. **55**(12), 119–125 (2017)
17. M. Firdaus, S. Rahmadika, K.H. Rhee, Decentralized trusted data sharing management on internet of vehicle edge computing (IoVEC) networks using consortium blockchain. Sensors **21**(7), 2410 (2021). https://doi.org/10.3390/s21072410
18. C. Li, Y. Fu, F.R. Yu, T.H. Luan, Y. Zhang, Vehicle position correction: a vehicular blockchain networks-based GPS error sharing framework. IEEE Trans. Intell. Transp. Syst. **22**(2), 898–912 (2021)
19. S.Y. Song, Y. Fu, F.R. Yu, L. Zhou, Blockchain-enabled internet of vehicles with cooperative positioning: a deep neural network approach. IEEE Internet Things J. **7**(4), 3485–3498 (2020)
20. S. Olariu, A survey of vehicular cloud research: trends, applications and challenges. IEEE Trans. Intell. Transp. Syst. **21**(6), 2648–2663 (2020)

21. S. Baselga, L. Garcia Asenjo, Global robust estimation and its application to GPS positioning. Comput. Math. Appl. **56**(3), 709–714 (2008)
22. I. Skog, P. Handel, In-car positioning and navigation technologies a survey. IEEE Trans. Intell. Transp. Syst. **10**(1), 4–21 (2009)
23. I.J. Goodfellow, Y. Bengio, A. Courville, *Deep Learning* (MIT Press, Cambridge, MA, USA, 2016). http://www.deeplearningbook.org
24. M.T. SHagan, H.B. Demuth, M.H. Beale, O. De Jesus, *Neural Network Design* (Martin Hagan, Cambridge, MA, USA, 2014)
25. J. Hu, F. Hua, W. Tian, Robot positioning error compensation method based on deep neural network. J. Phys. Conf. Ser. (Singapore, SINGAPORE) (2020)
26. Z. Li, J. Kang, R. Yu, D. Ye, Q. Deng, Y. Zhang, Consortium blockchain for secure energy trading in industrial internet of things. IEEE Trans. Ind. Inf. **14**, 3690C3700 (2018)
27. M. Castro, B. Liskov, Practical byzantine fault tolerance and proactive recovery. ACM Trans. Comput. Syst. **20**(4), 398–461 (2002)
28. W. Liu, A novel time delay estimation algorithm for 5G vehicle positioning in urban canyon environments. Sensors **20**(18) (2020)
29. N. Alam, A. Kealy, A.G. Dempster, An INS-aided tight integration approach for relative positioning enhancement in VANETs. IEEE Trans. Intell. Transp. Syst. **14**(4), 1992–1996 (2013)

Chapter 10
Cooperative Cloud-Edge Computing for Integrated Sensing and Communication in Internet of Vehicles

Shuyuan Zhao, Daoxun Li, and Yongdong Zhu

Abbreviations

CNN	Convolutional Neural Network
DNN	Deep Neural Network
ISAC	Integrated Sensing and Communication
MEC	Multi-access Edge Computing
OTFS	Orthogonal Time Frequency Space
RSU	Road Side Unit
V2I	Vehicle-to-Infrastructures
V2V	Vehicle-to-Vehicle

1 Introduction

Over the past few years, with the evolution of wireless communication technology and the popularity of intelligent vehicles, vehicular networks have attracted significant attention from academia and industry. The globally connected car market is projected to reach USD 56.3 billion by 2026, with a compound annual growth rate of 19.0%

S. Zhao · D. Li · Y. Zhu (✉)
Zhejiang Lab, Interdisciplinary Innovation Research Institute, Hangzhou, China
e-mail: zhuyd@zhejianglab.com

S. Zhao
e-mail: zhaosy@zhejianglab.com

D. Li
e-mail: lidx@zhejianglab.com

from 2021 to 2026. More than 125 million passenger cars sold between 2018 and 2022 will be equipped with embedded connectivity [1]. At the same time, with the development of autonomous driving technologies, a wide variety of valuable services (such as ultra-high-definition video, real-time navigation, and traffic safety-related services) have emerged and will bring the explosive growth of data traffic and colossal energy consumption in vehicular networks. Especially as the deep neural network (DNN) models gradually replace traditional algorithms in computation-intensive services, the demands of task latency and energy consumption in Internet of Vehicles become much higher. Therefore, collaborative cloud-edge computing, which splits and offloads computing tasks to cloud and edge servers, is considered to improve the service experience.

1.1 Motivation

Considering the power and resource-constrained vehicle ability for intelligent transportation systems services, pre-processing and offloading computing tasks are required. Under the intensive and massive data communication environment in Internet of Vehicles, it is hard to meet the real-time reliability of the computing tasks. Offloading the computing tasks to cloud or edge servers is a solution that can take advantage of the computing resources in the whole network to address the computation-intensive applications.

Presently, researchers have studied the computation offloading strategy, but most of the literature is a separate study on cloud computing offloading or edge computing offloading, and little attention is paid to the collaboration between cloud and edge computing. In addition, the literature often assumes that the directly connected multi-access edge computing (MEC) server has sufficient resources to meet all requirements. In fact, constrained by hardware and cost, the resources provided by the MEC server are relatively limited, especially in areas with dense vehicles. Relying only on edge computing may not be able to meet the requirements of massive data. In addition, due to the exponential growth of sensing information in Internet of Vehicles, the problem of scarce spectrum resources has become increasingly prominent. Therefore, integrated sensing and communication (ISAC) technology, which can jointly design the sensing and communication systems to share the same frequency band and hardware to reduce the latency, is considered to be used in Internet of Vehicles. Therefore, in this section, we focus on the critical technologies of integrated sensing and communication for cooperative cloud-edge computing in Internet of Vehicles.

1.2 Main Contributions

Main contributions to this chapter are as follows:

(1) This chapter introduces collaborative cloud-edge-end computing for deep learning training and inference in detail. Among them, this chapter focuses on early-exit strategy, model splitter and communication compression;

(2) This chapter establishes a general mathematical for the joint communication and computation resource allocation problem and solves it with a deep reinforcement learning-based solution.

(3) This chapter discusses the critical issues of cooperative cloud-edge computing to be solved in the future.

1.3 Structural Organization

Section 2 introduces the background and research status of the key technologies involved in cooperative cloud-edge computing, including intelligent computing and ISAC. Section 3 presents collaborative training and inference methods, joint communication and computation resource allocation strategy in detail. Section 4 looks forward to the future development of cooperative cloud-edge computing, and Sect. 5 summarizes the whole chapter.

2 Background

2.1 Intelligent Computing

2.1.1 Cloud Computing

With the rapid development of information and communication technology in the past few decades, cloud computing has been recognized by academics and industry as a compelling paradigm for managing and delivering services over the internet. The work [2] indicates that cloud computing can provide flexible computing resources and storage capabilities. However, As the application scale expands and the number of access devices increases, the network bandwidth will become a bottleneck, making it difficult for cloud computing to support large-scale real-time computing and data requests.

2.1.2   Edge Computing

Latency is almost inevitable in the cloud computing architecture, considering the multi-level routing of the backbone network that the data transmission process from the access point to the cloud computing center. Therefore, edge computing is proposed as an effective way to reduce latency by bringing some computing and storage resources closer to data sources. The work [3] defines edge computing as an enabling technology that allows computation to be performed at the edge of the network, with downstream data representing cloud services downstream and upstream data representing IoT services. Compared with cloud computing, edge computing can significantly reduce the bandwidth pressure and satisfy the low latency requirements of many emerging applications like AR/VR, image processing, etc. Edge computing also provides several additional benefits like security, privacy, and scalability. In addition, regarding deep learning with edge computing, the work [4] introduces the applications where deep learning is used at the network edge, and various approaches to quickly execute deep learning inference through edge computing. The work [5] proposes a distributed DNN computing system that utilizes available computation and communication resources to coordinate collaborative DNN inference on heterogeneous edge devices through a dynamic DNN model partition approach. However, Edge computing also has its own limitations, such as limited computing resources, higher costs, more incredible difficulty in prevention, etc.

2.1.3   Collaborative Computing

In recent years, with the widespread use of deep learning algorithms in computation-intensive applications, the constrained computing resources in vehicles are insufficient to satisfy the complex DNN inferences. Considering the advantages and disadvantages of cloud computing and edge computing, collaborative cloud-edge computing is proposed to improve the service quality of computation tasks. The work [6] also indicates that edge servers with caching and offloading to cloud servers are complementary to cloud computing, and they can reduce the total execution delay for users within the collaborative cloud-edge computing system. In this system, users' tasks are split into multiple parts and offloaded to edge and cloud, respectively. In this process, the delay caused by uplink/downlink transmission must be considered, which significantly impact the design of collaborative cloud-edge resource allocation strategy.

Furthermore, a flexible and reasonable task segmentation strategy is also an import part of the collaborative cloud-edge computing framework. The work [7, 8] proposes a DNN partitioning method to leverage hybrid computation resources in proximity for real-time DNN inference and jointly optimize model partitioning and early exit method to maximize the accuracy while not violating the latency deadline. Based on the previous work, the work [9] proposes a device-edge collaborative inference system based on early-exit policy in the task stream scenario. The work [10] further introduces a distributed inference system that co-optimizes the early-exit policy and convolutional neural network (CNN) splitting at the runtime and proposes an early-exit-aware cancellation mechanism that allows the interruption of the inference.

Cooperative cloud-edge computing is usually heterogeneous and unstable, at the same time, users' needs are significantly differentiated, dramatically increases the difficulty of allocating computing, storage, and network resources.

2.2 ISAC for Cooperative Cloud-Edge Computing

With the evolution of wireless communication technology and the popularity of intelligent vehicles, a wide variety of innovative vehicle applications requiring higher data transmission rates have emerged, such as real-time video transmission services, 3D high-definition maps and autonomous driving-related services. These computing-intensive applications will bring explosive growth of data traffic, which puts forward higher requirements of the performance indicators in Internet of Vehicles, such as bandwidth and delay. Due to the exponential growth of the amount of sensing information in Internet of Vehicles and the improvement of the performance requirements of the vehicle communication system, the problem of scarce spectrum resources has become increasingly prominent. Besides, the performance of single-vehicle sensing devices is limited as the electromagnetic environment is becoming more complex. Information transmission between sensing devices can be realized through communication networking technology, thereby achieving rapid fusion of a large amount of sensing data. As well as a more accurate characterization of the environment, communication functions also require the assistance of sensing functions to achieve better performance, such as more accurate sensing-assisted channel estimation and beam alignment. Therefore, the technology of ISAC in Internet of Vehicles has excellent research value and broad application prospects, which can jointly design the sensing and communication systems to share the same frequency band and hardware to improve service quality while reducing both hardware and signalling costs [11]. It is also the general trend of future intelligent transportation systems.

Currently, studies on ISAC in Internet of Vehicles mainly include sensing-assisted communication enhancement, communication-assisted cooperative perception, ISAC waveform and system design, etc. Based on these studies, the enhanced sensing ability can provide prior information for the optimal and fast scheduling of distributed computing tasks, and can also provide richer data sources for artificial intelligence services and applications to enhance the robustness of the DNN training models. The enhanced communication ability further improves the ubiquitous computing capability of the computing power network and realizes the instant allocation of computing power resources. So, ISAC can significantly improve the performance of cooperative cloud-edge computing in Internet of Vehicles.

2.2.1 Sensing-assisted Communication Enhancement

In this part, the research focuses on channel estimation and beam alignment. In [12], a channel estimation network named ChanEstNet is proposed, due to that the downlink channel estimation performance is limited by the fast time-varying and non-stationary characteristics in high-speed mobile scenarios. ChanEstNet uses convolutional neural networks to extract channel response feature vectors and recurrent neural networks for channel estimation, and can significantly improve the performance of channel estimation. Yuan et al. [13] designed a novel ISAC-assisted orthogonal time frequency space (OTFS) transmission scheme for vehicular networks. Based on the echoes reflected by the vehicles, the motion parameters of vehicles can be estimated at the road side unit (RSU), which are exploited for predicting the vehicle states. Benefiting from the slow time-varying DD domain channel coefficients, the channel parameters can be predicted.

Millimeter wave is regarded as one of the critical technologies to meet the various performance requirements of the future intelligent Internet of Vehicles. To compensate for the high path loss in the millimeter-wave frequency band, beamforming prediction and beam tracking have been extensively studied. Researchers in [14] investigate a radar-assisted predictive beamforming design for vehicle-to-infrastructures (V2I) communication by exploiting the dual-functional radar communication technique by establishing a novel extended Kalman filtering framework to track and predict the kinematic parameters of each vehicle. Side information derived from radar mounted on the infrastructure operating in a given mmWave band can be used to adapt the beams of the vehicular communication system operating in another millimeter wave band. In [15], a set of algorithms is proposed to perform the beam alignment task from extracting information, from the radar signal to the beams illuminating the different antennas in the vehicle. Since directional communications in Internet of Vehicles are severely hindered by beam pointing issues, a beam alignment procedure has to be periodically carried out to guarantee communication reliability. To speed up the process of beam sweeping when dealing with massive MIMO links, Brambilla et al. [16] proposes a method that exploits a priori information on array dynamics provided by an inertial sensor on transceivers to assist the beam alignment procedure.

2.2.2 Communication-assisted Cooperative Perception

By sharing sensing data, vehicles can expand the sensing range, improve the sensing reliability, and realize more complex decision-making planning, vehicle control and other functions. Researchers in [17] design a multi-vehicle cooperative driving system using cooperative perception along with experimental validation. By cooperative perception based on data-level fusion, vehicles can know the traffic situation even beyond line-of-sight or field-of-view. Although data-level fusion has the highest degree of data utilization, it has high requirements for communication bandwidth and is challenging to meet the real-time needs of Internet of Vehicles. Wang et al. [18]

proposes a vehicle-to-vehicle (V2V) approach for perception and prediction that transmits compressed intermediate representations of the perception and prediction neural network, achieving the best compromise between accuracy improvements and bandwidth requirements. Cui et al. [19] analyzes the performance of different information fusion methods and points out that the fusion of processed data can reduce the amount of transmitted data and significantly reduce the network communication load, while the fusion of raw sensor data has a better cooperative perception effect, but it will increase the communication burden and has a particular negative impact on the cooperative perception results.

2.2.3   ISAC Waveform and System Design

To support the high data rate demand of intelligent vehicular applications, the millimeter-wave automotive radar spectrum at 76–81 GHz can be utilized for communication. For this purpose, ISAC waveform and system design are becoming more important [20–23]. In [20], an adaptive OFDM waveform design problem for joint automotive radar and communication systems with given statistics about the extended target and signal-dependent clutter is studied to perform both functions of sensing and communication using the same waveform. Hieu et al. [21] introduces an intelligent real-time dual-functional radar-communication system for autonomous vehicles, which enables autonomous vehicles to perform both radar and data communications functions to maximize bandwidth utilization as well as significantly enhance safety.

3   Integrated Sensing and Communication for Cooperative Cloud-Edge Computing in Internet of Vehicles

3.1   Collaborative Cloud-Edge-End Computing for Deep Learning Training and Inference

3.1.1   Overall

Collaborative cloud-edge-end computing is an emerging technology to help mobile devices handle the massive computational demands of modern CNN. With the development of deep learning, CNN was widely used in intelligent vehicular networks and brought tremendous computational pressure to vehicles. To achieve the fast and reliable execution of CNN inference, the collaborative computing strategy must consider the real-time requirements of tasks, dynamics, instability of network connections, rational utilization of resources, etc. Several techniques and methods, such as early-exit policy, CNN splitting, and communication compression, are used to improve user satisfaction as well as system robustness.

3.1.2 Collaborative Training and Inference Methods

Collaborative training and inference methods always contain two phases: offline and online. The offline phase consists of introducing an early-exit mechanism in the CNN architecture, jointly training them using the training data set, and evaluating each early-exit point's accuracy and computational cost. In addition, the CNN model splitting method is also introduced to detect the candidate points in the model where computation can be split between device, edge and cloud. In the online phase, the task scheduler dynamically decides on the optimal split and early-exit policy by obtaining the initial parameters from the offline phase, target response time and network conditions. Then in the collaborative inference stage, communication data is compressed to reduce the bandwidth usage, and the inference process can exit early to satisfy the task response time requirement when facing network unavailable or remote server failure problems.

(1) Early-Exit Strategy

Early-exit strategy is initially designed to be an effective method to accelerate model inference and relies on multi-exit model architectures (Fig. 1). Unlike the single output structure of the general DNN, the multi-exit model adds multiple exit branches alone to the depth of the model architecture. This multi-exit model supports a wide range of latency budgets while being portable across cloud, edge and device, making it particularly suitable for collaborative computing scenarios. The work [8] first proposed an entropy-based early-exit, and the work [24] further proposed a confidence-based early exit model to solve the overthinking in inference. In this way, model inference can achieve high confidence by intermediate exits and complete early with an acceptable result. In collaborative cloud-edge-device computing, we need to place at least one early exit point at the device to ensure the task response time when the cloud/edge server is unavailable. And other exit points can be deployed to the device, edge or cloud according to the FLOPs requirement at the exit point, device capabilities, etc.

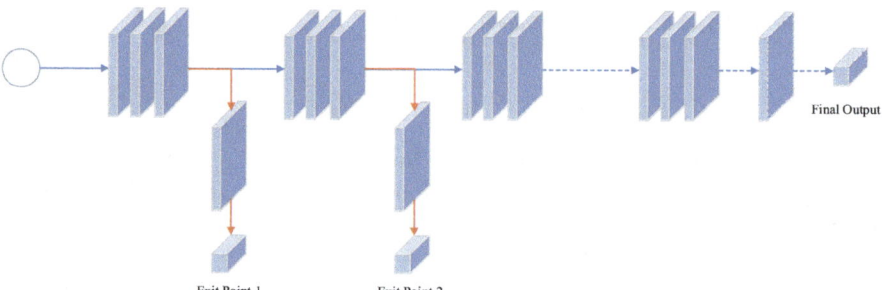

Fig. 1 Early-exit strategy

(2) **Model Splitter**

DNN model typically consists of a sequence of neuron layers and can be split at layer granularity. For a DNN model with layers, there are available points to split. Different split points yield varying trade-offs between device/edge/cloud computing and data transfer time. To adjust the split position dynamically during the online stage, all parameters of the candidate split points should be identified at the offline phase, considering: (1) the size of data to be transmitted at each split point, (2) compression ratio and compression time of transmitted data, (3) the average accuracy of the exit points before split point, (4) the estimated execution time on both sides of split points. At the online phase, the splitter first refines the relevant parameters according to the device/edge/cloud status and calculates the transmission delay based on the actual network bandwidth. Then according to the importance of the task goals, the iterative calculation is performed to find the optimization result that satisfies the most critical optimization objective. This process will iterate in a specific cycle until the task is completed.

(3) **Communication Compression**

The intermediate data of the DNN model usually contains many zero values, especially in the calculation results of the rectified linear unit (ReLU), which is the most widely used activation function. The neuron layers of the DNN model always generate a large amount of intermediate data, which leads to a massive pressure of network transmission. Reducing the amount of transmitted data through data compression is a natural solution. One solution is reducing the data precision, such as lower the 32-bit float down to a 16-bit float or 8-bit signed integer. However, this method may lead to a reduction in the accuracy of the model, so it must be reasonably evaluated in the offline phase so that the final result can meet the accuracy requirement of the task. The other is the compression method for sparse arrays. This method reduces data size by only preserving non-zero values. And it should be noted that the data compression and decompression times on different devices are diverse, which need to be estimated offline and considered by model scheduling.

3.2 *Joint Communication and Computation Resource Allocation in Cooperative Cloud-Edge Computing Internet of Vehicles*

In this part, a general mathematical resource allocation model in cooperative cloud-edge computing Internet of Vehicles is established based on the technology of ISAC and CNN splitting.

3.2.1   System Model

As shown in Fig. 2, a typical cooperative cloud-edge computing vehicular network with M RSUs and N vehicles is illustrated. Assume RSUs are evenly distributed and each RSU is configured with a MEC server. Denote C_i^m, F_i^m $(i = 1, 2, \ldots, M, M+1)$ as the total memory capacity and computing resource (CPU cycles/s) of each MEC server, respectively ($M + 1$ represents the cloud server). Computation data can be transmitted between MEC servers and between MEC and cloud servers through the fiber-wired link (with a fixed data rate of R_0), and the bandwidth of each RSU is B_i. Denote C_j^v, F_j^v $(j = 1, 2, \ldots, N)$ as the total memory capacity and computing resource of each vehicle, respectively. in time slot t, each vehicle j may generate a new computing task $\varphi_j^t = \{x_j^t, c_j^t, f_j^t, L_j^t\}$ with probability p, where x_j^t is the size of input data for the computation (in bits), c_j^t is the required memory of the task (in bits), f_j^t is the total number of CPU cycles required to accomplish the task, L_j^t is the delay threshold of the task. If the newly generated tasks cannot be accomplished locally on time, the vehicles will offload their computing tasks to the MEC or cloud servers through RSUs.

Assume all tasks can be partitioned continuously. Denote ω as the locally processing proportion of the task and $(1 - \omega)$ as the offloading proportion of the task. The data size, the required memory and the required number of CPU cycles of the offloading task are scaled by the same proportion for analytical tractability. So, ωx_j^t, ωc_j^t, ωf_j^t represent the data size, the required memory and the required number of CPU cycles of the locally processing proportion of the task.

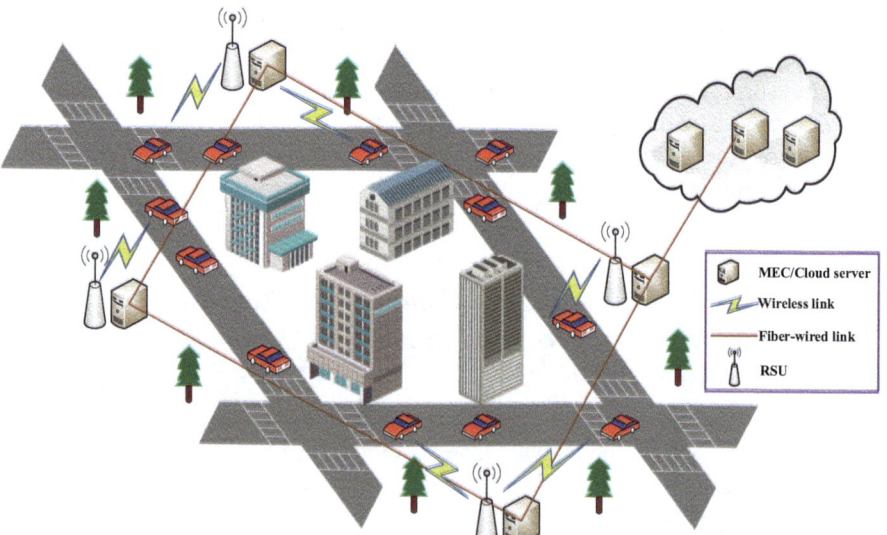

Fig. 2 Cooperative cloud-edge computing vehicular network

(1) **Communication Model**

Assume each vehicle can upload task data to the MEC servers through the V2I link and interference is neglected. Define the $\mathcal{H}(t)$ as the access matrix of all vehicles:

$$H(t) = \begin{bmatrix} h_1^1 & \cdots & h_1^M \\ \vdots & h_j^i & \vdots \\ h_N^1 & \cdots & h_N^M \end{bmatrix} \tag{1}$$

where $h_j^i(t) = 1$ denotes vehicle j is within the coverage of RSU i in time slot t, otherwise $h_j^i(t) = 0$. Define $\mathcal{B}(t) = \{\beta_1^t, \cdots, \beta_j^t, \cdots, \beta_N^t\}$ as the spectral resource allocation vector for vehicles, and β_j^t is the uplink bandwidth allocated by the directly connected RSU for vehicle j in time slot t. So according to the Shannon formula, the uplink data rate can be calculated as follows:

$$R_j^t = \beta_j^t log_2 \left(1 + \frac{P_j h r_j^{-\alpha}}{\sigma^2} \right) \tag{2}$$

where P_j is the transmission power of vehicle j, r_j is the distance between vehicle j and its connected RSU, α is the path loss factor, h is the small scale fading, σ^2 is the noise power.

Define $\mathcal{W}(t) = \left\{ w_1^t, \cdots, w_j^t, \cdots, w_N^t \right\}$ as the locally processing proportion vector of all the tasks in time slot t. Considering the offloading destination of different tasks, the transmission delay can be calculated as follows:

$$T_{w,j}^t = \begin{cases} \dfrac{\left(1-w_j^t\right)x_j^t}{R_j}, & offload\ to\ the\ directly\ connected\ RSU \\ \dfrac{\left(1-w_j^t\right)x_j^t}{R_j} + \dfrac{\left(1-w_j^t\right)x_j^t}{R_0}, & offload\ to\ other\ RSUs\ or\ cloud\ server \end{cases} \tag{3}$$

where $\frac{(1-w_j^t)x_j^t}{R_0}$ denotes the transmission delay between different MEC servers or between MEC servers and cloud servers. So the energy consumption of wireless transmission can be given by:

$$E_{w,j}^t = P_j \frac{\left(1-w_j^t\right)x_j^t}{R_j} \tag{4}$$

(2) **Computation Model**

Define $\mathcal{V}(t) = \left\{ v_1^t, \cdots, v_j^t, \cdots, v_N^t \right\}$ and $\mathcal{M}(t) = \left\{ m_1^t, \cdots, m_i^t, \cdots, m_M^t \right\}$ as the computation resource allocation vector at vehicles and MEC or cloud servers in time slot t, so the computing delay can be given by

$$\begin{cases} T_{v,j}^t = \frac{w_j^t f_j^t}{v_j^t}, & computig\ locally \\ T_{m,j}^t = \frac{\left(1-w_j^t\right)f_j^t}{m_j^t}, & computing\ at\ MEC\ or\ cloud\ servers \end{cases} \quad (5)$$

Define κ_v and κ_m as the energy consumption coefficient dependent on chip architectures [25], so the energy consumption can be given by

$$\begin{cases} E_{v,j}^t = \kappa_v \left(w_j^t f_j^t\right)\left(v_j^t\right)^2, & computig\ locally \\ E_{m,j}^t = \kappa_m \left[\left(1-w_j^t\right)f_j^t\right]\left(m_j^t\right)^2, & computing\ at\ MEC\ or\ cloud\ servers \end{cases} \quad (6)$$

3.2.2 Problem Formulation

In summary, the problem of joint communication and computation resource allocation in cooperative cloud-edge computing Internet of Vehicles can be summarized as: in time slot t, when a typical vehicle generates a new computing task, if the local computation resources are not enough to satisfy the task requirements, the vehicle can offload an appropriate portion of the task to the most appropriate MEC or cloud server for processing based on the status of the remaining resources of Internet of Vehicles. When multiple vehicles in the network simultaneously offload their tasks, an optimal strategy of communication and computation resource allocation should be proposed to decide the most appropriate offloading portion and offloading destinations of tasks for all vehicles and allocate the corresponding spectrum and computing resources for all tasks to reduce the latency and energy consumption as low as possible while the probability that tasks are executed successfully is as high as possible.

According to Eqs. (3) and (5), define the binary indicator variable δ_j^t as the task request status of vehicle j in time slot t ($\delta_j^t = 1$ means a new task request happens, otherwise $\delta_j^t = 0$). Denote $\}(t) = \left\{g_1^t, \cdots, g_j^t, \cdots, g_N^t\right\}$ as the target server number where vehicles offload their tasks ($g_j^t = 0$ means the whole task will be processed locally, $g_j^t = M + 1$ means the task will be offloaded to the cloud server). So the total processing time of tasks can be given by

$$\begin{aligned} T_j^t &= \delta_j^t \left(T_{v,j}^t + T_{w,j}^t + T_{m,j}^t\right) \\ &= \delta_j^t \left(\frac{w_j^t f_j^t}{v_j^t f} + \frac{\left(1-w_j^t\right)x_j^t}{R_j} + \gamma \frac{\left(1-w_j^t\right)x_j^t}{R_0} + \frac{\left(1-w_j^t\right)f_j^t}{m_j^t f}\right) \end{aligned} \quad (7)$$

where γ denotes whether the vehicle is connected to a non-directly connected MEC server. If $h_j^{g_j^t} = 1, \gamma = 0$, otherwise $\gamma = 0$. Note that the transmission delay and energy consumption for sending back the computation results from MEC or cloud servers to vehicles are neglected, due to the tiny data size and the higher downlink rate in most instances [26]. The total energy consumption of tasks can be given by:

$$E_j^t = \delta_j^t \left(E_{v,j}^t + E_{w,j}^t + E_{m,j}^t \right)$$

$$= \delta_j^t \left\{ \kappa_v \left(w_j^t f_j^t \right) \left(v_j^t \right)^2 + P_j \frac{\left(1-w_j^t\right)x_j^t}{R_j} + \kappa_m \left[\left(1-w_j^t\right)f_j^t \right] \left(m_j^t f\right)^2 \right\} \quad (8)$$

Define $c_i^m(t)$, $c_j^v(t)$, $f_i^m(t)$, $f_j^v(t)$, $b_i(t)$ as the remaining memory of MEC server i, the remaining memory of vehicle j, the remaining computation resource of MEC server i, the remaining computation resource of vehicle j, the remaining spectrum resource of RSU i in time slot t, respectively. When $t = 0$, the initial values of the above variables are all the upper limit. Considering the long-term performance based on the latency and energy consumption of Internet of Vehicles, the corresponding optimization problem can be formulated as:

$$\max_{\mathcal{W}(t),\mathcal{V}(t),\mathcal{M}(t),\mathcal{G}(t),\mathcal{B}(t)} \lim_{\tau \to \infty} \sum_{t=0}^{\tau} \left\{ \eta_1 \frac{\sum_{j=1}^{N} \delta_j^t \cdot 1\left(T_j^t < L_j^t\right)}{\sum_{j=1}^{N} \delta_j^t} + \eta_2 e^{-\sum_{j=1}^{N} \delta_j^t T_j^t} + \eta_3 e^{-\sum_{j=1}^{N} \delta_j^t E_j^t} \right\}$$

$$(9)$$

$$s.t. C1 : v_j^t \le f_j^v(t), \forall j$$

$$C2 : \sum_{j \in \left\{j | g_j^t = i\right\}} m_j^t \le f_i^m(t), \forall i$$

$$C3 : \sum_{j \in \left\{j | h_j^t(t)=1\right\}} \beta_j^t \le b_i(t), \forall i$$

$$C4 : c_j^t \le c_j^v(t), \forall j$$

$$C5 : \sum_{j \in \left\{j | g_j^t = i\right\}} c_j^t \le c_i^m(t), \forall i$$

where C1 is the constraint of total computation resource of vehicles, C2 is the constraint of total computation resource of MEC or cloud servers, C3 is the constraint of total spectrum resource of RSUs, C4 is the constraint of total memory of vehicles and C5 is the constraint of total memory of MEC or cloud servers.

The above long-term optimization problem involves the sequential decision problem in continuous coherent time slots, and the network environment and user requests change dynamically. So the problem is difficult to be solved with traditional optimization methods. As a sequential decision-making method, reinforcement learning can have specific advantages in solving such problems by performing continuous trial-and-error learning in the target environment and modifying the policy based on the feedback results. Therefore, the MEC-based dynamic task allocation and resource scheduling problem will be modeled as a Markov decision process and solved with deep reinforcement learning algorithms in the next section.

3.2.3 Deep Reinforcement Learning-Based Solution

In this section, we first define the state space, action space and reward function for the proposed problem. Then, the utility maximization problem is formulated as a Markov decision process. To avoid the curse of dimensionality, a deep reinforcement learning-based solution that uses deep neural networks to estimate the action-value function of Q-learning is proposed.

(1) State Space

The state of the MEC-assisted Internet of Vehicles in time slot t can be determined by the state of all vehicles, MEC servers and the cloud server.

$$s_t = \{[\delta_1^t, x_1^t, f_1^t, c_1^t, D_1^t, f_1^v(t), c_1^v(t)], \cdots, [\delta_N^t, x_N^t, f_N^t, c_N^t, D_N^t, f_N^v(t), c_N^v(t)],$$

$$[f_1^m(t), \cdots, f_M^m(t), f_{M+1}^m(t)],$$
$$[c_1^m(t), \cdots, c_M^m(t), c_{M+1}^m(t)], [b_1(t), \cdots, b_M(t)], \mathcal{H}(t)\} \tag{10}$$

(2) Action Space

In time slot t, the agent selects the action a_t from the action space A at state s_t. a_t should decide the offloading proportion of tasks, the offloading destination, the locally-required computation resource, the computation resource required at MEC or cloud servers, and the required spectrum resource for each vehicle.

$$a_t = \{[w_1^t, g_1^t, v_1^t, m_1^t, \beta_1^t], \cdots, [w_N^t, g_N^t, v_N^t, m_N^t, \beta_N^t]\} \tag{11}$$

(3) Reward Function

The reward function is the objective function of problem (9) in time slot t.

$$r_t = \eta_1 \frac{\sum_{j=1}^N \delta_j^t \cdot 1\left(T_j^t < L_j^t\right)}{\sum_{j=1}^N \delta_j^t} + \eta_2 e^{-\sum_{j=1}^N \delta_j^t T_j^t} + \eta_3 e^{-\sum_{j=1}^N \delta_j^t E_j^t} \tag{12}$$

Deep Q-learning is an effective value-based reinforcement learning method and widely used in the literature about communications and networking. DQN enables the agent to adaptively learn the optimal behavior separately within a specific context in each time slot [27]. In each time iteration, an agent t observes its current state s_t, takes an action a_t based on its policy $\pi(s)$, and receives its immediate reward r_t together with its new state s_{t+1}. The objective is to find the optimal policy π^* that maximizes the discounted cumulative reward $R = \sum_{t=0}^{\tau} \gamma \cdot r_t$, where $\gamma \in [0, 1)$ is the discounted factor. The optimal action-value function can be given by

$$Q^*(s_t, a_t) = r(s_t, a_t) + \gamma \max_{a_{t+1} \in A} Q^*(s_{t+1}, a_{t+1}) \tag{13}$$

If the state-action space becomes huge, classic Q-learning algorithm cannot be applied since many states are rarely visited. Therefore, DQN improves Q-learning by approximating the Q-function with a DNN of weights θ as a Q-network. The Q-network updates its weights at each iteration on a data replay buffer to minimize the following loss function derived from a target network with the same architecture as the Q-network but with old weights.

$$L(\theta) = \mathop{E}_{(s,a,r,s' \in D)} \left[(y_t - Q(s_{t+1}, a_{t+1}|\theta))^2 \right] \tag{14}$$

$$y_t = r_t + \gamma \max_{a_{t+1} \in A} Q^*(s_{t+1}, a_{t+1}|\theta') \tag{15}$$

4   Future Directions

4.1   Robustness, Stability and Security

The development of collaborative computing will bring about a rapid increase in the number of edge computing devices, which will get great challenges to subsequent operation and maintenance. As a distributed system, the crash of each node or network congestion may lead to the failure of the task. Moreover, edge servers are usually deployed separately and heterogeneously, unlike cloud computing, which leads to extremely high maintenance costs.

In addition, due to the characteristics of close to users in edge computing, some edge devices are deployed in open areas in a distributed manner, exposed to unsafe conditions, and are more vulnerable to attack, resulting in device security and data security issues. To deal with these problems, it is necessary to research and design the cloud-edge safety and maintenance scheme of the whole life cycle to improve safety and reduce the difficulty of maintenance.

4.2   Multi-agent Reinforcement Learning

With the increase of vehicles and MEC servers in Internet of Vehicles, the multi-user cooperation and competition problem described will become very complicated, making the state and action space of the traditional single-agent reinforcement learning algorithm very large. In addition, obtaining the current status information in a practical application environment is difficult, and the delay due to data collection may be intolerable. Therefore, the distributed cooperative scheme based on the multi-agent reinforcement learning method to manage the multi-dimensional resources is a promising research work. Some literature has now been studied on the multi-agent reinforcement learning method used in Internet of Vehicles. But, some important problems, such as excessive complexity, poor convergence performance, and excessive state and action space, need to be further optimized and improved.

5   Conclusion

In this chapter, cooperative cloud-edge computing in Internet of Vehicles is discussed based on intelligent computing and integrated sensing and communication. Collaborative cloud-edge-end computing technologies for deep learning training and inference, such as early-exit policy, CNN splitting, and communication compression, are introduced to improve user satisfaction and system robustness. Also, this chapter analyses the joint communication and computation resource allocation strategy, proposes a DQN-based solution and summarizes future directions for cooperative cloud-edge computing in Internet of Vehicles.

References

1. 5GAA, White Paper, MNO network expansion mechanisms to fulfil connected vehicle requirements. https://5gaa.org/wp-content/uploads/2020/06
2. R. Buyya, C.S. Yeo, S. Venugopal, J. Broberg, I. Brandic, Cloud computing and emerging IT platforms: vision, hype, and reality for delivering computing as the 5th utility. Futur. Gener. Comput. Syst. **25**(6), 599–616 (2009)
3. W. Shi, J. Cao, Q. Zhang, Y. Li, L. Xu, Edge computing: vision and challenges. IEEE Internet Things J. **3**(5), 637–646 (2016)
4. J. Chen, X. Ran, Deep learning with edge computing: a review. Proc. IEEE **107**(8), 1655–1674 (2019)
5. L. Zeng, X. Chen, Z. Zhou, L. Yang, J. Zhang, CoEdge: cooperative DNN inference with adaptive workload partitioning over heterogeneous edge devices. IEEE/ACM Trans. Netw. **29**(2), 595–608 (2021)
6. X. Yang, Z. Fei, J. Zheng, N. Zhang, A. Anpalagan, Joint multi-user computation offloading and data caching for hybrid mobile cloud/edge computing. IEEE Trans. Veh. Technol. **68**(11), 11018–11030 (2019)
7. E. Li, Z. Zhou, X. Chen, Edge intelligence, in *Proceedings of the 2018 Workshop on Mobile Edge Communications*, pp. 31–36 (2018)
8. S. Teerapittayanon, B. McDanel, H.T. Kung, BranchyNet: fast inference via early exiting from deep neural networks. in *2016 23rd International Conference on Pattern Recognition (ICPR)*, pp. 2464–2469 (2016)
9. W. Shi, S. Zhou, Z. Niu, M. Jiang, L. Geng, Multi-user co-inference with batch processing capable edge server. IEEE Trans. Wirel. Commun. 1–1 (2022)

10. S. Laskaridis, S. I. Venieris, M. Almeida, I. Leontiadis, N. D. Lane, SPINN: synergistic progressive inference of neural networks over device and cloud. in *Proceedings of the 26th Annual International Conference on Mobile Computing and Networking*, pp. 1–15 (2020)
11. A. Liu et al., A survey on fundamental limits of integrated sensing and communication. IEEE Commun. Surv. & Tutor. **24**(2), 994–1034 (2022)
12. W. Yuan, F. Liu, C. Masouros, J. Yuan, D.W.K. Ng, N. Gonzalez-Prelcic, Bayesian predictive beamforming for vehicular networks: a low-overhead joint radar-communication approach. IEEE Trans. Wireless Commun. **20**(3), 1442–1456 (2021)
13. W. Yuan, Z. Wei, S. Li, J. Yuan, D.W.K. Ng, Integrated sensing and communication-assisted orthogonal time frequency space transmission for vehicular networks. IEEE J. Sel. Top. Signal Process. **15**(6), 1515–1528 (2021)
14. F. Liu, W. Yuan, C. Masouros, J. Yuan, Radar-assisted predictive beamforming for vehicular links: communication served by sensing. IEEE Trans. Wireless Commun. **19**(11), 7704–7719 (2020)
15. N. Gonzalez-Prelcic, R. Mendez-Rial, R.W. Heath, Radar aided beam alignment in MmWave V2I communications supporting antenna diversity. in *2016 Information Theory and Applications Workshop (ITA)*, pp. 1–7 (2016)
16. M. Brambilla, M. Nicoli, S. Savaresi, U. Spagnolini, Inertial sensor aided mmWave beam tracking to support cooperative autonomous driving. in *2019 IEEE International Conference on Communications Workshops (ICC Workshops)*, pp. 1–6 (2019)
17. S.-W. Kim et al., Multivehicle cooperative driving using cooperative perception: design and experimental validation. IEEE Trans. Intell. Transp. Syst. **16**(2), 663–680 (2015)
18. T.-H. Wang, S. Manivasagam, M. Liang, B. Yang, W. Zeng, U. Raquel, V2vnet: Vehicle-to-vehicle communication for joint perception and prediction. ECCV (2020)
19. G. Cui, W. Zhang, Y. Xiao, L. Yao, Z. Fang, Cooperative perception technology of autonomous driving in the internet of vehicles environment: a review. Sensors **22**(15), 5535 (2022)
20. C.D. Ozkaptan, E. Ekici, O. Altintas, Enabling communication via automotive radars: an adaptive joint waveform design approach. in *IEEE INFOCOM 2020-IEEE Conference on Computer Communications*, pp. 1409–1418 (2020)
21. N.Q. Hieu, D.T. Hoang, N.C. Luong, D. Niyato, iRDRC: an intelligent real-time dual-functional radar-communication system for automotive vehicles. IEEE Wirel. Commun. Lett. **9**(12), 2140–2143 (2020)
22. D. Ma, N. Shlezinger, T. Huang, Y. Liu, Y.C. Eldar, Joint radar-communication strategies for autonomous vehicles: combining two key automotive technologies. IEEE Signal Process. Mag. **37**(4), 85–97 (2020)
23. S. Sun, A.P. Petropulu, H.V. Poor, MIMO radar for advanced driver-assistance systems and autonomous driving: advantages and challenges. IEEE Signal Process. Mag. **37**(4), 98–117 (2020)
24. Y. Kaya, S. Hong, T. Dumitras, Shallow-deep networks: understanding and mitigating network overthinking. in *Proceedings of the 36th International Conference on Machine Learning*, vol. 97, pp. 3301–3310 (2019)
25. J. Bi, H. Yuan, S. Duanmu, M. Zhou, A. Abusorrah, Energy-optimized partial computation offloading in mobile-edge computing with genetic simulated-annealing-based particle swarm optimization. IEEE Internet Things J. **8**(5), 3774–3785 (2021)
26. H. Zhou, K. Jiang, X. Liu, X. Li, V.C.M. Leung, Deep reinforcement learning for energy-efficient computation offloading in mobile-edge computing. IEEE Internet Things J. **9**(2), 1517–1530 (2022)
27. N.C. Luong et al., Applications of deep reinforcement learning in communications and networking: a survey. IEEE Commun. Surv. & Tutor. **21**(4), 3133–3174 (2019)

Chapter 11
Big Data for Internet of Vehicles and Smart Transportation

Qi Liu, Chaowei Wang, Jie Gao, Chuanxin Zeng, Hong Zhu, Shangguan Wei, Feng Yin, Huanlai Xing, and Lexi Xu

Q. Liu
China Unicom Smart City Research Institute and Beijing Jiaotong University, Beijing, China
e-mail: liuqi49@chinaunicom.cn

C. Wang
School of Electronic Engineering, Beijing University of Posts and Telecommunications, Beijing, China
e-mail: wangchaowei@bupt.edu.cn

J. Gao
China Unicom Research Institute, Beijing, China
e-mail: gaojie49@chinaunicom.cn

C. Zeng
China Unicom Smart City Research Institute, Beijing, China
e-mail: zengcx12@chinaunicom.cn

H. Zhu
China National Institute of Standardization, Beijing, China
e-mail: zhuhong@cnis.ac.cn

S. Wei
School of Electronic and Information Engineering, Beijing Jiaotong University, Beijing, China
e-mail: wshg@bjtu.edu.cn

F. Yin (✉)
School of Science and Engineering, The Chinese University of Hong Kong, Shenzhen, China
e-mail: yinfeng@cuhk.edu.cn

H. Xing
School of Computing and Artificial Intelligence, Southwest Jiaotong University, Chengdu, China
e-mail: hxx@home.swjtu.edu.cn

L. Xu (✉)
China Unicom Research Institute and Beijing University of Posts and Telecommunications,

Y. Zhu et al. (eds.), *Communication, Computation and Perception Technologies for Internet of Vehicles*, https://doi.org/10.1007/978-981-99-5439-1_11

201

Abbreviations

HDFS   Hadoop Distributed File System.
HQL   SQL-like Query Language.
IoV   Internet of Vehicles.
NDFS   Nutch Distributed File System.
NFS   Network File System.
VRIT   Virtual-Real Interactive Testing.

1 Introduction

Due to the rapid progress of computer and information technology, as well as the acceleration of vehicle and transportation industry, the vehicle and transportation industry experiences the data explosion. Vehicle and transportation big data has 4 V features, including volume, velocity, variety, value [1]. Vehicle and transportation big data has attracted the attention from both the academics and industries [2].

In order to seek the value of big data, Hadoop ecology has been built and widely used in recent years. Meanwhile, data mining technologies are employed for big data analysis and mining.

Employing big data resources of vehicle and transportation can lead to the digital revolution of vehicle and transportation industry. Therefore, this chapter studies big data, as well as both software and hardware infrastructure of big data. Then, in this chapter, we design a novel vehicle and transportation big data platform, which consists of six layers. Furthermore, we apply this big data platform in vehicle and transportation industry.

Main contributions to this chapter are as follows:

1. This chapter introduces big data, including big data origin, big data features, the popularity of data mining, big data mining technology. Furthermore, this chapter introduces the big data software infrastructure–Hadoop, including HDFS, MapReduce, etc. In addition, this chapter also introduces big data hardware infrastructure.
2. This chapter designs vehicle and transportation big data platform. The designed platform includes 6 layers, and this chapter elaborates on vehicle and transportation data source layer, vehicle and transportation data aggregation layer, vehicle and transportation data storage layer, vehicle and transportation data processing layer, vehicle and transportation data analysis and management layer, vehicle and transportation data application layer.

Beijing, China
e-mail: davidlexi@hotmail.com

3. This chapter elaborates on the application of vehicle and transportation big data in the IoV, especially IoV system for Beijing 2022 Winter Olympics. Furthermore, this chapter introduces digital twin IoV and intelligent transportation.

This chapter is organized as follows: Sect. 2 introduces big data origin and features, as well as software and hardware infrastructure of big data. Section 3 introduces the designed vehicle and transportation big data platform. Section 4 elaborates on application of big data in transportation industry. Section 5 summarizes the whole chapter.

2 Background of Big Data

2.1 Big Data Origin/Features and Data Mining

2.1.1 Big Data Origin and Features

With the rapid progress of society and the widespread application of information and computer technology, the era of big data has arrived [1]. The term "big data" was proposed by McKinsey, which is a world famous enterprise consulting company. In recent years, big data has been widely mentioned and used in various industries. The term "big data" has been also explained in detail on the White House government website. Big data has a profound impact on various industries, just as Gary King, who is a professor of sociology at Harvard University, said, "The arrival of the era of big data is a new quantitative revolution among various fields."

Initially, big data itself is not a specific concept, as the name implies, it is the scale of the data scale is very huge, whilst in the past, there are also nouns that represent the scale of data, such as "large-scale data", "super-large-scale data", etc. [3]. It can also be understood as follows: Big data refers to massive complex data sets with high value that are difficult to complete analysis and calculation in a short time by using existing theories, methods, technologies and tools.

In order to achieve the value of big data, data mining technology has been employed in many industries. With the in-depth research of big data and data mining technology, the overall productivity and efficiency of industries can be improved significantly [2].

2.1.2 The Popularity of Data Mining Technology

For data mining technology, compared with the traditional data model, the popularity of data has been significantly improved [4]. Meanwhile, with the improvement of data popularity, the efficiency of data analysis, data extraction and data storage has also been increased, and hence, the overall efficiency of data operation has been improved. In the era of big data, due to the surge of the overall amount of data, this put high requirements of high speed data mining. However, for many traditional data mining technologies/algorithms, the data flow performance is poor, which cannot meet the technical requirements of big data analysis and mining.

2.1.3 Description of Data Mining Technology

In recent years, people gradually rely on smart terminals/devices and Internet. Meanwhile, the total amount of social data, which is generated by various industries and People's daily life, also surges dramatically [1]. This also results in new types of data sources, which put new requirements of data mining technology. Accordingly, the distribution density of data with relatively high potential value decreases continuously, and the average value of the data decreases gradually. Therefore, it is difficult for traditional data mining technology to achieve effective analysis and prediction. In the era of big data, data mining technology also experiences fast development, in order to address above-mentioned challenges and process/mine huge amount of data resources [1, 2].

2.2 Software Infrastructure of Big Data

Massive data storage is not new challenge. It has appeared for long time. In some industries, companies have stored and processed large amount of data volume via Network File System (NFS). The disadvantage of NFS is that it is difficult to make full use of multiple computers to analyze massive data simultaneously. Therefore, traditionally NFS and methods cannot effective the scenario, where the amount of data is huge (e.g., a log file with a few of GB data).

Hadoop was officially introduced by the Apache Software Foundation in 2005 [5]. It is inspired by MapReduce and Google File System In 2006, MapReduce and Nutch Distributed File System (NDFS) were incorporated into a project named as Hadoop. Hadoop is a widely-used tool for categorizing search keywords on the Internet, meanwhile, it is also with high scalability. For example, what happens when greping a large 10 TB file? On traditional systems, this would take a long time. Hadoop is designed to address these challenges with high efficiency.

Nowadays, Hadoop becomes Apache top project. As described in Fig. 1, Hadoop is a flexible and highly-available framework, and this framework is suitable for huge scale computation and data processing [5].

The core of Hadoop is HDFS and MapReduce, meanwhile, Hadoop2.0 also includes YARN. HDFS is the basis of data storage and management in Hadoop system. HDFS simplifies the file consistency model and provides high-capacity data access for applications with large data sets through streaming data access [6].

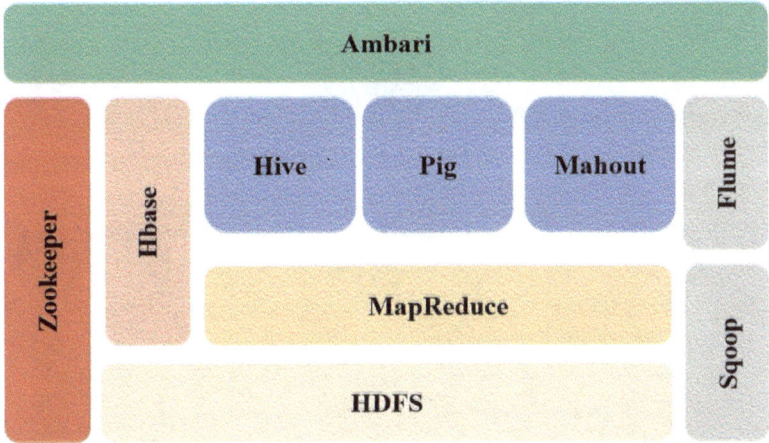

Fig. 1 Framework of hadoop technology

In the architecture of HDFS shown in Fig. 2, NameNode is the Master node. NameNode manages HDFS namespace and data block mapping information, configures replica policies, and processes client requests [6]. DataNode is the Slave node that stores actual data and then reports the storage information to NameNode [6].

The Secondary NameNode shares the workload. Merge fsimage and fsedits periodically and push to NameNode. In an emergency, the Secondary NameNode can be used to recover the NameNode.

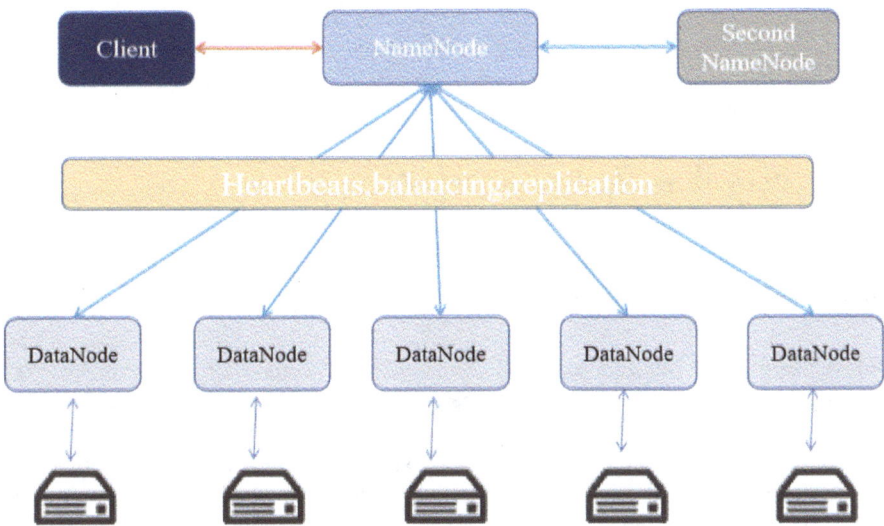

Fig. 2 Architecture of HDFS

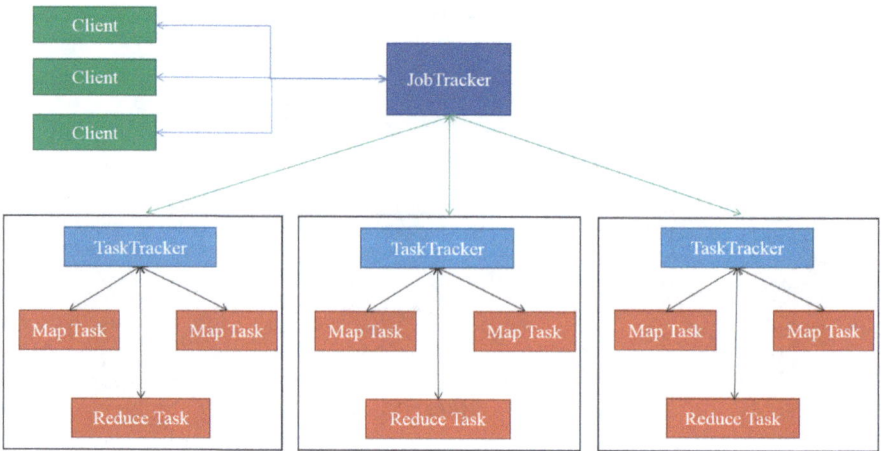

Fig. 3 Framework of mapReduce

The whole big data cluster is calculated using MapReduce computing framework, which is described in Fig. 3. MapReduce is a calculation model used to compute the huge scale of data.

Job Tracker is the Master node, and it is to manage all jobs, job/task monitoring, error handling, etc. [5]. It divides the task into many tasks and then assigns these tasks to the Task Tracker. Task Tracker is the Slave node that runs both Map Task and Reduce Task. In addition, Task Tracker also interacts with the Job Tracker to report the status of the task [5].

Map Task will parse each data record, send it to the Map written by the user, execute it, and write the output result to the local disk [5]. Reducer Task remotely reads input data, sorts the data, and transfers the data in groups to the Reduce function written by the user for execution.

In addition to HDFS, Hadoop includes several other big data computing tools, for example, HIVE, HBase, Zookeeper, Sqoop, Mahout [1, 5, 6].

2.3 Hardware Infrastructure of Big Data

Before designing a big data platform, we should first consider the required storage capacity, computing power, real-time analysis, data storage cycle and other factors. Then we design the platform architecture according to these requirements.

Meanwhile, the robustness of the platform should be considered. For example, the failure of any node will not impact the ordinary use of the platform, and the damage of any disk will not lead to data loss. The basic requirement of Hadoop big data platform infrastructure is to ensure high availability of management nodes, such as NameNode and ResourceManager.

Therefore, ResourceManager should be implemented on these nodes. In addition, to ensure the security of HDFS data, an appropriate number of copies for Hadoop block storage should be set. For example, if you set three replicas, no data will be lost when any two DataNodes in the cluster fail. In terms of network, it is recommended that the server of each node be bound with two NICs, set the network to redundant mode, and bind with switches to ensure the normal operation of the node network if a single NIC fails or a single switch fails.

The planning of storage and computing resources on a big data platform should consider the actual application requirements, such as the amount of existing and future growing data, the data storage period, the amount of intermediate result data of daily computing tasks, and the data redundancy space.

The hardware planning principle of big data platform is as follows: If the exactly the storage and computing resource requirements are known, then the big data platform can be configured according to the requirements. However, if storage and computing resource requirements cannot be accurately predicted, it is important to leave appropriate space for expansion, such as enough cabinet locations, network interfaces, disk interfaces, etc. In practical applications, the storage capacity is generally easy to estimate, however, the computing resources are difficult to estimate, hence, leaving enough interfaces for expansion is a problem that should be considered.

For computing resource planning, it depends on the applications that will be run. For memory-hungry big data components such as Spark, HBase, and ElasticSearch, it is recommended that the server memory of the compute node be large (e.g., 64 GB, 128 GB, 256 GB) [1, 5]. Since it is difficult to evaluate the computing resource requirements in the early stage, both CPU and memory can be configured based on above principles. If a computing performance bottleneck exists, horizontal expansion can be performed in the later stage.

3 Vehicle and Transportation Big Data Platform

3.1 Overview of Vehicle and Transportation Big Data Platform

The vehicle and transportation big data platform integrates a series of data resources, in the aspects of drivers, vehicles, roads, etc. Meanwhile, the data of this platform covers urban roads, expressways, parking lots, charging sites and other transportation infrastructure [2, 7]. The utilization of vehicle and transportation data resources can help optimize the adaptation and control of transportation resources. Therefore, this big data platform tries to help the city transportation management and operation, the decision support of the government, and help the development of "smart cities".

In order to realize above-mentioned professional transportation data management and improve the efficiency of transportation data, the vehicle and transportation big data platform is designed according to four principles:

(1) Open architecture: The big data platform adopts an open architecture, which can support third-party algorithms, massive data storage, parallel computing, on-demand use, dynamic distribution, distributed deployment, cloud storage, dual-live data center, etc. The open structure can effectively improving the efficiency of transportation data integration and the intelligent level of the platform.

(2) Vehicle/transportation data processing ability: The big data platform has high requirements of timeliness transportation data processing, and the diverse dimensions data [1]. This platform can employ data cleaning and data conversion as well as distributed parallel computing technologies to produce high-quality vehicle/transportation data, as well as extract the potential high value of transportation information.

(3) Transportation data mining capability: Vehicle/transportation data mining capability aims at achieving the in-depth of massive transportation data mining and application, on the basis of statistics, machine learning algorithms, deep learning algorithms etc. Typical algorithms include transportation data classification, clustering, association, prediction, and hidden transportation data knowledge discovery [2].

(4) Meeting the requirements of future intelligent transportation management: This platform should support diversified services as well as assist the future intelligent transportation management, for example, transportation data analysis model, intelligent transportation statistical analysis, intelligent transportation signal control, dynamic road monitoring and early warning, intelligent parking and other functions [7].

Based on above four principles, the vehicle and transportation big data platform includes six layers, as shown in Fig. 4.

3.2 Vehicle and Transportation Data Source Layer

In the vehicle and transportation big data platform, the major data source can be generally categorized into the following four categories:

The first category is vehicle data source [2]. Specifically, this big data platform collects vehicle data, including vehicle monitoring platform database, vehicle operation database, road monitoring system database, vehicle terminal data, etc.

The second category is transportation administration department data source [7]. Typical data source includes taxi management system database from transportation administration commission, transportation induction system database, transportation signal control system database, license plate and driving license management system database, electronic toll system database, transportation accident management system database, etc.

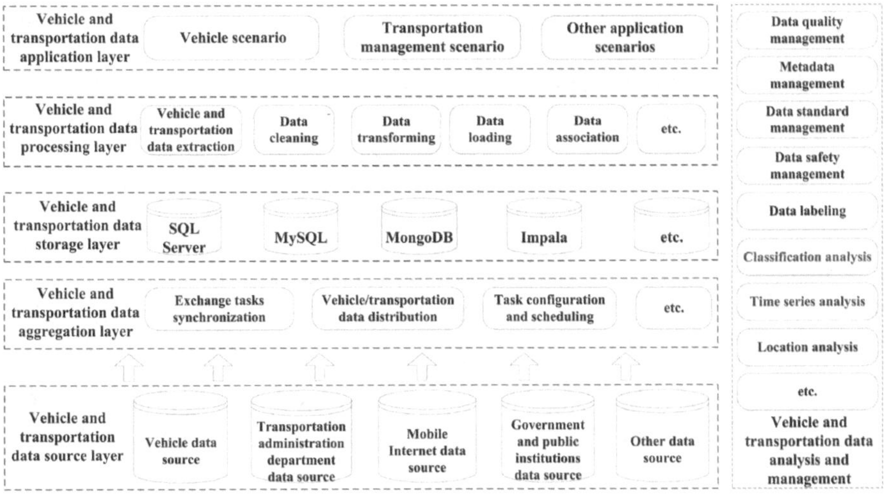

Fig. 4 Architecture of vehicle and transportation big data platform

The third category is mobile Internet data source. Specifically, this platform tries to collect data from Internet travel platform. Meanwhile, this platform tries to collect Internet map system (e.g., Baidu Map, Gaide Map) data, for example, real-time transportation information, real-time congestion alarm information [8].

The fourth category is government and public institutions data source. Specifically, this platform tries to connect city residents and household registration management system database, city planning system database, city police database. In addition, this big data platform also tries to connect to government departments, for example, city administration department database, environmental protection department database, emergency rescue department database, etc.

3.3 Vehicle and Transportation Data Aggregation Layer

The aim of vehicle and transportation data aggregation layer is to connect and access above-mentioned four categories of data sources, including vehicle data source, transportation administration department data source, mobile Internet data source, government and public institutions data source. The vehicle and transportation data aggregation layer is the foundation of the platform, because it connects multi-source heterogeneous transportation data sources via a series of methods, including database collection, interface collection, file collection, and media collection, etc.

The data aggregation layer can provide the interface/channel for vehicle and transportation data extraction and convergence. During this process, this layer supports continuous transmission from the interrupted point. In addition, this data aggregation layer has the following functions: the exchanging tasks synchronization of vehicle/

transportation data, vehicle/transportation data distribution, task configuration and scheduling, etc. [9].

Exchanging tasks synchronization function of vehicle/transportation data: This layer employs the background coding, trigger mechanism, release/subscribe mode, SQL Job mode, service broker message queue mode. In this way, the data aggregation layer tries to realize the data exchange tasks synchronization, together with various vehicle/transportation data sources.

Vehicle/transportation data distribution function: This layer employs a series of technologies to empower the vehicle/transportation data distribution, including pipeline, memory, multi-thread technology, GUI, etc. In this way, the data aggregation layer supports data distribution, as well as support the capability of cross-Hadoop, MPP, RDBMS and other platforms/database.

Task configuration and scheduling: This layer employs Azkaban to realize the fast distributed task configuration and scheduling function with extendible capability.

3.4 Vehicle and Transportation Data Storage Layer

From above sub-sections, the vehicle and transportation big data platform connects to a wide range of data sources, including vehicle data source, transportation administration department data source, mobile Internet data source, government and public institutions data source. Meanwhile, above data sources have various data category, for example, text data, figure data, video data, GIS data, digital information [1, 7]. The aim of data storage layer is to effectively storage above multi-source heterogeneous vehicle/transportation data.

The vehicle and transportation data storage layer employs SQL Server database and MySQL database to store the structured vehicle/transportation data, as shown in Fig. 5. In addition, this layer employs MongoDB and Impala to storage the non-structured vehicle/transportation data.

For vehicle and transportation data storage, the data storage layer sets heat storage and cold storage, according to its usage frequency and level. Specifically, the data

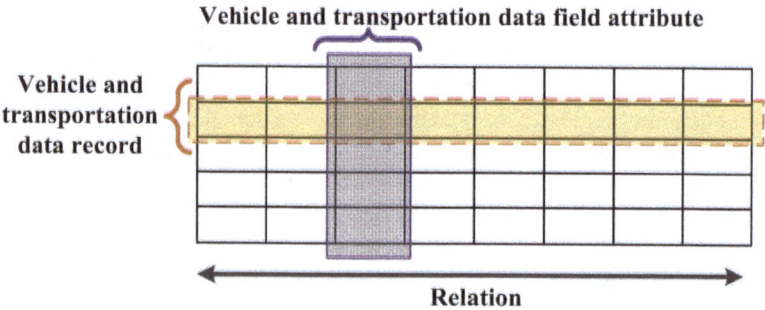

Fig. 5 Structured vehicle/transportation data

storage layer employs heat storage for frequently used data and important data. The data storage layer employs cold storage for non-frequently used data as well as non-significant data. In addition, the data storage layer employs cloud storage, dual-live data center, in order to keep the safely storage of the heterogeneous vehicle and transportation data.

3.5 Vehicle and Transportation Data Processing Layer

The objective of data processing layer is to improve the vehicle and transportation heterogeneous data quality as well as address the following problems, including vehicle and transportation data missing, low/timeliness data update, data jumbling, data processing delay, etc.

Effective data processing layer can assist vehicle and transportation big data platform, in this way to assist accurate V2X management, timely transportation accident and escape analysis, timely transportation 'monitoring, accurate road congestion management, effective car parking, etc.

In order to reach above-mentioned objective, the data processing layer designs the following module:

Vehicle and transportation data extraction module is designed to extract the original multi-source vehicle and transportation data, and its key is extraction process and extraction rules. This module employs both full extraction technology and increment extraction technology to extract different types of multi-source heterogeneous data.

Vehicle and transportation data cleaning module is designed to delete the dirty data and duplicated data. Meanwhile, this module also tries to correct the error data and checks the data consistency, as well as transfer the format. After above data cleaning process, this module can effectively improve the vehicle and transportation data quality.

Vehicle and transportation data transformation module is designed to address the data inconsistency problem, and its typical methods includes unified name, unified units, data granularity transformation, business rule transformation, data format transformation, etc. Meanwhile, this module also generates new data via data segmentation, data recombination, etc.

Vehicle and transportation data loading module is designed to load the clean data into the database tables, on the basis of the pre-defined database system model. Effective vehicle and transportation data loading can save the database storing space, as well as improve the efficiency of the data analysis and application.

Vehicle and transportation data association module is designed to improve both the data processing efficiency and data value for its further analysis and application. The data association module associates key fields from multiple vehicle related systems and transportation relevant platforms.

3.6 Vehicle and Transportation Data Analysis and Management Layer

Vehicle and transportation data analysis and management layer is to analyse various vehicle and transportation data, and effectively manage above data sources as well as improve the data quality for data application. Therefore, this layer includes the following modules.

Data quality management module includes data quality rule management, data quality comprehensive inquiry, data quality problem discovery, data quality problem analysis, data inspection, etc.

Metadata management module includes meta-model management, metadata collection management, metadata basic management, metadata analysis services, etc.

Data standard management module is designed to manage the data basic standards, data technical standards, data labelling standards, and data evaluation standards, etc. This module can guarantee the unified/comprehensive data standard.

Data safety management module employs the data secret level classification model to classify various vehicle and transportation data. Furthermore, this module takes the relevant data encrypting according to data secret level. In addition, this module also supports desensitization encryption rule, desensitization encryption algorithm, management of desensitization encryption task.

Data labeling module is designed to label vehicle and transportation data, on the basis of statistical processing, model mining, combination definition, etc. Then the marked labels can assist clustering and statistics in the data application layer.

This vehicle and transportation data analysis and management layer also supports a series of data analysis algorithm, including correlation analysis, classification analysis, clustering analysis, time series analysis, location analysis, neural networks based analysis, machine learning and other data mining methods. In addition, this layer supports the visualization of data analysis results.

3.7 Vehicle and Transportation Data Application Layer

Following above-mentioned five layers, vehicle and transportation application layer can provide a variety of vehicle and transportation data services to meet the requirements of various vehicle and transportation application scenarios. The following Sect. 4 will introduce the vehicle and transportation application.

4 Application of Big Data in Vehicle and Transportation Industry

As discussed in above sections, big data has been widely applied in the transportation industry. Big data can bring new methods and ideas to address traditional transportation problems. In recent years, the popular Internet of Vehicles (IoV) and digital twin are typical big data application in the transportation industry. Big data technology endows transportation perception, interconnection, analysis, prediction and control capabilities to a certain extent. Therefore, big data technology and application can improve the road safety, and alleviate the transportation congestion, as well as improve transportation services.

4.1 Application of Big Data in Internet of Vehicles

IoV is an important development direction in the transportation and vehicle industry. The IoV system refers to the collection, storage and transmission of all working conditions and static and dynamic information of the vehicle by installing onboard terminal equipment on the vehicle dashboard. Generally, IoV system has real-time and real scene function, as well as uses the networks to realize human-vehicle interaction.

In the IoV system, accurate detection, tracking and recognition of transportation targets is the basis for transportation applications. This sub-section will focus on the transportation data acquisition and multi-sensor data fusion perception methods by taking the typical application of "Science and Technology Winter Olympics" as a case.

4.1.1 Typical Application Cases of "Science and Technology Winter Olympics"

Facing the complex environment of the Beijing Shougang Winter Olympics Park, a 5G smart car networking business system was built to meet the core needs of efficient and safe transportation of intensive personnel, materials and equipment during the Winter Olympics 2022. The architecture is shown in Fig. 6, including one platform, two networks, more than five car models connections, and ten major business scenarios.

The deployment of the 5G intelligent IoV system in the Winter Olympic Park includes three modules: multi-source fusion perception, human-vehicle–road collaboration, and intelligent driving monitoring. In this system, more than 180 devices (such as 5G base stations, RSUs (Road Side Unit), sensing equipment) are connected, more than 30 unmanned vehicles of more than 5 types of vehicles (e.g., Robotaxi, retail vehicles, retail vehicles, distribution vehicles, sweepers) are connected. In the Beijing Shougang Winter Olympic Park, more than 100 vehicle–road-cloud joint

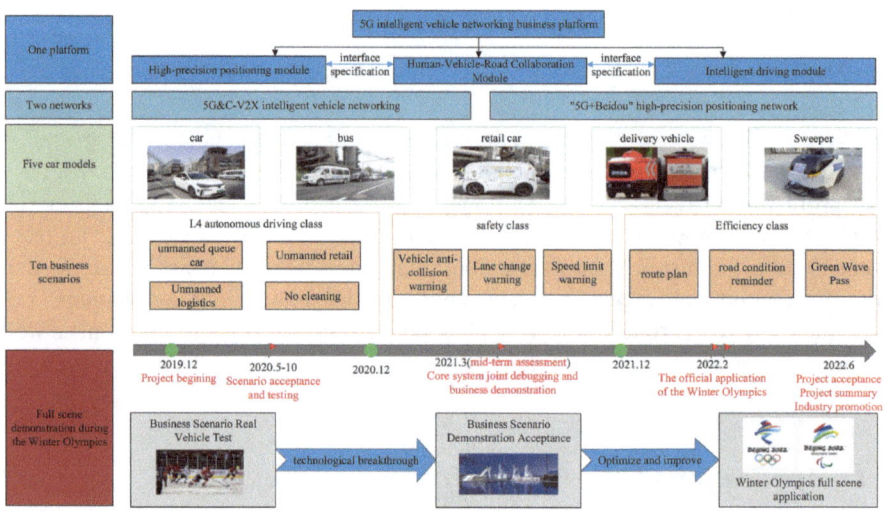

Fig. 6 Winter olympics 5G intelligent car networking system architecture

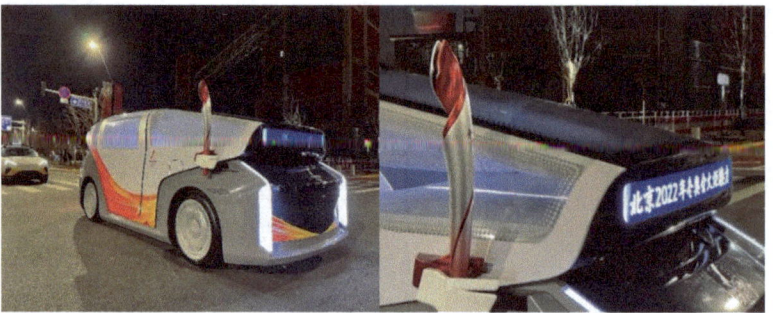

Fig. 7 5G intelligent vehicle network helps realize the first unmanned car torch relay

debugging tests have been conduct for autonomous driving, unmanned retail and other business vehicles. The system has completed demonstration and verification in 10 major events (such as "Meet in Beijing", torch relay), as exemplified in Figs. 7 and 8.

4.1.2 Data Collection Based on Internet of Vehicles System

In the IoV system, the transportation data and information comes from multiple sources. It includes raw transportation data, perceived transportation data, and semantic transportation information provided by transportation systems, V2X systems, and autonomous vehicles.

Fig. 8 5G unmanned sweeper helps to build 5G + smart winter olympic village

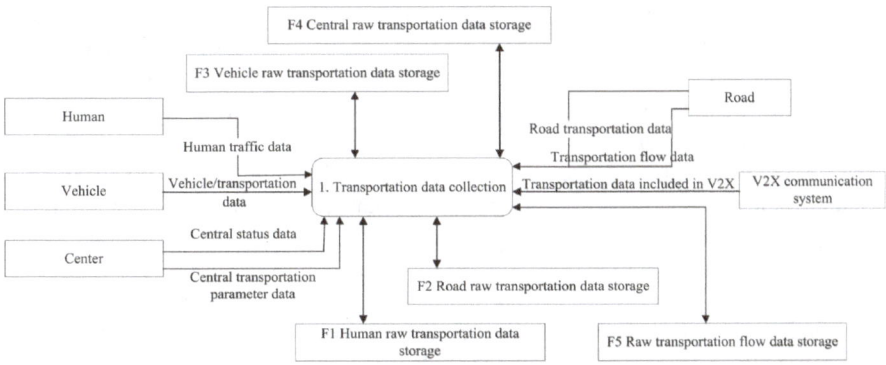

Fig. 9 Transportation data collection

In the IoV system, the objects of transportation data collection are the transportation participants in the transportation system, including people, vehicles, roads and centers, as well as the data of transportation subjects indirectly acquired through the IoV system, as shown in Fig. 9. People, vehicles, roads and centers all need access to describe their transportation state. The center itself also has parameters that describe its system state. In addition, the road is the transportation carrier, and the transportation flow data is the collection target of transportation situational awareness. The collected raw transportation parameter data needs to be persisted to four different data storage of F1-F4. Transportation flow data is persisted in the F5 transportation flow data storage.

4.1.3 Perception Data Fusion of Internet of Vehicles System

In the IoV system, multi-sensor fusion perception can expand the perception capability, optimize the perception effect, and improve the reliability of perception. This sub-section will focus on the multi-sensor data fusion technology of the IoV system.

In the IoV system, transportation situational awareness, prediction and control strategy are all based on transportation information collection. Recently, domestic and foreign scholars observe and count vehicles through various sensing methods, such as cameras, geomagnetism, millimeter-wave radar, and V2X communication. For sensing methods, a single sensor has inherent defects and is difficult to guarantee the reliable and stable environmental sensing/perception capability. Therefore, the multi-sensor collaborative fusion perception scheme is adopted to improve the reliability of perception and eliminate the perception blind area.

Multi-sensor collaborative fusion perception includes three aspects: fusion perception of multiple types of sensors, fusion perception of multiple sensors of the same type at the same location, and sensor fusion perception of multiple time scales.

The first aspect is the fusion perception of multiple types of sensors. As listed in Table 1, although the video data is easy to identify the target type, it is not as good as radar in detecting target status and anti-interference ability. The advantage of millimeter wave radar in finding the target status can make up for the weakness of video data. The excellent detection ability and detection accuracy of radar can also perceive the environment more reliably. Therefore, the multi-sensor fusion method based on video data is adopted to project the radar information onto the recognized image data, and the candidate frame provided by the radar data can also enhance the detection ability of the video data.

Due to the large range of radar detection, sometimes the data detected by radar cannot be timely detected by video for object type detection. For the object approaching the monitoring point, the trajectory is tracked firstly, and after the corresponding object can be detected in the video data, the types of bodies correspond to trajectories, thereby expanding the perception space.

The second aspect is the fusion perception of multiple sensors of the same type at the same location, as shown in Fig. 10. Although the same observation point can identify transportation participants within the field of view, in the actual scene, the blind area of the field of vision formed by the mutual occlusion of objects will bring difficulties to the recognition, and the insufficient perception will also bring potential dangers. The multi-point observation in the vehicle–road cooperative environment can address this problem. Multiple sensors cooperate with each other to form a global transportation situation.

Table 1 Comparison of application effects of multiple types of data

	Video data	Millimeter wave radar data	Lidar data
Target detection	Bad	Bad	Excellent
Target recognition	Excellent	Bad	Good
Precision	Bad	Excellent	Excellent
Detection distance	Good	Excellent	Good
Lighting anti-interference	Good	Excellent	Excellent
Atmospheric anti-interference	Bad	Excellent	Good

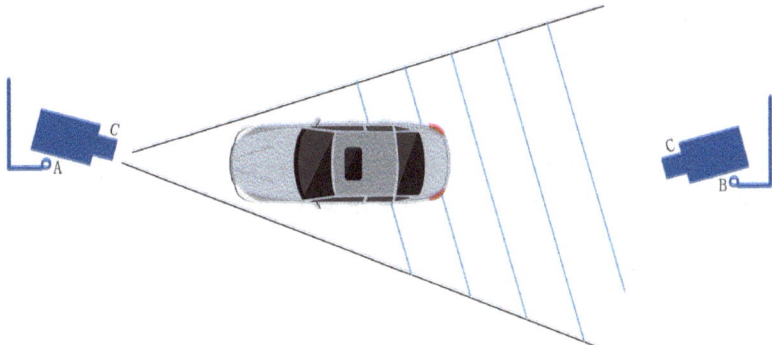

Fig. 10 Schematic diagram of multi-point sensors of the same type

The third aspect is multi-time scale sensor fusion perception, as shown in Fig. 11. On some road sections or with extremely small transportation forks or even no forks, although the sensor does not cover the sensing range of the entire road section, the transportation flow on the road can still be estimated in the control center. The estimation is through the passing data at both ends of the road section, combined with the information of vehicles that have passed and not passed. Furthermore, combined with the passing time and the travel time of the road, it is also possible to roughly estimate the transportation distribution status of the road section, and then make corresponding modifications to the control devices (e.g., traffic lights) to achieve fusion perception beyond the monitoring range.

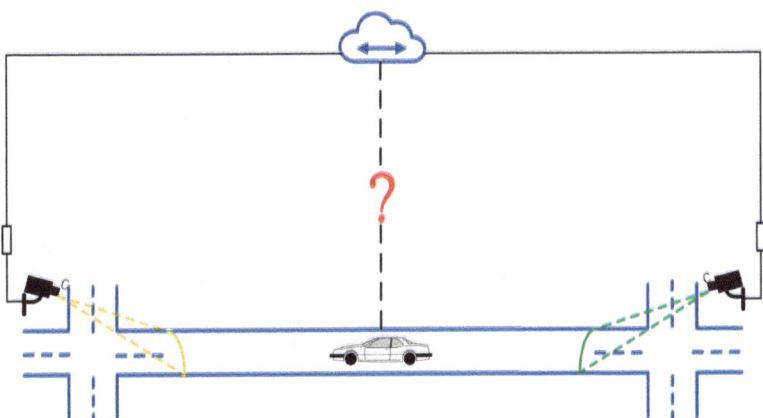

Fig. 11 Multi-time scale fusion perception

4.2 Digital Twin Applications for Internet of Vehicles and Intelligent Transportation

The development of digital transportation in China should employ data as the key element. The new generation of artificial intelligence and big data technology accelerates the process of "Internet of Everything" in the transportation industry. Intelligent transportation is mapped from the physical world to the digital world with the four elements of "vehicle-people-road-environment". The digital twin combines data mining, big data visualization and other technologies to display road conditions, transportation situation, potential risks in real-time. They are used to build an efficient and safe digital transportation system. This sub-section will take a large-scale mixed transportation group collaborative behavior simulation platform as an example, and focus on the hardware architecture and testing environment of the virtual-real combined transportation simulation system, as well as two testing methods that conform to real application scenarios.

4.2.1 Collaborative Behavior Simulation Platform for Large-Scale Hybrid Transportation Groups

The hardware of the intelligent collaborative behavior simulation system for hybrid transportation groups mainly includes three types of modules: virtual simulation server cluster, human–computer interaction equipment, and hardware-in-the-loop testing machine. The overall system hardware architecture is shown in Fig. 12.

To meet the requirement of rapid testing of vehicle–road collaborative group control efficiency, a virtual and real transportation simulation platform architecture was designed. The simulation platform is mainly composed of three systems: the vehicle–road group operating environment data simulation/collection and fusion processing system (data generation system), vehicle–road group collaborative decision-making and optimal control system (algorithm testing system), vehicle road group testing scene sequence generation and performance analysis system (scene verification system). In terms of real vehicle road testing equipment, it integrates support item perception, interaction, and decision-making function equipment, and maps it to the virtual testing environment through the state synchronization drive mechanism. In terms of virtual transportation environment driving, simulation tools are used to simulate and generate transportation environment perception data, and a simulation sequence of mixed transportation scenarios is constructed. Combined with the development of specific functions, the virtual-real combination simulation and testing for intelligent cooperative behavior of heterogeneous transportation groups is realized (Fig. 13).

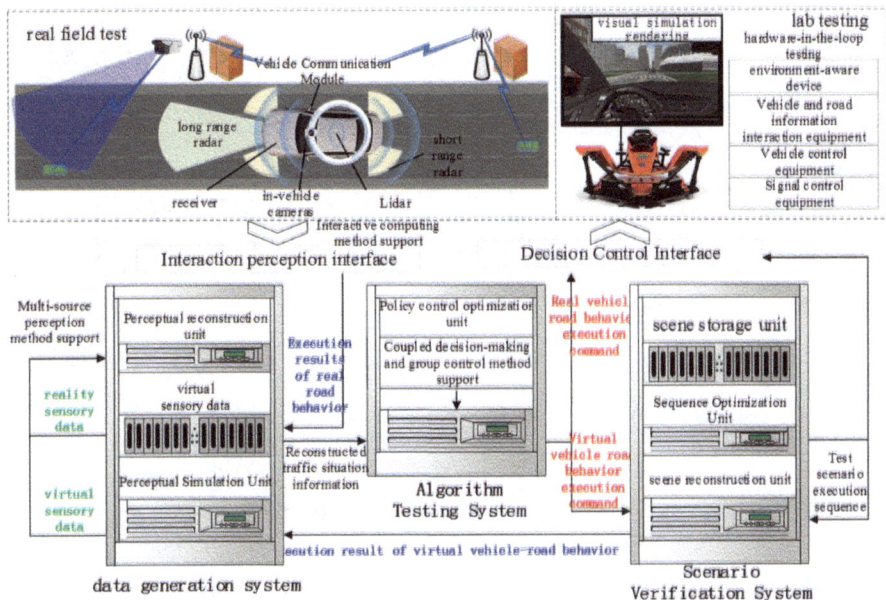

Fig. 12 Hardware architecture of intelligent collaborative behavior simulation system for hybrid transportation group

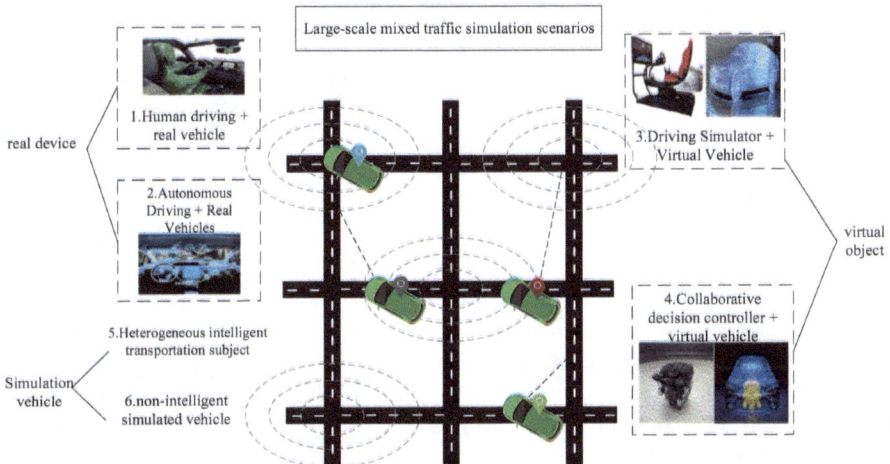

Fig. 13 Large-scale virtual-real combined test environment for mixed transportation

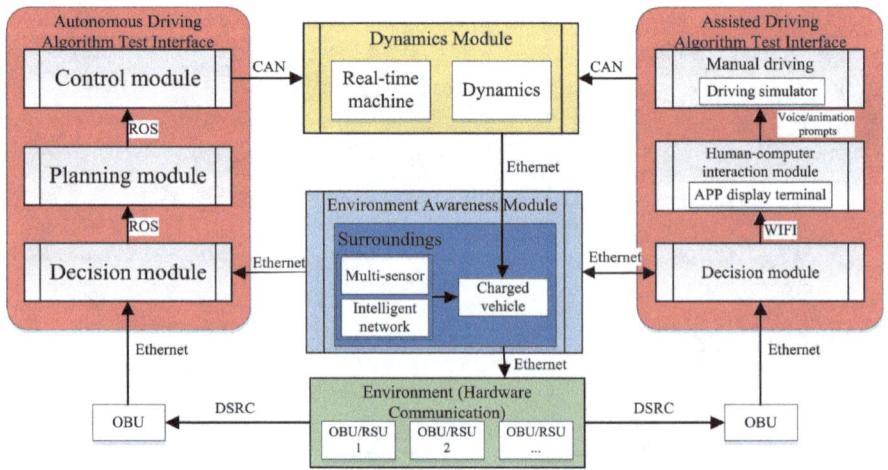

Fig. 14 Multi-component integrated co-simulation testing framework

4.2.2 Multi-component Integrated Co-simulation Testing Method

We consider the testing requirements of the decision-making, planning and control functions of intelligent vehicles in intelligent transportation. Therefore, the simulation testing system should include functional modules, such as scene modeling, decision-making planning, intelligent network connection, various sensors, dynamics, display warning, data analysis, etc. The designed multi-component integrated co-simulation testing framework is shown in Fig. 14 [10].

Figure 15 shows the overall framework of the data flow channel design of the simulation system. Among the transportation flow simulation, communication simulation and sensor simulation modules, all simulation objects (e.g., vehicles, traffic lights) continuously interact in real-time, and various simulation tasks are completed synchronously. The main principle framework of data synchronization is shown in Fig. 16. While the traffic state is synchronized, the communication simulation and the sensor simulation are continuously working according to their pre-defined rules and protocols. The communication simulation results and sensor simulation results are defined as corresponding standard data frames, which are sent to the decision-making module through Ethernet, and are parsed by the decision-making module into the data structure required by the decision-making algorithm.

4.2.3 Virtual-Real Interactive Testing Method

Recently, the virtual-real interactive testing (VRIT) scheme represented by digital twins has been proposed [11, 12]. VRIT is an effective test method to evaluate the security and intelligence of intelligent network connection. In the work of [11],

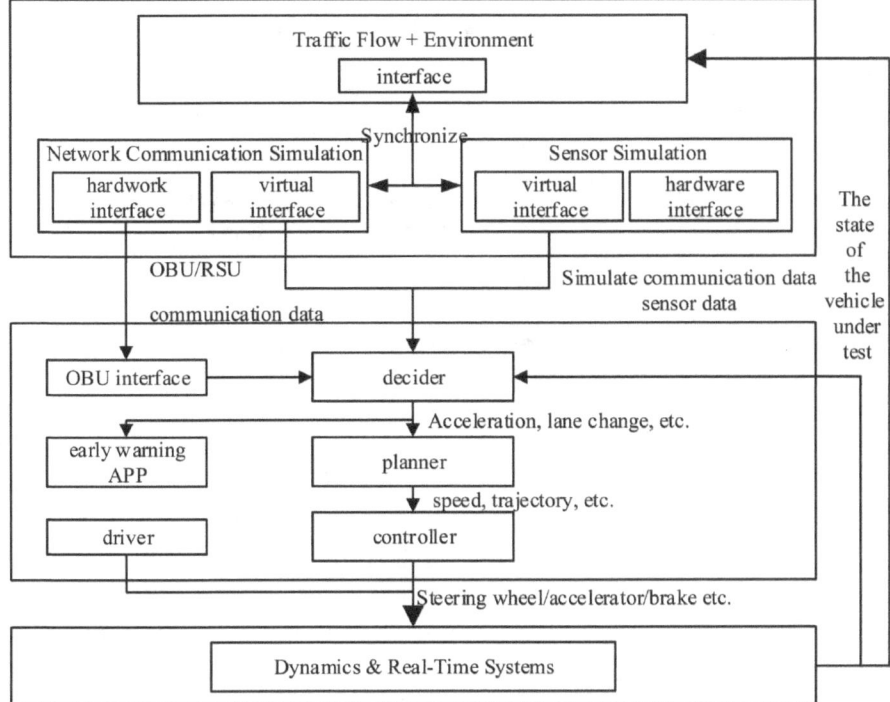

Fig. 15 Data interaction framework

VRIT can provide an interactive channel connecting the physical space and the virtual space, and VRIT can also add a synchronization space to ensure the state consistency of the physical space and the virtual space. These extra spaces generate reference trajectories for mirrored objects in each operation cycle.

As described in Fig. 17, in the physical space of VRIT, the operation of the object under test is related to its controller. Physical space sends data to virtual space. Virtual space simulates transportation flow by redrawing actual vehicles. Heterogeneous data from multiple spaces is divided into view layer, execution layer and command layer. View layers change the position and pose of mirrored objects. The execution layer changes the object state in the view layer including motion data, such as speed and yaw rate. The real-time operation of the command layer directly affects the execution layer. Data in different spaces of each layer will be assigned the corresponding priority. Heterogeneous data with different priorities will affect the generated reference trajectory.

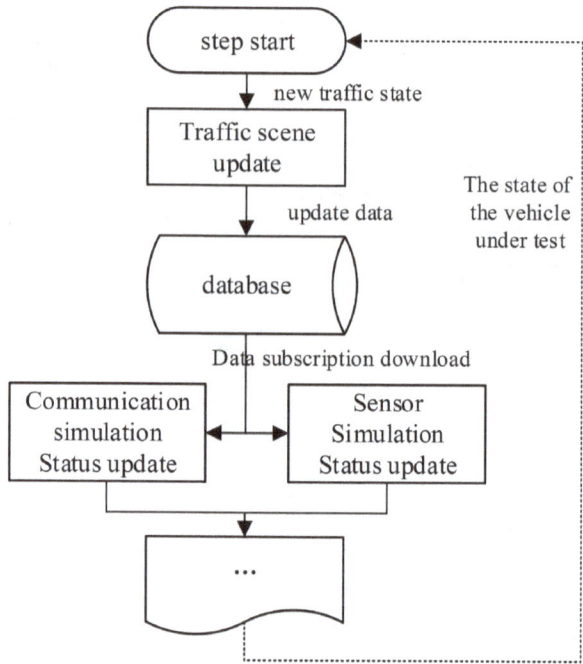

Fig. 16 Simulation data update logic

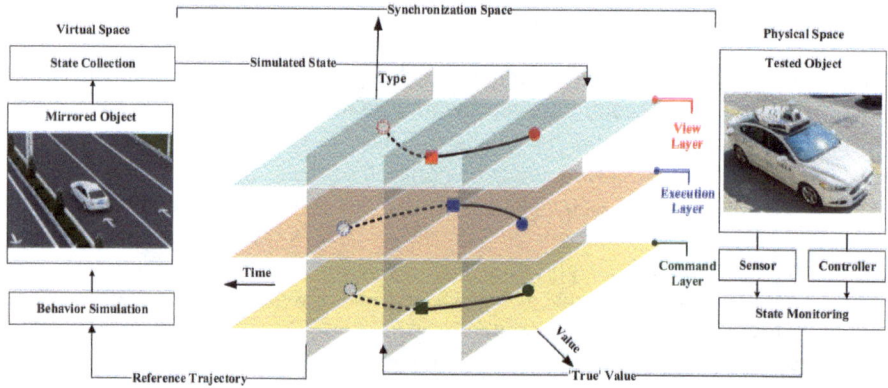

Fig. 17 Model framework for the synchronization of physical space and virtual space trajectories

5 Conclusion

In recent years, big data accelerates the digital revolution of vehicle and transportation industry. In this chapter, initially, we study big data origin and features, data mining technology, as well as both software and hardware infrastructure of big data. Then, this chapter designs a novel vehicle and transportation big data platform, which consists of six layers. Based on the vehicle and transportation big data platform, this chapter introduces our application in IoV, as well as digital twin IoV and intelligent transportation.

References

1. https://en.wikipedia.org/wiki/Big_data
2. H. Zhang, L. Xu, X. Cheng, W. Chen, X. Zhao, Big data research on driving behavior model and auto insurance pricing factors based on UBI. in *3rd International Conference on Signal and Information Processing, Networking and Computers (ICSINC)*, (Chongqing, China, 2017), pp. 404–411
3. J. Gao, X. Cheng, L. Xu, H. Ye, An interference management algorithm using big data analytics in LTE cellular networks. in *2016 16th International Symposium on Communications and Information Technologies (ISCIT)*, (Qingdao, China, 2016), pp. 246–251
4. J. Gao, X. Cheng, L. Xu, L. Cao, K. Chao, A coverage self-optimization algorithm using big data analytics in WCDMA cellular networks. in *1st International Conference on Signal and Information Processing, Networking and Computers (ICSINC)*, (Beijing, China, 2015)
5. http://hadoop.apache.org/
6. https://hadoop.apache.org/docs/r1.2.1/hdfs_design.html
7. L. Xu, Y. Luan, Y. Cao, H. Liu, X. Cheng, R. Dong, H. Zhang, Framework and design discussion on city smart transportation management big data platform. Des. Tech. Posts Telecommun. **5**, 7–12 (2020)
8. L. Xu, Y. Cao, H. Yang, C. Sun, T. Zhang, B. Wen, X. Cheng, C. Song, X. He, Research on telecom big data platform of LTE/5G mobile networks. in *18th IEEE International Conference on Ubiquitous Computing and Communications (IEEE IUCC)*, (Shenyang, China, 2019), pp. 756–761
9. https://en.wikipedia.org/wiki/Extract,_transform,_load
10. W. ShangGuan, L.G. Chai, D.F. Chu, S.F. Zheng, F. Hui, *Simulation and Evaluation Method for Intelligent Cooperative Behavior of Mixed Traffic Group.* (China Communications Press, 2022), pp. 164–180
11. W. Qiu, S. Wei, B. Cai, L. Chai, Heterogeneous data-based spatiotemporal trajectory synchronization for virtual–real interactive testing. Computer-Aided Civ. Infrastruct. Eng. (2022)
12. T. Luo, M. Zhang, Z. Pan, Z. Li, N. Cai, J. Miao, Y. Chen, M. Xu, Dream-experiment: a MR user interface with natural multi-channel interaction for virtual experiments. IEEE Trans. Visual Comput. Graphics **26**(12), 3524–3534 (2020)

Chapter 12
Noval Enabling Technology for V2X Network: Blockchain

Yuntao Liu, Kainan Zhu, Wei Hua, and Yongdong Zhu

Abbreviations

AVs Autonomous Vehicles
CAV Connected Autonomous Vehicles
DAG Directed Acyclic Graph
DLT Distributed Ledger Technology
DPoS Delegated Proof of Stake
DRL Deep Reinforcement Learning
DSA Digital Signature Algorithm
ECDSA Elliptic Curve Digital Signature Algorithm
IoV Internet of Vehicles
LBS Location-based service
MEC Multi-access Edge Computing
OBU On-Board Unit
PBFT Practical Byzantine Fault Tolerance
PoS Proof of Stake
PoW Proof of Work
P2P Peer-to-Peer
RSA Rivest-Shamir-Adleman

Y. Liu · K. Zhu · W. Hua · Y. Zhu (✉)
Zhejiang Lab, Interdisciplinary Innovation Research Institute, Hangzhou, China
e-mail: zhuyd@zhejianglab.com

Y. Liu
e-mail: liuyt@zhejianglab.com

K. Zhu
e-mail: zhukainan@zhejianglab.com

W. Hua
e-mail: huawei@zhejianglab.com

RSU   Road-Side Unit
TA   Traffic Administration
V2X   Vehicle-to-Everything

1 Introduction

With the continuous advancement of information technology, vehicles are becoming more intelligent than ever. Vehicle-to-Everything (V2X) communication enables interconnection between vehicles and other entities, including roadside infrastructure, pedestrians, and cloud servers. On the one hand, vehicles can improve driving safety and traffic efficiency by using information from V2X networks. On the other hand, by exchanging information about traffic conditions, the adverse effects of environmental pollution, accident rate, and traffic congestion on the entire traffic system can be alleviated while improving road traffic's convenience, comfort, and safety.

1.1 Motivation

V2X network incorporates numerous entities, such as vehicles, wireless base stations, roadside units (RSUs), multi access edge computing (MEC) servers, and cloud providers. The powerful capabilities of the V2X network are based on the interaction between vehicles and other entities. Unfortunately, in practical application scenarios, it is difficult for a vehicle to freely interact with other entities due to interactive information's sensitivity and privacy issues. For example, since malicious entities (vehicles or others) can easily publish false information or tamper with shared data, vehicles are prone to have security issues.

Blockchain has attracted extensive attention from both academia and industry since it was first introduced in Bitcoin as a distributed ledger technology [1]. Blockchain can protect data from being tampered with due to the use of chained data structures and cryptography. Therefore, any participant can trust the data on the blockchain without any doubt. Blockchain enables secure and private interactions between untrusted participants without relying on a central agent. Furthermore, because data are replicated on multiple distributed nodes, blockchain can solve the single point of failure problem. Therefore, many applications have adopted blockchain by taking advantage of the above attractive properties.

With the help of blockchain technology, a trusted environment for vehicles to communicate with other entities can be created. In addition, the blockchain can also ensure the integrity and immutability of the core information of vehicles, enabling many innovative services based on the V2X to be realized. Therefore, integrating blockchain technology into the V2X network can bring considerable benefits to vehicles.

This chapter conducts a comprehensive survey to shed light on the application of blockchain in V2X-assisted autonomous driving.

1.2 Main Contributions

There is a lot of survey research work on the application of blockchain in V2X networks [2–5]. In contrast to these surveys, our survey focuses on blockchain integration in V2X network scenarios. The contributions of this chapter are summarized as follows.

(1) We discuss the basic principles and key technologies of blockchain.
(2) We survey the state-of-the-art research work of integrating blockchain into V2X and discuss the adoption of blockchain for V2X scenarios.
(3) We highlight typical applications and highlight the specific challenges blockchain can address.
(4) Based on the survey, we present future research directions and opportunities in this field.

1.3 Structural Organization

Section 2 presents an overview of blockchain technology and introduces the V2X network. Section 3 discusses the existing application of blockchain in V2X networks. The future research directions related to blockchain are highlighted in Sect. 4. Finally, in Sect. 5, the concluding remarks of this chapter are presented.

2 Background

2.1 Blockchain Technology

2.1.1 Overview of Blockchain

Blockchain is a peer-to-peer distributed ledger technology that uses a decentralized consensus mechanism to maintain a complete, distributed, immutable ledger database to build trust and consensus in a decentralized network. Due to the immutability of ledger data, distributed participants in a blockchain system without trust relationships can establish consensus and trust.

The basic unit of a traditional blockchain system is the block. Figure 1 shows the connection structure of a traditional blockchain. All blocks are linked together to form a chain. Newly generated blocks are attached to the chain to form a ledger, which

Fig. 1 Block in traditional blockchain

is maintained by multiple distributed participants in the system. Hence, blockchain technology is sometimes known as distributed ledger technology (DLT).

Blockchain has the characteristics of decentralization, immutability, auditability, fault tolerance, and non-repudiation guarantees. These characteristics make it a potential solution to enhance the security of various distributed systems, such as the financial industry [6], food industry [7], energy industry [8], supply chain [9], and others [10].

Blockchain relies on key technologies such as structure, cryptography, consensus, and smart contracts. These key technologies are briefly discussed below.

2.1.2 Structure of Blockchain

Blockchain is a combination of block and chain. Therefore, block and chain are essential parts of the blockchain system.

The typically linear structure of a blockchain system consists of a series of blocks that are used to store information related to transactions occurring on the system. Except for the genesis block, each block contains the hash information of the previous block. All blocks are layered, one on top of the other, with the genesis block being the foundation, and they grow in height until the end of the blockchain is reached and the sequence is complete. A chain, one after another, connects all blocks.

Although linear structures are common, another type of structure called a directed acyclic graph (DAG) can also be used [11]. The basic structure of the DAG blockchain is shown in Fig. 2. Unlike a linear structure, a block is connected to its multiple predecessors in a DAG blockchain. Therefore, a DAG blockchain has many chains. In addition, some DAG blockchain systems use transactions as the smallest unit (some DAG blockchain systems can also support the smallest unit as a block). Since there is neither a block nor a chain, theoretically, a DAG DTL is not a blockchain. However, its concurrency mechanism can significantly improve blockchain performance. Hence, DAG is an emerging change for blockchain-based ecosystems. Figure 2 shows the structure of the DAG blockchain.

2.1.3 Cryptography

Cryptography is widely used in blockchain systems to ensure security, privacy, and anonymity. The hash function and digital signature are two typical applications of cryptography in the blockchain.

Fig. 2 DAG blockchain

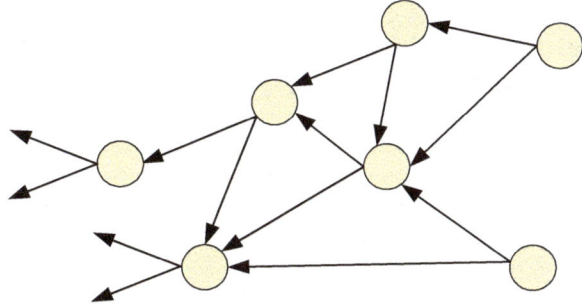

A hash function can encode variable-length input data into data of a specific length. [12]. SHA-3, SHA-256, and SHA-512 are three hash functions that are widely used in blockchain. A hash function has many properties: (1) Irreversibility. The process of hash function processing is one-way. It is almost impossible to calculate the original input data from the processed output data; (2) Determinism. The output data has a fixed size regardless of the input data size; (3) Collision resistance. Even if the two input data differ by only one byte, the output value will significantly differ. Figure 3 illustrates the properties of the hash function.

Each block in the blockchain contains the previous block's hash value, and the hash value connects each block to the previous block to build a chain. In this way, owing to the use of hashing function, any attempt to tamper with block data will cause a drastic change in its hash value and other blocks of the entire blockchain. Therefore, the hash functions guarantee the integrity and tamper-proof of the blockchain system. Data on the blockchain is credible, and any participant can trust the blockchain without any

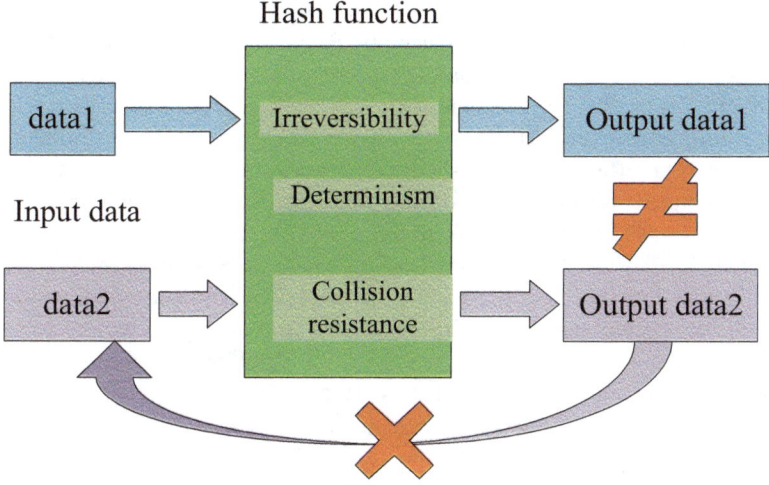

Fig. 3 Properties of Hash function

doubt, enabling secure and private interactions among untrusted participants without relying on a central agent. In addition, another application of the hash function is the Merkle hash tree [13], which uses very little hash data to represent the transactions in the block. The Merkle hash tree can quickly check blocks' integrity and the transactions' authenticity.

The hash function realizes that the transaction data cannot be tampered with, but it cannot guarantee the non-repudiation of the transaction data. Blockchain uses the digital signature to ensure the non-repudiation of data. Digital signature mainly includes two processes signature and verification. The signer uses its private key to sign the data during the signing process. The verifier uses the signer's public key to verify the data in the verification part. RSA (Rivest-Shamir-Adleman), DSA (Digital Signature Algorithm), and ECDSA (Elliptic Curve Digital Signature Algorithm) are commonly used as digital signature algorithms. For example, ECDSA is widely adopted in many blockchain systems, such as Bitcoin, Litecoin, and Ethereum, to achieve non-repudiation of transaction data.

2.1.4 Consensus Mechanism

The maintenance and management of the blockchain system are done by a set of participants in a peer-to-peer network who do not trust each other. Therefore, a consensus mechanism must exist to make all participants agree on the blockchain system's transactions (state). The consensus mechanism is the core of the blockchain system, in which all or some participants execute to maintain and verify transactions in the blockchain system. Some of the notable consensus mechanisms are Proof of Work (PoW) [1], Proof of Stake (PoS) [14], Delegated Proof Of Stake (DPoS) [15], and practical byzantine fault tolerance (PBFT) [16].

The PoW is one of the most widely used consensus mechanisms in existing blockchain systems. It is firstly introduced by Bitcoin, also known as the Nakamoto consensus. We take PoW as an example to elaborate on the consensus mechanism. The core of PoW mechanisms includes block publisher selection and chain selection.

Every node in Bitcoin has the right to pack blocks, but only one node packs a block that can be effectively published to the network during an interval. During the packaging process of each block to be published, a random number needs to be calculated and added to the block so that the block's hash value meets a particular requirement. Due to the randomness of the hash value, notes (miners) need to perform many trial calculations to get the random number. The first node that computes to get the random number can publish its block. Therefore, nodes with more computing power have a better chance of being selected for the publishers.

Furthermore, the longest chain rule is used in the Nakamoto consensus. The blockchain system is distributed, and there is no unified time synchronization. At the same time, there are many forked chains in the blockchain ledgers due to network transmission delays and other reasons. The node takes the longest chain as the legal chain and appends new blocks after this chain.

2.1.5   Smart Contracts

The smart contract was first proposed by cryptographer Nick Szabo in 1995 [17]. A smart contract is a set of digitally defined commitments, including agreements on which contract participants can execute those commitments.

A smart contract can be a piece of executable code or script in a specific implementation. It can be executed on all nodes. Thus, the definition of smart contracts must have precise and deterministic conventions to avoid inconsistencies. Hence, the smart contract is an autonomous actor that behaves transparently and predictably. In addition, a smart contract's rules and execution logic must also be clear and explicit. Essentially, a smart contract works like an if–then statement of a computer program. The smart contract can automatically execute the corresponding agreed terms when a preset condition or state in the blockchain system is met or triggered. In a sense, a smart contract can realize the automatic digital execution of a paper contract.

Traditional blockchain systems, such as Bitcoin, is just a system that supports financial transfer functions. A smart contract can extend the functions of the blockchain system from simple financial transfers to various complex services and make the blockchain system a highly autonomous system that can provide efficient and consistent services without relying on any trusted entity.

2.2   V2X

The Internet of Vehicles (IoV), also known as V2X, has evolved from traditional vehicular ad hoc networks (VANET) and involves entities such as humans, infrastructures, and other heterogeneous entities. As the name implies, V2X enables the exchange of information between vehicles and all entities.

V2X is crucial for achieving intelligent transportation systems and is expected to improve traffic safety, mitigate congestion, and reduce fuel consumption and pollution, which are the significant challenges of modern transportation. V2X can bring various reliable communication services to connected vehicles and expand the functions of vehicles and improve the safe driving of vehicles. For example, for a CAV, through V2X communication, information about the surrounding environment, other vehicles, and roads can be collected to achieve safer driving.

However, at the same time, V2X also brings some challenges to CAVs. These problems may be similar to those faced in traditional ad hoc networks. Since some inherent characteristics of connected vehicles, such as different vehicle capabilities, speed of vehicles, intermittent connections, and others, the seriousness of traditional problems is magnified, and new issues are introduced in V2X [18].

3   Blockchain for V2X Scenarios

Since V2X networks contain various sensitive information and involve many
distrusted entities, the blockchain can be naturally applied in V2X networks. There
have been many research works on how blockchain technology can be applied in
V2X. This section comprehensively analyzes and summarizes the incorporation of
blockchain technology and V2X application scenarios from the following seven
aspects.

3.1   Trust Management

In IoV, vehicles can communicate with other vehicles, sensors, pedestrians, RSUs,
infrastructure, cloud servers, and others through OBUs. Communication with other
entities can improve traffic safety, mitigate congestion, and reduce fuel consumption
and pollution. However, in the actual V2X environment, it is hard to ensure that each
entity in the network is safe and reliable. The potential for dishonest and misbehaving
peers in the V2X leads to various security and privacy threats. Therefore, establishing
trust among these potentially untrusted entities is one of the significant challenges
in V2X.

Given the possibility of misbehaviors vehicles in V2X, the messages shared
between vehicles are not always reliable. Zhang et al. [19] propose a blockchain-
based trust management system for this issue. In this system, they design a complete
reputation value update algorithm to effectively detect those vehicles that send false
messages and then reduce their reputation values for punishment. The reputation
values are recorded into the blockchain to prevent tampering by attackers. Additionally, they devise a consensus mechanism that combines PoW and PoS to ensure that
vehicles with higher reputation value can be prioritized for updates on the blockchain.
The simulation results show that the trust management system has obvious restrictions on malicious vehicles, which improves the accuracy of the vehicle's judgment
of events according to the received messages.

In the same area, Singh et al. [20] use smart contracts to build a blockchain-based
decentralized trust management scheme for V2X. In this scheme, the core information of interactions between vehicles is called events. And for each event, the
vehicle exchanges many messages with its peers and checks the integrity and authenticity of each message received. If any inconsistencies in the messages are detected,
the vehicle reports to RSUs for action. RSUs are edge nodes capable of running
a distributed consensus of blockchain for trust management. Based on the smart
contract logic deployed, RSUs execute the management and maintain the trust values
of vehicles depending on their behavior in a decentralized manner. In addition, by
introducing sharing technology, the transaction load on the entire blockchain network
is divided into multiple subnets. Each subnet maintains a set of localized transactions,
thereby reducing the load on the main blockchain and improving the entire network

transaction throughput performance. Furthermore, an incentive strategy for vehicle participation event detection is designed. The system can reward vehicles that detect actual events and report them accurately. Conversely, vehicles can be revoked from the system if they do not perform well.

Some studies construct trusted environments from the perspective of vehicle identity authentication. A blockchain-based vehicle registration management scheme is introduced to create trust management for V2X [21]. They design a reliable framework for managing vehicle registrations via blockchain and smart contracts. Specifically, each vehicle is assigned a unique identity using the physical unclonable function (PUF), and certifications are issued to each vehicle by RSUs to protect vehicles' privacy. Furthermore, a dynamic consensus algorithm, dPoW, is devised, which can scale and change its operation according to the incoming traffic generated by vehicles.

V2X promotes the burst of a large number of innovative services. These innovative services also depend on trust among interacting entities. For example, the complex computing tasks of the vehicle can be processed efficiently in the nearby edge infrastructures, thus reducing the computing and storage burden of the vehicle. Compared with traditional cloud computing, it can also reduce the delay of computing tasks. In [22], an efficient privacy-preserving authentication protocol is presented. Specifically, edge computing and consortium blockchains are combined to support efficient computation and storage capabilities with low communication delay while providing data availability. The core data of the computing offloading task in edge computing is audited and verified in the alliance chain, and the verified data is permanently, openly, and transparently stored in the blockchain system. First, the pseudonym-based mechanism is used to hide the vehicle identification number (VIN) code, enabling privacy protection of the vehicle. Second, identity-based cryptography is adopted to fulfill data integrity between the cluster leader and vehicles, abandoning traditional PKI-based certificates and providing effective data integrity for vehicles to transmit messages. Third, a cluster selection algorithm and a key update algorithm based on the Chinese remainder theorem are given to ensure the forward and backward security of the transmitted information. Finally, security analysis and experimental evaluation of communication and computational costs demonstrate the effectiveness of the proposed protocol.

3.2 Data Sharing

In V2X, connected vehicles, especially connected autonomous vehicles, are equipped with many sensing devices. Connected autonomous vehicles continuously generate various types of data, such as vehicle state data, vehicle surrounding perception data, and road traffic data. The generated data of connected autonomous vehicles can be shared with other entities in V2X. For example, sharing road traffic data can improve and enhance the driving experience and vehicle safety. However, despite apparent benefits, such data sharing is difficult to apply in practice due to the lack of trust between vehicles and privacy concerns.

Chai et al. offer a hierarchical blockchain-based sensing information (SI) message cooperation mechanism [23]. In this cooperation mechanism, vehicles can communicate with roadside units (RSUs) to share their sensing information (SI). Vehicles can access SIs saved in surrounding nodes and cooperatively sense traffic conditions and road information to improve road safety. According to the SI influence range, SI is divided into low-scope SI and high-scope SI. For low-scope SIs, the audit process is conducted by a small scale of peers. High-scope SIs are audited on a small scale and then verified by higher-level peers on a larger scale in different layers of ledgers. Through layered blockchain, the overall consense is achieved. Correspondingly, mining nodes achieve consensus in different layers and cache various SI latency can be significantly reduced to accommodate V2X applications with high-speed transmission and low-latency requirements.

In [24], a hybrid blockchain PermiDAG is developed for secure data sharing. The PermiDAG consists of a main blockchain ledger maintained by RSU nodes and a local DAG ledger run by selected vehicle nodes. The main blockchain ledger records all data sharing events between vehicles, including providers, consumers, data profiles, and transaction summaries in the local DAG ledger. Microtransactions in data sharing events are stored in the local DAG ledger to improve the efficiency of the blockchain in latency-sensitive V2V data sharing. Data sharing related microtransactions in the local DAG ledger are synchronized periodically to the main blockchain. To ensure the quality of shared data, a two-stage verification mechanism is devised. The selected vehicle nodes perform the first-stage quality verification in local DAG, and then RSUs execute the regular validation of transactions and blocks in the main blockchain.

3.3 Resource Sharing

In V2X, except the data sharing, there can also be other resource sharing scenarios, such as computing resource sharing, storage resource sharing, learning knowledge sharing.

A consortium blockchain-based framework for secure computing resource sharing in V2X is designed in [25]. In this framework, vehicles provide computing resources, and the computing resource requester can be vehicles or other entities in V2X. RSUs collect transaction records, verify data integrity and maintain a consortium blockchain. In order to defend against malicious behavior by service requesters and service providers, multi-step smart contracts are designed to enable secure resource sharing. When the computing resource sharing task is completed, RSUs verify the related transactions. According to the verification results, the smart contract automatically performs corresponding operations on the providers and requesters of computing resources, rewarding honest participants and punishing dishonest ones. The Byzantine Fault Tolerance-based Proof of Stake (BFT-based PoS) consensus protocol is applied in the consortium blockchain to achieve consensus efficiently.

Furthermore, a contract-based incentive mechanism is designed to incentivize vehicles to share their computing resources with service requesters in asymmetric information scenarios.

Since connected vehicles generate a large amount of data and multimedia content, caching these data and content in other devices (other vehicles or RSU, BS, MEC) near vehicles can reduce the caching overhead of connected vehicles. Dai, Y integrates deep reinforcement learning (DRL) and blockchains into in-vehicle networks for smart and secure content caching [26]. This article proposes a blockchain-empowered distributed and secure content caching framework, where vehicles act as cache requesters and cache providers, and BSs act as verifiers to maintain the blockchain system. Due to the high mobility of vehicles, a DRL-based method is utilized to learn dynamic network topology and time-varying wireless channel conditions to achieve the best matching between cache requesters and cache providers. Moreover, a new DPoS-based blockchain consensus mechanism, Proof of Utility (PoU), is designed. The PoU can speed up the selection of block verifiers and achieve fast verification of blocks.

Knowledge sharing allows vehicles to exchange their learned experiences. It can speed up the learning process of vehicles and improve the intelligence ability of vehicles. For example, vehicles can collaborate to train a traffic perception model. Since the data collected by each vehicle has a certain limitation, learning the models trained by other vehicles helps the vehicle to obtain a comprehensive model and improve the accuracy of the model. The knowledge shared among vehicles is the parameter information of the model. In a sense, sharing knowledge is a manifestation of swarm intelligence.

A hierarchical blockchain-enabled federated learning framework is provided for V2X knowledge sharing [27]. In this framework, knowledge is shared in the form of learning parameters during the federated learning process. The proposed framework consists of one Top Chain (TC) in the top layer and multiple Ground Chains (GCs) in the ground chain layer. In the ground chain layer, vehicles are responsible for collecting their surrounding environmental data as the federated learning training set. The learning results are sent to nearby RSU in the form of transactions. RSUs are responsible for collecting the transactions within their communication ranges and packaging transactions as candidate blocks. According to the variety of locations and communication ranges of RSUs, the ground chain layer contains multiple GC chains. The consensus process is implemented among RSUs located in one identical GC chain, and the published blocks reveal the federated learning results, which are also the representatives of vehicular knowledge. In the top layer, RSUs integrate the learning results from GC chain ledgers and their local training results into the transactions of the TC chain, and then the transactions are collected by BSs and recorded in the TC ledger. In the TC chain, the ledger assembles the sharing knowledge from both vehicles and RSUs, which can be used for traffic analysis from a global perspective.

3.4 Trading Application

Electric vehicles are a primary form of CAVs and have gradually gained popularity. Electric vehicles, as the mobile storage facility of distributed energy, are an essential part of the Energy Internet. They make vehicle-to-grid (V2G) energy trading possible due to their flexible mobility, which is conducive to addressing the growing energy demand and extremely uneven energy distribution. V2G is a trading mode of peer-to-peer (P2P) in which mobile vehicles share the remaining energy of a node with other nodes to effectively achieve peak-load shifting, improve energy efficiency, and reduce transmission loss.

A vehicle-to-vehicle (V2V) energy trading system is developed based on a consortium blockchain [28]. In this system, energy transactions are conducted between vehicles. Fog computing nodes are deployed in social hotspots, verify and save energy transactions, and maintain the blockchain network through a consensus mechanism. Vehicles can access the energy blockchain history from fog nodes and query records of interest. Combining the PBFT algorithm and DPoS algorithm, a more efficient consensus algorithm DPOSP is designed as the consensus mechanism to reduce the verification latency and support fast transaction payments. Xia et al. also propose a similar idea to introduce blockchain to ensure the trustworthiness, security, and reliability of vehicle-to-vehicle (V2V) energy transactions [29].

With the development and emergence of various network services in V2X, a large amount of data is collected and stored. Data will be one of the most critical assets in future V2X systems. Data transactions can be traced based on the blockchain to protect the rights of both parties. In [30], a blockchain-based V2X data transaction framework is presented to ensure the security and authenticity of data transactions. Vehicles play different roles in the data transaction process, including data sellers collecting data, data buyers requesting data, and idle nodes that neither selling nor buying data. The role of each node can be switched according to its requirements. The edge server acts as a broker for data transactions and uses smart contracts to manage the data transaction process. Each data buyer sends its data requirements to the nearest edge server, which will conduct an iterative double auction among vehicles and utilize smart contracts to match pairs of data transactions. Buyers and sellers broadcast data transactions to brokers for verification and auditing. The broker records the transaction data within a certain period, encrypts and digitally signs it, and packs it into blocks.

CAVs can be viewed as dynamic computing resource transporters. Therefore, in V2X, vehicles can trade their idle computing resources in a peer-to-peer (P2P) manner to meet smart cities' dynamic computing resource demands. Lin et al. [31] provide a blockchain-based peer-to-peer (P2P) computing resource trading system. Intelligent vehicles can share their idle computing resources with the computing resource trading system and obtain corresponding rewards. An intelligent vehicle can be a computing resource seller, a computing resource buyer, or an idle node according to its computing resource status. The MEC nodes are the managers of the local P2P trading system. They can store and manage the mobility profiles and computing resource status of intelligent vehicles in their area and verify and package computing resource transactions into blocks. At the same time, MEC nodes are miners of the blockchain system, realizing the consensus process of transaction data in various regions. The blockchain consensus protocol adopts the hybrid consensus protocol PoW-PBFT, combining the traditional PoW protocol with the PBFT protocol, which can improve transaction performance to ensure transaction security.

In addition, many other commercial trading application scenarios have been proposed. For example, a blockchain-based vehicle networking data transaction mechanism is designed in [32].

3.5 Collaboration Application

Location-based service (LBS) is an essential service provided by the Internet of Vehicles. However, geolocation information is sensitive. Location-based service providers (LSPs) can collect and analyze users' sensitive information based on location query requests. In addition, external attackers may also analyze the user's privacy by stealing information from the user's location query request process. Regarding privacy issues in practical location-based services, Li et al. [33] offer a blockchain-based LBS security preserving trust scheme. Specifically, the distributed k-anonymity algorithm is used to hide users' actual location so that users can avoid personal privacy leakage when requesting LBS. In addition, a trust management environment is constructed based on the blockchain to solve the trust problem among users in the distributed k-anonymity algorithm. For the blockchain environment, RSUs are used to build k-anonymous regions, detect malicious behavior, and remove malicious users from the system. Conflux, a DAG consensus protocol, is adopted for the blockchain. Compared with traditional Proof-of-X (PoX: such as PoW and PoS) and PBFT consensus mechanisms, Conflux can achieve greater system throughput without sacrificing security performance. Extensive experiments show that the proposed scheme is feasible and outperforms some state-of-the-art privacy-preserving methods.

In the traditional urban traffic monitoring system, the traffic administration (TA) needs to install cameras and other sensing devices on various roads to collect road information and realize real-time monitoring of traffic conditions. This approach is costly and vulnerable to damage from natural and human activities. With the help of the Internet of Vehicles, the task of gathering traffic information can be crowdsourced to vehicles on the road. In [34], a blockchain-based real-time traffic monitoring (BRTM) system is proposed to realize a reliable and efficient information transaction process between traffic managers and vehicles. There are two types of nodes in BRTM's blockchain network: full and lightweight nodes. Traffic administrations in different cities are full nodes. Full nodes perform a consensus process to validate candidate blocks and store the entire blocks. Vehicles participating in the blockchain system are lightweight nodes due to insufficient storage and computing power. Vehicle nodes only store block headers that are convenient for them to check and verify the transactions. For the consensus mechanism, a reputation-based DPoS consensus mechanism is designed in BRTM, which can reduce the confirmation time and improve the throughput of the system. Each full node in the BRTM system has a reputation value, which reflects its behavior in previous consensus rounds. A full node with a higher reputation value means it has a higher voting weight and is more likely to be the leader of a future consensus round, thus ensuring the reliability and efficiency of the consensus process.

3.6 Transportation Management

Based on the traffic data, accident events, road information, and other information of V2X, some new transportation applications can be generated, such as dynamic traffic control and management, vehicle driving methods, and parking management services.

In [35], a blockchain-based semi-centralized traffic signal control (SCTSC) system is introduced. The system dynamically manages and controls traffic lights through data collected by vehicles to improve traffic efficiency. Vehicles and infrastructure (such as traffic signal controllers) act as blockchain nodes in the system. Nodes are divided into simple nodes and consensus nodes. Simple nodes generate new messages and have the right to read the blockchain ledger data. Consensus nodes participate in the verification of messages, perform the consensus process, and write verified blocks into the ledger. Nodes have attribute sets and are grouped according to attributes. For example, vehicles traveling in the same direction on the same road are grouped. Vehicles in the same group can temporarily agree on signal timing change through voting. The temporary agreement and messages of agreement rounds are recorded on the blockchain ledger.

Compared with a single-vehicle driving mode, a platooning model can improve the traffic capacity of a city, ease traffic congestion, and reduce energy consumption. In a platooning model, a platoon head (PH) leads the platoon members (PMs) to drive together, and all PMs have the same interests or goals and keep a certain distance. A blockchain-based platooning solution is proposed in [36]. The solution designs a PH selection mechanism based on reputation value, allowing more experienced and credible vehicles to serve as platoon leaders. The PH selection mechanism can not only incentivize the honest behavior and contribution of vehicles but also suppress the selfish behavior of vehicles. At the same time, using smart contracts to realize the automatic payment mechanism can prevent false payments or malicious defaults. At the same time, using smart contracts realizes an automatic payment mechanism for platoon services, preventing false payments or malicious arrears.

Finding suitable parking spots for vehicles has become an important issue in crowded metropolises and urban areas. At the same time, many private parking spots are idle. Owners of private parking spaces have no desire to share their parking spots due to privacy concerns. In order to improve the utilization rate of private parking spots and alleviate the parking issue, Zhang et al. present a blockchain-based intelligent parking scheme with fairness, reliability, and privacy protection in [37]. This scheme adopts a group signature to realize anonymous authentication of drivers and parking plot owners and can trace malicious users. Based on the smart contract, drivers can get the correct parking plot matching result after paying, and the owner of the parking plot can receive the corresponding payment from the driver who rented the parking plot. The scheme achieves fairness in intelligent parking by rewarding well-behaved users while penalizing misbehaved users.

3.7   Evidence Service

The core data generated in V2X can be stored in the blockchain, and these data can be used as digital proof for audit evidence.

A blockchain-based connected vehicle forensic framework, Block4Forensic (B4F) is proposed in [38]. B4F collects and stores vehicle accident-related data from vehicles, maintenance centers, manufacturers, and insurance companies in a blockchain ledger, enabling automatic forensics after an accident. There are three types of data in B4F. The first is the incident data, which is generated when predefined events are triggered. The second is diagnostic data, produced by the vehicle periodically or in the case of a malfunction. The last one is maintenance data, which contains information about maintenance reports and is kept jointly by the maintenance center and the user. B4F designs four types of nodes in the system: leaders, validators, monitors, and clients. The client provides a signed transaction to B4F. A leader is randomly selected among the validator nodes (for example, manufacturers, maintenance centers, and insurance companies) for the package verification of a block and to execute the Byzantine agreement protocol. The monitors are law enforcement authorities who are not directly involved in the transaction verification process but keep a copy of the transaction ledger to participate in post-incident disputes.

An intelligent digital forensics system is developed based on blockchain for CAVs [39]. In this system, the blockchain is used to save the relevant data of the incident. The short random signature is used to hide the identity of the verification witness and protect the privacy of witnesses. Fine-grained access control based on ciphertext policy attribute encryption for evidence access is designed to provide more privacy protection for CAVs.

A blockchain-enabled certificate-based authentication scheme for vehicle accident detection and notification is designed [40]. In this scheme, the transactions related to vehicle accident detection and notification are stored in the blockchain ledger, which is maintained by cloud servers. After an authenticated vehicle detects an accident of itself or adjacent vehicles on the road, it sends the relevant accident information to its nearby cluster head (CH) as a transaction. The cluster head forwards the transaction to the corresponding RSUs and edge servers. Edge servers are responsible for preparing partial blocks containing transactions and Merkle tree roots, as well as digital signatures for this information, and then forwards to the blocks to associated cloud servers. Finally, the cloud servers execute a consensus process to complete the block creation, verification, and addition.

4 Future Directions

Blockchain application in the V2X still has many open and challenging problems, which can be regarded as future research directions and opportunities. In this section, we discuss such potential future research directions to resolve these problems.

4.1 Blockchain Performance

On the one hand, distributed ledgers and consensus protocols of blockchain systems tend to consume a lot of storage and computing resources. For a long time, the excessive resource overhead of the blockchain system has been criticized. Not all entities in V2X can afford huge resource overhead.

On the other hand, the performance of blockchain systems, such as throughput (transactions per second), transaction latency, and network capacity, still has some gaps with the requirements of real production environments.

Therefore, it is of great practical significance to study a lightweight and efficient blockchain system, which can reduce the overhead of system resources and make the processing of transactions more efficient and effective.

There has been much research on blockchain systems' lightweight and high efficiency [41]. However, there is not yet a blockchain system that has been widely proven in practice.

4.2 Quantum Attacks

A blockchain system relies heavily on its cryptographic technology and the computing power of its participants. However, with the advent of quantum computing, some of the technologies that current blockchains use have become less secure. Quantum computers offer powerful computing capabilities that radically differ from today's devices. Some encryption algorithms widely used in current blockchain systems are easily cracked by quantum computers, and the great computing capability of quantum computers can easily make blockchain systems subject to attack [42]. Therefore, it is urgent to research the post-quantum blockchain system for V2X.

4.3   Heterogeneous Service

Since each application has its characteristics, the current research direction is to create an exclusive blockchain system for a specific application. However, in real V2X scenarios, vehicles often have multiple applications simultaneously, which all need to rely on the blockchain system. For example, a vehicle can provide data sharing services with other vehicles or RSUs. Simultaneously, it receives context and computing offload services from a MEC. Traditional research scheme to build a dedicated blockchain system for each application is unrealistic in practical applications. In particular, considering the compatibility and interactivity among multiple blockchain systems and the corresponding overhead of energy and computing resources, the traditional scheme is difficult to apply in practical scenarios. Using only one blockchain system to support many application services is more flexible.

4.4   Incentive Mechanism

Blockchain transforms a traditional central control system into a distributed system controlled by multiple decentralized participants. Therefore, incentivizing these decentralized participants to maintain and manage the blockchain system is the key to the successful application of blockchain. The incentive mechanism includes positive reward and negative punishment. For example, in V2X business application scenarios, it is necessary to study how to simultaneously distribute benefits among participants such as MECs, service providers, vehicles, brokers, and penalty malicious and dishonest entities. If all participants can benefit from their contributions and suppress their malicious behavior, then the blockchain system can be truly integrated into industrial scenarios. Currently, most of the research on incentive mechanisms is carried out from the theoretical aspect, lacking optimization and verification in the business environment.

4.5   Integration with Emerging Technologies

With the development of big data, artificial intelligence, and other emerging technologies, it is a trend to integrate them in V2X applications based on the blockchain.

Big data technology can provide many data analysis methods to maximize the value of the information in the massive V2X data. In addition, big data technology can improve the quality of data in blockchain systems, including raw transaction-related data and data stored in the ledger.

In a blockchain system, each participant can be considered as an agent, and the optimization of the system can be achieved through cooperation and competition.

In a sense, a blockchain system can also be considered a type of swarm intelligence. Therefore, we can learn to optimize the blockchain from artificial intelligence technology, especially swarm intelligence and multi-agent reinforcement learning.

5 Conclusion

Blockchain technology is a promising emerging technology that can effectively solve some problems in V2X fields. Based on the introduction of the basic principles of V2X and blockchain, this chapter extensively discusses blockchain integration into V2X. From the application perspective, the application scenarios are classified into trust management, data sharing, resource sharing, trading application, collaboration application, transportation management and forensics application. Finally, future research directions are presented.

References

1. S. Nakamoto, Bitcoin: a peer-to-peer electronic cash system. Decentralized Bus. Rev. 21260 (2008)
2. C. Wang, X. Cheng, J. Li, Y. He, K. Xiao, A survey: applications of blockchain in the internet of vehicles. EURASIP J. Wirel. Commun. Netw. **2021**(1), 1–16 (2021)
3. M.B. Mollah et al., Blockchain for the internet of vehicles towards intelligent transportation systems: a survey. IEEE Internet Things J. **8**(6), 4157–4185 (2020)
4. R. Jabbar et al., Blockchain technology for intelligent transportation systems: a systematic literature review. IEEE Access (2022)
5. T. Alladi, V. Chamola, N. Sahu, V. Venkatesh, A. Goyal, M. Guizani, A comprehensive survey on the applications of blockchain for securing vehicular networks. IEEE Commun. Surv. Tutor. (2022)
6. Y. Guo, C. Liang, Blockchain application and outlook in the banking industry. Financ. Innovation **2**(1), 1–12 (2016)
7. M. Tripoli, J. Schmidhuber, Emerging opportunities for the application of blockchain in the agri-food industry (2018)
8. J. Wu, N.K. Tran, Application of blockchain technology in sustainable energy systems: an overview. Sustainability **10**(9), 3067 (2018)
9. J.-S. Kim, N. Shin, The impact of blockchain technology application on supply chain partnership and performance. Sustainability **11**(21), 6181 (2019)
10. B.K. Mohanta, D. Jena, S.S. Panda, S. Sobhanayak, Blockchain technology: a survey on applications and security privacy challenges. Internet of Things **8**, 100107 (2019)
11. Q. Wang, J. Yu, S. Chen, Y. Xiang, SoK: diving into DAG-based blockchain systems (2020). arXiv preprint arXiv:2012.06128
12. M. Naor, M. Yung, Universal one-way hash functions and their cryptographic applications, inn *Proceedings of the Twenty-First Annual ACM Symposium on Theory of Computing* (1989), pp. 33–43
13. R.C. Merkle, A digital signature based on a conventional encryption function, in *Conference on the theory and application of cryptographic techniques.* (Springer, 1987), pp.369–378
14. S. King, S. Nadal, Ppcoin: peer-to-peer crypto-currency with proof-of-stake. Self-published paper **19**(1) (2012)

15. D. Larimer, Delegated proof-of-stake (dpos). Bitshare whitepaper **81**, 85 (2014)
16. M. Castro, B. Liskov, Practical byzantine fault tolerance. OsDI **1999**(99), 173–186 (1999)
17. N. Szabo, Formalizing and securing relationships on public networks. First Monday (1997)
18. H.K. Verma, K.P. Sharma, Evolution of VANETS to IoV: applications and challenges. Tehnički glasnik **15**(1), 143–149 (2021)
19. H. Zhang, J. Liu, H. Zhao, P. Wang, N. Kato, Blockchain-based trust management for internet of vehicles. IEEE Trans. Emerg. Top. Comput. **9**(3), 1397–1409 (2020)
20. P.K. Singh, R. Singh, S.K. Nandi, K.Z. Ghafoor, D.B. Rawat, S. Nandi, Blockchain-based adaptive trust management in internet of vehicles using smart contract. IEEE Trans. Intell. Transp. Syst. **22**(6), 3616–3630 (2020)
21. U. Javaid, M.N. Aman, B. Sikdar, A scalable protocol for driving trust management in internet of vehicles with blockchain. IEEE Internet Things J. **7**(12), 11815–11829 (2020)
22. Q. Mei, H. Xiong, Y. Zhao, K.-H. Yeh, Toward blockchain-enabled IoV with edge computing: efficient and privacy-preserving vehicular communication and dynamic updating," in *2021 IEEE Conference on Dependable and Secure Computing (DSC)* (IEEE, 2021), pp. 1–8
23. H. Chai, S. Leng, M. Zeng, H. Liang, A hierarchical blockchain aided proactive caching scheme for internet of vehicles, in *ICC 2019–2019 IEEE International Conference on Communications (ICC)* (IEEE, 2019), pp. 1–6
24. Y. Lu, X. Huang, K. Zhang, S. Maharjan, Y. Zhang, Blockchain empowered asynchronous federated learning for secure data sharing in internet of vehicles. IEEE Trans. Veh. Technol. **69**(4), 4298–4311 (2020)
25. S. Wang, D. Ye, X. Huang, R. Yu, Y. Wang, Y. Zhang, Consortium blockchain for secure resource sharing in vehicular edge computing: a contract-based approach. IEEE Trans. Netw. Sci. Eng. **8**(2), 1189–1201 (2020)
26. Y. Dai, D. Xu, K. Zhang, S. Maharjan, Y. Zhang, Deep reinforcement learning and permissioned blockchain for content caching in vehicular edge computing and networks. IEEE Trans. Veh. Technol. **69**(4), 4312–4324 (2020)
27. H. Chai, S. Leng, Y. Chen, K. Zhang, A hierarchical blockchain-enabled federated learning algorithm for knowledge sharing in internet of vehicles. IEEE Trans. Intell. Transp. Syst. **22**(7), 3975–3986 (2020)
28. G. Sun, M. Dai, F. Zhang, H. Yu, X. Du, M. Guizani, Blockchain-enhanced high-confidence energy sharing in internet of electric vehicles. IEEE Internet Things J. **7**(9), 7868–7882 (2020)
29. S. Xia, F. Lin, Z. Chen, C. Tang, Y. Ma, X. Yu, A Bayesian game based vehicle-to-vehicle electricity trading scheme for blockchain-enabled internet of vehicles. IEEE Trans. Veh. Technol. **69**(7), 6856–6868 (2020)
30. C. Chen, J. Wu, H. Lin, W. Chen, Z. Zheng, A secure and efficient blockchain-based data trading approach for internet of vehicles. IEEE Trans. Veh. Technol. **68**(9), 9110–9121 (2019)
31. X. Lin, J. Wu, S. Mumtaz, S. Garg, J. Li, M. Guizani, Blockchain-based on-demand computing resource trading in IoV-assisted smart city. IEEE Trans. Emerg. Top. Comput. **9**(3), 1373–1385 (2020)
32. K. Liu, W. Chen, Z. Zheng, Z. Li, W. Liang, A novel debt-credit mechanism for blockchain-based data-trading in internet of vehicles. IEEE Internet Things J. **6**(5), 9098–9111 (2019)
33. B. Li, R. Liang, W. Zhou, H. Yin, H. Gao, K. Cai, LBS meets blockchain: an efficient method with security preserving trust in SAGIN. IEEE Internet Things J. **9**(8), 5932–5942 (2021)
34. J. Guo, X. Ding, W. Wu, Reliable traffic monitoring mechanisms based on blockchain in vehicular networks. IEEE Trans. Reliab. (2021)
35. L. Cheng et al., SCTSC: A semicentralized traffic signal control mode with attribute-based blockchain in IoVs. IEEE Trans. Comput. Soc. Syst. **6**(6), 1373–1385 (2019)
36. Chen et al., Smart-contract-based economical platooning in blockchain-enabled urban internet of vehicles. IEEE Trans. Indus. Inform. **16**(6), 4122–4133 (2020)
37. C. Zhang, L. Zhu, C. Xu, C. Zhang, H. Westermann, BSFP: blockchain-enabled smart parking with fairness, reliability and privacy protection. IEEE Trans. Veh. Technol. **99**, 1–1 (2020)
38. M. Cebe, E. Erdin, K. Akkaya, H. Aksu, S. Uluagac, Block4Forensic: an integrated lightweight blockchain framework for forensics applications of connected vehicles (IEEE, 2018), pp. 50–57

39. R. Tyagi, S. Sharma, S. Mohan, Blockchain enabled intelligent digital forensics system for autonomous connected vehicles, in *2022 International Conference on Communication, Computing and Internet of Things (IC3IoT)* (IEEE, 2022), pp. 1–6
40. A. Vangala, B. Bera, S. Saha, A.K. Das, N. Kumar, Y. Park, Blockchain-enabled certificate-based authentication for vehicle accident detection and notification in intelligent transportation systems. IEEE Sens. J. 21–14 (2021)
41. M. Kamal, G. Srivastava, M. Tariq, Blockchain-based lightweight and secured V2V communication in the internet of vehicles. IEEE Trans. Intell. Transp. Syst. **99** (2020)
42. A.K. Fedorov, E.O. Kiktenko, A.I. Lvovsky, *Quantum Computers Put Blockchain Security at Risk* (Nature Publishing Group, 2018)

Chapter 13
An Introduction to Trust Management in Internet of Vehicles

Yu'ang Zhang, Chenchen Lv, Chaklam Cheong, and Yue Cao

Abbreviations

IoV	Internet of Vehicle
ITS	Intelligent Transportation System
RSU	Road Side Unit
PKI	Public Key Infrastructure
BSM	Basic Safety Message
TA	Trusted Authority
DDoS	Distributed Denial of Service
DTMS	Distributed Trust Management System

1 Introduction

Today's society has been deeply integrated with the Internet. In the field of transportation, Internet of Vehicles (IoVs) have also received extensive attention and development. In China, the "14th Five-Year Information and Communication Industry Development Plan" pointed out that it is necessary to promote the research and development

Y. Zhang (✉) · C. Lv · C. Cheong · Y. Cao
School of Cyber Science and Engineering, Wuhan University, Wuhan, China
e-mail: yuang.zhang@whu.edu.cn

C. Lv
e-mail: lvbuer@whu.edu.cn

C. Cheong
e-mail: cheongchaklam@whu.edu.cn

Y. Cao
e-mail: yue.cao@whu.edu.cn

© The Author(s), under exclusive license to Springer Nature Singapore Pte Ltd. 2023   245
Y. Zhu et al. (eds.), *Communication, Computation and Perception Technologies for Internet of Vehicles*, https://doi.org/10.1007/978-981-99-5439-1_13

of IoVs, and speed up the testing and verification of key technologies. As shown in Fig. 1, IoVs are used for the communication between vehicles and other traffic participants, including vehicle-to-vehicle (V2V), vehicle-to-infrastructure (V2I), vehicle-to-pedestrian (V2P), and vehicle-to-cloud network interconnection (V2N). Vehicles use their GPS equipment and various sensors to monitor the speed, location, surrounding traffic conditions, and communicate with neighboring vehicles. With IoVs as the enabling technology, various applications can be developed for intelligent transportation systems (ITSs), such as real-time traffic flow monitoring, collision avoidance, navigation, and even advanced autonomous driving. For example, the group standard issued by the China Society of Automotive Engineers [1] clarifies 17 Day 1 applications, including forward and intersection collision warning, green wave speed guidance, congestion notification, near-field payment, etc. These applications are generally categorized into safety applications and efficiency applications. The large-scale deployment of these applications will bring significant improvement to travel efficiency and traffic safety.

Due to the close relationship between IoVs and the physical world, the intrusion of IoVs may cause extremely serious consequences. However, car manufacturers have yet to implement robust in-vehicle intrusion detection systems, which makes commercial vehicles vulnerable to attackers. In 2018, security researchers at Tencent found 14 common security vulnerabilities affecting in-vehicle communication modules, gateways, and infotainment systems, after analyzing the electronic control units of various BMW models. They obtained the ability to send commands from the vehicle gateway to different CAN buses to affect vehicle functions. In 2020, a cybersecurity firm named Context Information Security discovered that the tire pressure monitoring system in one Ford model could be intercepted. Attackers could send false messages to conceal the status of tires when the tires are flat. The above examples show that it is possible for attackers to invade the on-board system and force the vehicle to behave maliciously. For IoVs, if vehicles participating in the communication are controlled by attackers, they can launch various attacks to

Fig. 1 The participants in IoVs

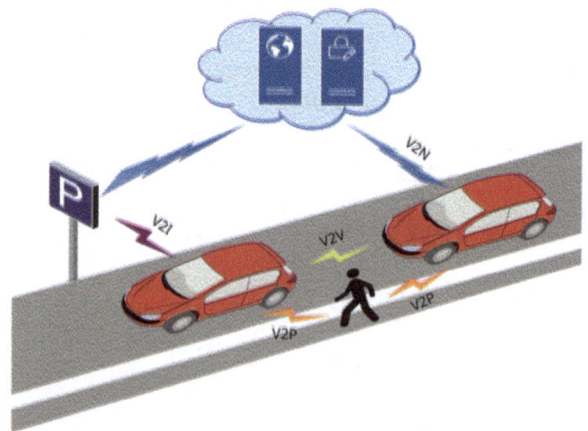

undermine the performance of the network. Therefore, security solutions must be deployed in the communication process.

Cryptography-based public key infrastructure, certificates, signatures, and other methods have been effectively applied in IoVs. However, an invaded vehicle can have a valid certificate while still behaving maliciously. Cryptography-based methods can solve the problem of identity trust, but it is difficult for them to conduct behavioral trust assessments and detect authenticated attackers. Therefore, the trust management becomes an emerging technique in the security of IoVs.

In computer science, the term "trust" means the "estimated subjective probability that an entity exhibits reliable behavior for particular operations under a situation with potential risks" [2]. The trust management in IoVs aims to create a trusted environment for vehicles and other entities to exchange messages. In general, the trust management process focuses on the behavior of vehicles and can be divided into three stages: the trust evidence collection, local trust evaluation, and trust aggregation. During the evidence collection stage, vehicles obtain trust-related information from neighboring vehicles and the surrounding environment. Then, each vehicle locally assigns trust values to others based on a trust model. The trust authority (TA) aggregates these local trust values to obtain a more comprehensive view of the security situation in the network. Practically, the trust management is usually combined with cryptography techniques to ensure the security requirements of IoVs.

2 Security Problems in IoVs

2.1 Characteristics of IoVs

Compared to traditional networks, IoVs have the following unique characteristics [3]:

(1) The mobility of vehicles: The positions of vehicles in the network change frequently, and the moving speeds of different vehicles in different environments vary significantly. In addition, due to restrictions on the road network (e.g., traffic regulations), the movements of vehicles may contain a certain pattern.

(2) Intermittent connections: In IoVs, due to the movement of vehicles, the network topology changes rapidly. The duration of each interaction between vehicles is short, and the wireless communication link may be interrupted frequently. Therefore, vehicles cannot form a stable connection like in traditional networks.

(3) Acentric structure: The role of each vehicle in the network is equal, functioning as both an end user of the network and a router. In addition to this, IoVs are free to access. Once the vehicle enters the communication range of other vehicles, the network connection is automatically established, and as the vehicle moves, the connection is automatically disconnected.

(4) The limitation of resources. The computing resources, storage resources, and power resources of vehicles are limited. Although modern connected cars tend to have powerful on-board chips, they have to strike a balance between performance and power consumption. Specifically, when a vehicle is running out of battery, some equipment (e.g., sensors and cameras) may not function properly.

2.2 Security Requirements of IoVs

The security of IoVs mainly includes the following aspects.

(1) Confidentiality: This is to ensure that the information in the network is not obtained or used by unauthorized entities. In IoVs, a large portion of messages are sent via broadcast, so it is particularly important to ensure the confidentiality of messages.
(2) Availability: The availability guarantees that the information can be used normally by authorized entities when needed, especially under harsh conditions or in the presence of malicious vehicles.
(3) Data integrity: The purpose of ensuring data integrity is to make it impossible for attackers to tamper with or forge messages. The network should be able to detect the anomaly when integrity is compromised. For example, if there is a vehicle sending falsified traffic accident data, the traffic efficiency can be greatly reduced.
(4) Accountability: The accountability refers to the ability to determine the creator of a message. In IoVs, different message creators may have different priorities. For example, information created by government-managed traffic signs should have a higher priority. If the accountability of messages cannot be guaranteed, then messages with a high priority may not be processed properly.
(5) Non-repudiation: It means that in the process of message transmission, neither the sender nor the receiver can deny their participation, especially in case of a dispute.

2.3 Common Malicious Attacks in IoVs

The malicious attacks in IoVs are usually divided into two categories: active attacks and passive attacks. Active attacks mean that attackers actively interfere with the message forwarding process, such as dropping messages arbitrarily or sending false warnings. Passive attacks mean that attackers do not interact with other vehicles and only eavesdrop on their communication. Moreover, attacks can be launched by both authenticated vehicles (i.e., internal attacks) and unauthenticated vehicles without valid certificates (i.e., external attacks). External attacks can be stopped by cryptography-based methods, while the trust management systems mainly focus on internal attacks. Some typical examples of internal attacks in IoVs are listed below.

(1) Black/gray hole attack: Vehicles that implement the black hole attack always drop their received messages, compromising the availability of the network. Similarly, gray hole attackers discard part of the messages they received according to a certain strategy or a certain probability.

(2) Sybil attack: Sybil attack means that the attacker simulates fake identities of vehicles to carry out malicious behavior. At the same time, it may still use its real identity to behave as a normal vehicle. The fake identities can be both generated and the stolen identities of other vehicles.

(3) Selfish vehicles: Selfish vehicles can be divided into independent selfish nodes and social selfish vehicles. The independent selfish vehicles only send messages generated by themselves, and do not forward messages of other vehicles. Social selfish vehicles preferentially send the messages generated by vehicles with close social relations with themselves, and the messages of other vehicles will be discarded.

(4) Sinkhole attack: The sinkhole attack refers to the malicious vehicle publishing false information about the routing path according to the routing algorithm. Attackers can forge the quality metrics for a specific path, so that their neighbor vehicles tend to send messages to attackers, thereby gaining control of the network.

(5) Man-in-the-middle attack: Malicious vehicles can eavesdrop on two or more benign vehicles, or even tamper with the content of their messages, without the attacked vehicles being aware of it.

(6) Message tampering attack: The messages in IoVs often contain various information required by other vehicles. The attacker tampered with their received messages before relaying them, so as to mislead other vehicles about the situation of surrounding traffic.

(7) Distributed Denial-of-Service (DDoS) attack: Similar to DDoS attacks against cloud servers, malicious vehicles inject into the network a large number of useless messages to consume the resources of network. Jamming is a typical technique to launch DDoS attacks. Attackers send stronger signals in the wireless channel to hinder the normal communication.

(8) False data injection: Vehicles often need to send their perceived event information to other vehicles that interact with them. For example, when a vehicle detects a traffic accident, it will send messages to inform other vehicles of the accident, including the time and location of accident. Malicious vehicles may send bogus event messages, such as the occurrence of fake traffic accidents, to disrupt the normal operation of other vehicles. Besides, attackers can generate fake GPS signals to spoof GPS devices, making vehicles unable to detect their real locations.

(9) Replay attack: The attacker may continuously resend previously received messages into the network, which may undermine the authentication mechanism and result in unexpected behaviors of IoV applications.

Table 1 presents a summary of the above attacks.

Table 1 Attacks in IoVs

Attack method	Internal/external	Violated requirement
Black/gray hole	Internal	Availability
Sybil	Internal	Accountability
Selfish vehicles	Internal	Availability
Sinkhole	Internal	Availability
Man-in-the-middle	Internal	Confidentiality
Message tampering	Internal	Data integrity
DDoS	Internal and external	Availability
False data injection	Internal	Data integrity
Replay	Internal and external	Data integrity

3 Taxonomy of Trust Management in IoVs

The concept of trust originates from the sociology and psychology research, and is considered the basis for long-term collaborations. In IoVs, since the problem of identity trust can be solved by enforcing a cryptography-based solution (e.g., the public key infrastructure), the trust management mainly focuses on the behaviors of authenticated entities with the permission to participate in the communication process. The high mobility and distributed nature pose great challenges to the design of trust models in IoVs. According to the evaluation targets, trust models are usually divided into three categories: entity-centric trust models, data-centric trust models, and hybrid trust models.

3.1 Entity-Centric Trust Model

The entity-centric trust model evaluates the trust of each vehicle in terms of two aspects: the interactions between vehicles and recommendations from neighboring vehicles. The evaluation is based on a distributed manner, in which vehicles individually compute trust values for one another. Other techniques, such as vehicle clustering and road side unit (RSU) based trust computing, are also integrated to improve the accuracy of trust evaluation and reduce the overhead.

Typical models include the distributed trust management system (DTMS) based on the watchdog mechanism proposed by Khan et al. [4]. In this method, the network first elects a leader vehicle in a vehicle group to be responsible for the trust evaluation. Then the watchdog mechanism is used to make other vehicles send reports to the leader about malicious behaviors around them. Minhas et al. [5] proposed a multi-faceted trust model, which combines different factors (e.g., the experience of sending vehicle, the role and priority of vehicle, and the opinion of majority of vehicles) to evaluate the trust. In this method, the role-based trust calculation is done using the public key infrastructure. A trust model based on provenance information was proposed in [6]. When a vehicle interacts with other vehicles, it will generate and store an interaction record. Each record includes the ID of two vehicles, the number of sent and received messages, and the number of messages generated by each vehicle. Based on these data, indicators such as the packet loss rate can be calculated to deal with black hole attacks. Wu et al. [7] proposed a trust model for IoVs based on the Bayesian theory. In this model, the idea of Bayesian inference is applied to the trust evaluation, where the trust value of vehicles is described by the Beta distribution. The trust evaluation is combined with the GeoDTN+Nav routing protocol [8] to secure the message relay. Zhang et al. [9] aimed at the problem of selfish vehicles in the network refusing to cooperate, and designed the observation protocol to monitor the behavior of vehicles. On this basis, the trust value is calculated by combining the direct trust value and the recommended trust value, and the incentive scheme is applied to punish the detected selfish vehicles.

The entity-centric trust model maintains the long-term trust relationship between vehicles, which can be reused in later cooperation. The main concern is data sparsity. If vehicles are scattered in a relatively large area with sparse interactions, the entity-centric model may not have sufficient information to carry out accurate trust evaluation.

3.2   Data-Centric Trust Model

The data-centric trust model is of paramount importance since messages in IoVs tend to contain critical information such as emergency announcements. Undetected false alarms can significantly undermine the efficiency of road traffic. Correspondingly, the data-centric model usually relies on the information from the real world, such as the geographic information and traffic conditions. Considering the timeliness of messages, data-centric trust will not be stored for a long time, and is usually computed on a per message basis.

Different from the entity-centric trust model, the data-centric trust model mainly focuses on evaluating the trustworthiness of the received messages. For example, in the model proposed by Raya et al. [10], vehicles give reports for a certain event message. The reports from different vehicles are assigned different weights according to the time and location of the event. Then the decision-making module derives the trustworthiness of message from these reports using theories such as Bayesian inference. Huang et al. [11] proposed a voting mechanism with variable weights for the data trust evaluation. This model sets the weight according to the number of vehicles between the scene of an accident and message senders. The vehicle that directly observes the accident has the highest weight on the messages it sends. In the process of message transmission, vehicles with more hops away from the accident-observing vehicle are given smaller weights. Wu et al. [12] proposed an infrastructure-assisted trust model. In this model, when a vehicle is informed of the occurrence of an event, it judges the trust level of the event according to the distance from the vehicle to the event, and sends this information to the RSU. The RSU updates the event trust list, and sends the trust value back to the vehicle. Zhang et al. [13] utilized geographic information to calculate the data trust, including the Manhattan distance from the vehicle location to the event, and the number of obstacles in the path to the event.

In addition to aforementioned trust models based on geographic and spatio-temporal information, statistical-based methods are also used in the evaluation of data-centric trust. In IoVs, vehicles continuously send basic safety messages (BSMs). BSMs usually include the vehicle's position, speed, acceleration, and other operating data, which can be forged by attackers. Aiming at location forging attacks, Le et al. [14] designed three features for machine learning models, including the movement plausibility, minimum distance to trajectories, and minimum translation distance to trajectories. Two statistical learning models, namely KNN and SVM, are utilized to judge the trustworthiness of position data in BSMs. To deal with attackers sending fake speed and traffic flow information, Zaidi et al. [15] adopted the Greenshield model to estimate the surrounding traffic flow with respect to the speed. Then the hypothesis test is utilized to determine whether the message was injected by an attacker.

3.3   Hybrid Trust Model

The hybrid trust model calculates the trust value of both vehicles and messages, in which the trust of messages is usually calculated on the basis of entity trust. Compared to the entity-centric and data-centric models, the hybrid trust model can be relatively complicated and may introduce considerable communication overheads.

Rawat et al. [16] proposed a trust model that combines probabilistic and deterministic methods. Specifically, the probabilistic method is used for the calculation of entity trust, and the deterministic method is used for the calculation of message trust. The trust of vehicles is calculated by the Bayesian theory, based on whether multiple copies of the same message received from different vehicles are consistent. The calculation of message trust combines the two factors of vehicle coordinates and received signal strength (RSS). In the model proposed by Shrestha et al. [17], the clustering algorithm is used to distinguish normal and malicious vehicles. Then the trust value of message received by the vehicle is calculated based on the random walk algorithm. Li et al. [18] proposed an attack-resistant trust management scheme (ART). In this model, the entity trust consists of two aspects, including the functional trust (i.e., whether the vehicle can realize its function), and recommendation trust (i.e., whether the recommended trust value given by the vehicle is trustworthy). Then the message trust is jointly determined by the information collected from multiple neighboring vehicles.

4 Simulation Tools for IoVs

The evaluation of the trust models and applications in IoVs requires powerful simulation tools. The simulation of IoVs consists of two parts: the mobility simulation and network simulation. Currently, there exist widely-adopted solutions for both of them. For example, the Simulation of Urban Mobility (SUMO) [29] is designed for mobility simulation, and the ns-3 network simulator [30] is utilized for network simulation. Therefore, efforts have been made to integrate these two kinds of simulators to create simulators for IoVs. Notably, Veins [19] and Eclipse MOSAIC [20] are two well-designed simulators of this kind. There are also simulators that offer the inherent capability of mobility simulation, such as the Opportunistic Network Environment (ONE) simulator [21].

4.1 Veins

Veins is one of the most popular open source frameworks for running vehicular network simulations. It was initially released by Christoph Sommer et al. [19] in 2011 and has been undergoing active development since then. Veins is based on two well-established simulators: OMNeT++ [31], an event-based network simulator, and SUMO, a road traffic simulator. The two simulators are connected via a TCP socket, communicating through the standardized Traffic Control Interface (TraCI) protocol. This allows bidirectionally-coupled simulation of road traffic and network communications. Veins also inherits the powerful graphical interface from OMNET++, as shown in Fig. 2.

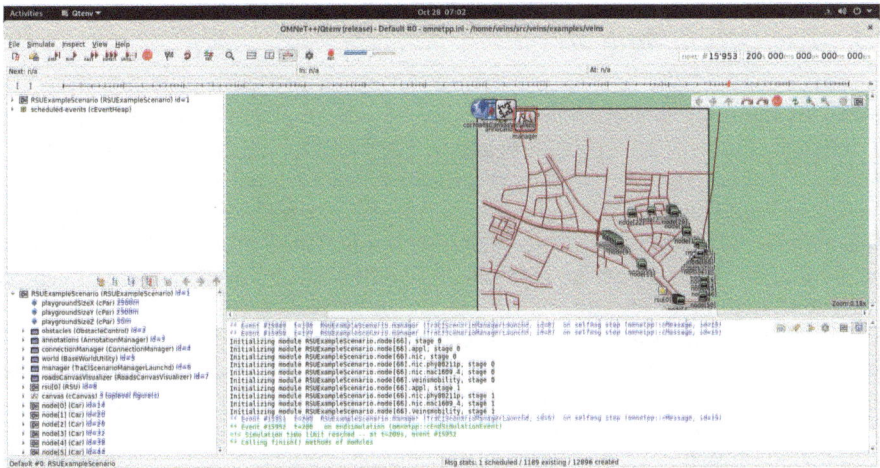

Fig. 2 The GUI of Veins

Veins provides a comprehensive suite of modules that serve as a foundation for simulating applications. This modular design enables users to tailor the simulator to their specific needs. For example, to model physical layer effects accurately, the MiXiM module is integrated into Veins. For wireless communication technologies adopted in IoVs, Veins offers modules that simulate IEEE 802.11p and IEEE 1609.4 DSRC/WAVE standards, as well as cellular networking such as C-V2X and 5G NR-V2X. For cybersecurity research, F^2MD (Framework For Misbehavior Detection) [32] extends Veins with a rich set of attack models and detection schemes. It also enables Veins to utilize external machine learning modules for advanced misbehavior detection.

4.2 Eclipse MOSAIC

Eclipse MOSAIC, formerly known as VSimRTI [20], was initially developed by Fraunhofer FOKUS and DCAITI (Daimler Center for Automotive IT Innovations). It employs the concept of Federates and Ambassadors to couple individual simulators to the whole environment. Similar to Veins, it adopts SUMO as the mobility simulator. For the network simulation, it offers the choice of two external simulators (i.e., OMNET++, ns-3) and two built-in simulators (i.e., Simple Network Simulator, Cell). Eclipse MOSAIC mainly focuses on application simulation. It provides abstractions of various ITS-related functionalities, such as the on-board operating system, RSU, perception of traffic signs, and environment sensors. A commercial edition is also available, which extends the open source version with the ability to simulate electric mobility, including battery models and charging station simulations.

4.3 The ONE Simulator

The ONE simulator is written in Java by Ari Kernen et al. [21] from the University of Helsinki, Finland. This project began in 2008 and the latest version v1.6.0 was released in 2015. The ONE simulator provides a user-friendly simulation environment. It is built from scratch and does not rely on external simulators such as ns-3. Although the ONE simulator lacks the realistic simulation of network protocols, in the simulation centering on node behavior, it offers strong ease of use and can be easily extended for various scenarios.

The ONE simulator is based on a discrete event engine, which divides the continuous time into multiple simulation rounds according to the preset time step. The ONE simulator consists of comprehensive functional packages, providing simulations of core network entities, routing algorithms, the application layer, node movement models, etc. It also provides a graphical user interface as shown in Fig. 3. The interface displays basic information such as maps, nodes, and event logs. The status of nodes can be shown on the map, such as the communication range of each node, the connections between nodes, etc. The current routing information of each node (e.g., message buffer) can also be displayed.

For mobility simulations, apart from importing maps and trajectories from external sources, the ONE simulator can also utilize programmed movement models. This function is implemented in the movement package. During the simulation, the package is responsible for calculating the future path for each node based on the movement model. The commonly used movement models such as random movement, map-based movement, shortest path movement, and working day movement model have been implemented as examples.

Moreover, to facilitate the result analysis, a report generation package is also implemented. When events such as message creation, delivery, and deletion occur during simulation, the report package obtains these data through event listeners, and

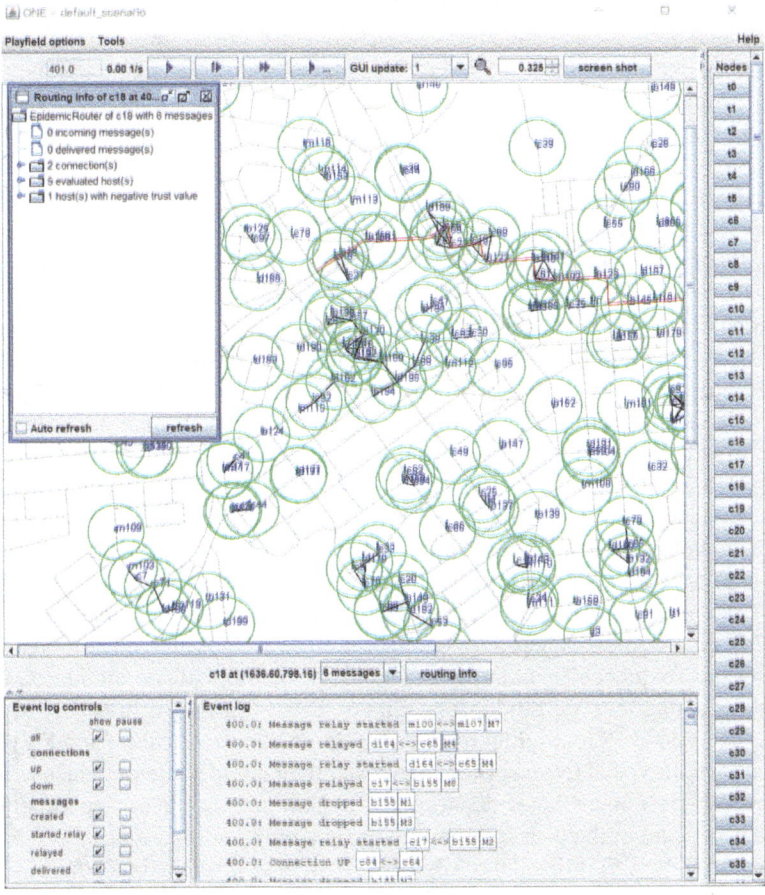

Fig. 3 The GUI of ONE simulator

performs corresponding statistical calculations. The simulator also provides utilities for post-processing such as drawing statistical graphs.

5 Future Research Opportunities

5.1 Social Relationships in IoVs

The trust evaluation in general consists of two aspects: the individual trust evaluation and relational trust evaluation [2]. The individual trust evaluation is based on previous interactions and experience with an entity, while the relational trust evaluation takes into account the similarity, centrality, and importance of entities. In IoVs, derived from the mobility patterns, vehicles also exhibit social relationships similar to those of people in their movement and interaction behaviors. For example, if the owners of two vehicles live and work at the same place, their vehicles are likely to take similar commute routes during workdays. In addition, vehicle owners may have established close relationships in real world, which is also worth consideration for the trust evaluation.

Specifically, the interaction history between vehicles, vehicle trajectories, and geographic information can be used to describe the social relationship between vehicles. In order to reduce the overhead of Epidemic routing algorithm, Li et al. [22] utilized the encounter history to measure the similarity between vehicles. Then virtual connections between vehicles are constructed according to the similarity between vehicles, and dynamic communities are formed on this basis. The algorithm divides vehicles into communities with frequent internal interactions. Clustering algorithms can also be used for the calculation of social relationships. Yang et al. [23] applied the affinity propagation (AP) algorithm to form clusters of vehicles. Then the cluster head is selected, and the trust model is designed based on the intra-cluster consensus. The indicator for clustering can be designed differently to adapt to actual scenarios.

These data (trajectories, interaction histories, etc.) can be detected and recorded within the network. Besides, with the development of Online Social Networks (OSNs), it is also possible to integrate real social relationship data into IoVs.

In [24], each vehicle categorizes others as known friends and anonymous friends based on the frequency of email interactions. Vehicles with no email interaction available are categorized as random encounters. Then, vehicles in different categories are assigned with different initial trust values. In order to connect the mail exchange information with VANTEs, users need to register their email addresses and usernames with the vehicle management department, and the management department then issues a certificate to the user for communication and identification.

Kerrache et al. [25] utilized the Advogato Trust Metric to calculate the trust value of vehicle owners themselves according to their interactions in social networks. Owners are assigned different "Honesty Human Factor" (HFF) according to the trust calculation. When calculating the trust value of vehicles in IoVs, the HFF is also taken into account. Jedari et al. [26] proposed a watchdog system for detecting selfish vehicles. They claim that the social relationships such as family members and colleagues are relatively stable, and the corresponding vehicles are more inclined to cooperate with each other on message forwarding. Therefore, the real-world social relationship data are utilized to assist the trust calculation.

The above-mentioned trust models based on OSNs require users to provide their social accounts, thus facing privacy protection issues.

5.2 Development of Standard Evaluation Procedures and Datasets

The design and improvement of trust models require proper evaluation, both by simulations and in field tests. Usually, simulations are carried out before field tests. Moreover, the combination of machine learning techniques and trust models is attracting lots of research efforts. However, the lack of standard evaluation procedures makes it hard to reproduce and compare different simulation results. The simulation-based evaluation needs to be standardized in two aspects: the traffic scenario and communication datasets. There are well-developed traffic scenarios based on real-world traffic, such as the Luxembourg SUMO Traffic (LuST) [27]. Unfortunately, they are not widely adopted due to the lack of specific support for trust model simulations.

The IoVs operate in a highly diversified environment, especially with the development of 5G-related technologies. Therefore, representative datasets with the characteristics of IoV communication are required. There have been preliminary efforts in the development of datasets. Heijden et al. [28] utilized the Veins simulator to generate the Vehicular Reference Misbehavior Dataset (VeReMi), which contains malicious communication data with different types of position falsification attacks. Kamel et al. [29] extended the VeReMi dataset to provide more realistic sensor error models and more sophisticated position falsification strategies. However, these datasets are not comprehensive, considering the large-scale, heterogeneous, and complex communications in IoVs. In addition, at this stage, most of the datasets come from simulation experiments. The construction of datasets from real field tests is still a challenging problem.

5.3 Practical Issues of IoV Trust Models

The trust models may raise privacy concerns since they often rely on sensitive information such as vehicle trajectories. The cryptography-based privacy protection mechanisms have to be implemented alongside trust models. These mechanisms may require vehicles to frequently change their identities (or pseudonyms), which conflicts with the long-term identity requirements of the trust models. On the other hand, the frequent change of pseudonyms also significantly increases the difficulty of tracking the origin of messages and correlating messages from the same vehicle, adding additional barriers to trust models based on interaction histories.

In addition, the cryptography mechanism as well as the trust model itself may introduce greater computational and communication overhead, resulting in nontrivial delays in the message delivery. This is not suitable for many latency-sensitive applications in IoVs, especially safety applications. Therefore, the cost of authentication, trust calculation, and privacy protection must be fully evaluated, and such security solutions have to be tailored to the specific requirements of IoVs.

6 Conclusion

This chapter first introduces the basic characteristics and security issues in IoVs. Then, the security requirements and common attacks are analyzed. The trust models are detailed in three categories: the entity-centric, data-centric, and hybrid trust model. Moreover, simulation tools used for the evaluation of trust models in IoVs are introduced, including Veins, Eclipse MOSAIC, and the ONE simulator. Finally, the problems in current research and future opportunities are identified.

References

1. China Society of Automotive Engineers, in *Cooperative Intelligent Transportation System-Vehicular Communication Application Layer Specification and Data Exchange Standard (Phase I)* CSAE 53-2020 (Beijing, 2020)
2. J.-H. Cho, K.S. Chan, S. Adali, A survey on trust modeling. ACM Comput. Surv. (CSUR) **48**, 1–40 (2015)
3. Z. Lu, G. Qu, Z. Liu, A survey on recent advances in vehicular network security, trust, and privacy. IEEE Trans. Intell. Transp. Syst. **20**, 760–776 (2019). https://doi.org/10.1109/TITS.2018.2818888
4. U. Khan, S. Agrawal, S. Silakari, Detection of malicious nodes (DMN) in vehicular ad-hoc networks. Procedia Comput. Sci. **46**, 965–972 (2015). https://doi.org/10.1016/j.procs.2015.01.006
5. U.F. Minhas, J. Zhang, T. Tran, R. Cohen, A multifaceted approach to modeling agent trust for effective communication in the application of mobile ad hoc vehicular networks. IEEE Trans. Syst., Man, Cybern. Part C (Appl. Rev.) **41**, 407–420 (2011). https://doi.org/10.1109/TSMCC.2010.2084571
6. C. Ge, L. Zhou, G.P. Hancke, C. Su, A Provenance-aware distributed trust model for resilient unmanned aerial vehicle networks. IEEE Internet Things J. **8**, 12481–12489 (2021). https://doi.org/10.1109/JIOT.2020.3014947
7. Q. Wu, Q. Liu, L. Zhang, Z. Zhang, A trusted routing protocol based on GeoDTN+Nav in VANET. China Commun. **11**, 166–174 (2014). https://doi.org/10.1109/CC.2014.7085617
8. P. Cheng, K.C. Lee, M. Gerla, J. Härri, GeoDTN+Nav: Geographic DTN routing with navigator prediction for urban vehicular environments. Mob. Netw. Appl. **15**, 61–82 (2010). https://doi.org/10.1007/s11036-009-0181-6

9. L. Zhang, X. Zhang, C. An, C. Tang, A reputation-based incentive scheme for delay tolerant networks. Acta Electron. Sin. **42**, 1738–1743 (2014). https://doi.org/10.3969/j.issn.0372-2112.2014.09.012

10. M. Raya, P. Papadimitratos, V.D. Gligor, J.-P. Hubaux, On data-centric trust establishment in ephemeral ad hoc networks, in *IEEE INFOCOM 2008—The 27th Conference on Computer Communications* (2008), pp. 1238–1246

11. Z. Huang, S. Ruj, M.A. Cavenaghi, M. Stojmenovic, A. Nayak, A social network approach to trust management in VANETs. Peer-To-Peer Netw. Appl. **7**, 229–242 (2014). https://doi.org/10.1007/s12083-012-0136-8

12. A. Wu, J. Ma, S. Zhang, RATE: a RSU-aided scheme for data-centric trust establishment in VANETs. In: *2011 7th International Conference on Wireless Communications, Networking and Mobile Computing* (IEEE, Wuhan, China, 2011), pp. 1–6

13. X. Zhang, R. Li, W. Hou, J. Shi, Research on Manhattan distance based trust management in vehicular ad hoc network. Secur. Commun. Netw. **2021**, 1–13 (2021). https://doi.org/10.1155/2021/9967829

14. A. Le, C. Maple, Shadows don't lie: n-sequence trajectory inspection for misbehaviour detection and classification in VANETs. In: *2019 IEEE 90th Vehicular Technology Conference (VTC2019-Fall)* (2019), pp. 1–6

15. K. Zaidi, M.B. Milojevic, V. Rakocevic, A. Nallanathan, M. Rajarajan, Host-based intrusion detection for VANETs: a statistical approach to rogue node detection. IEEE Trans. Veh. Technol. **65**, 6703–6714 (2016). https://doi.org/10.1109/TVT.2015.2480244

16. D. Rawat, G. Yan, B.B. Bista, M. Weigle, *Trust on the Security of Wireless Vehicular Ad-Hoc Networking* (Ad Hoc & Sensor Wireless Networks, 2015)

17. R. Shrestha, S.Y. Nam, Trustworthy event-information dissemination in vehicular ad hoc networks. Mob. Inf. Syst. **2017**, 1–16 (2017). https://doi.org/10.1155/2017/9050787

18. W. Li, H. Song, ART: an attack-resistant trust management scheme for securing vehicular ad hoc networks. IEEE Trans. Intell. Transp. Syst. **17**, 960–969 (2016). https://doi.org/10.1109/TITS.2015.2494017

19. C. Sommer, R. German, F. Dressler, Bidirectionally Coupled network and road traffic simulation for improved IVC analysis. IEEE Trans. Mob. Comput. (TMC) **10**, 3–15 (2011). https://doi.org/10.1109/TMC.2010.133

20. B. Hilt, M. Berbineau, A. Vinel, A. Pirovano, simulation of convergent networks for intelligent transport systems with VSimRTI, in *Networking Simulation for Intelligent Transportation Systems: High Mobile Wireless Nodes* (2017), pp. 1–28

21. A. Keränen, J. Ott, T. Kärkkäinen, The ONE simulator for DTN protocol evaluation, in *Proceedings of the 2nd International Conference on Simulation Tools and Techniques. ICST (Institute for Computer Sciences, Social-Informatics and Telecommunications Engineering)* (BEL, Brussels, 2009), pp. 1–10

22. F. Li, J. Wu, LocalCom: a community-based epidemic forwarding scheme in disruption-tolerant networks, in *2009 6th Annual IEEE Communications Society Conference on Sensor, Mesh and Ad Hoc Communications and Networks* (2009), pp. 1–9

23. S. Yang, J. Li, Z. Liu, S. Wang, Managing trust for intelligence vehicles: a cluster consensus approach, in *Proceedings of the Second International Conference on Internet of Vehicles—Safe and Intelligent Mobility*, vol. 9502 (Springer-Verlag, Berlin, Heidelberg, 2015), pp. 210–220

24. R. Hussain, W. Nawaz, J. Lee, J. Son, J.T. Seo, A Hybrid Trust Management Framework for Vehicular Social Networks, in H.T. Nguyen, V. Snasel (eds) *Computational Social Networks* (Springer International Publishing, Cham, 2016), pp. 214–225

25. C.A. Kerrache, N. Lagraa, R. Hussain, S.H. Ahmed, A. Benslimane, C.T. Calafate, J.-C. Cano, A.M. Vegni, TACASHI: trust-aware communication architecture for social internet of vehicles. IEEE Internet Things J. **6**, 5870–5877 (2019). https://doi.org/10.1109/JIOT.2018.2880332

26. B. Jedari, F. Xia, H. Chen, S.K. Das, A. Tolba, Z. AL-Makhadmeh, A social-based watchdog system to detect selfish nodes in opportunistic mobile networks. Futur. Gener. Comput. Syst. 92:777–788 (2019).https://doi.org/10.1016/j.future.2017.10.049

27. L. Codeca, R. Frank, S. Faye, T. Engel, Luxembourg SUMO traffic (LuST) scenario: traffic demand evaluation. IEEE Intell. Transp. Syst. Mag. **9**, 52–63 (2017). https://doi.org/10.1109/MITS.2017.2666585

28. R.W. van der Heijden, T. Lukaseder, F. Kargl, VeReMi: A dataset for comparable evaluation of misbehavior detection in VANETs, in R. Beyah, B. Chang, Y. Li, S. Zhu (eds) *Security and Privacy in Communication Networks. SecureComm 2018. Lecture Notes of the Institute for Computer Sciences, Social Informatics and Telecommunications Engineering*, vol. 254 (Springer, Cham, 2018). https://doi.org/10.1007/978-3-030-01701-9_18

29. P.A. Lopez, M. Behrisch, L. Bieker-Walz, J. Erdmann, Y.-P. Flötteröd, R. Hilbrich, L. Lücken, J. Rummel, P. Wagner, E. Wießner, Microscopic traffic simulation using SUMO, in *The 21st IEEE International Conference on Intelligent Transportation Systems* (IEEE, 2018)

30. T.R. Henderson, M. Lacage, G.F. Riley, C. Dowell, J. Kopena, Network simulations with the ns-3 simulator. SIGCOMM demonstration **14**, 527 (2008)

31. A. Varga, OMNeT++, in *Modeling and Tools for Network Simulation* (Springer, 2010), pp. 35–59

32. J. Kamel, M.R. Ansari, J. Petit, A. Kaiser, I.B. Jemaa, P. Urien, Simulation framework for misbehavior detection in vehicular networks. IEEE Trans. Veh. Technol. **69**, 6631–6643 (2020)

Chapter 14
Misbehaviour Detection Mechanisms in Internet of Vehicles

Chenchen Lv, Yu'ang Zhang, Yue Cao, and Chakkaphong Suthaputchakun

Abbreviations

CA	Certificate Authority
CAMs	Cooperative Awareness Messages
IDS	Intrusion Detection System
OBD	On-Board Diagnostics
RSSI	Received Signal Strength Indication
RSU	Road Side Unit
TTP	Trusted Third-Party

1 Introduction

With the gradual improvement of human requirements for road traffic safety and efficiency, Internet of Vehicles emerges as the times require. The expansion of wireless networks allows Internet of Vehicles to bring many benefits to road safety, traffic efficiency, and local services [1]. Internet of Vehicles is designed to improve driving safety and travel efficiency, including improving road safety, improving traffic management for green driving, partially supporting autonomous vehicles, and infotainment services. In Internet of Vehicles, the vehicle acts as both a router node and a terminal node. Internet of Vehicles also includes other infrastructure components, called Road Side Unit (RSU), which are sparsely placed along the road. The

C. Lv (✉) · Y. Zhang · Y. Cao
School of Cyber Science and Engineering, Wuhan University, Wuhan, China
e-mail: lvbuer@whu.edu.cn

C. Suthaputchakun
Electrical and Computer Engineering Department, Bangkok University, Bangkok, Thailand

© The Author(s), under exclusive license to Springer Nature Singapore Pte Ltd. 2023 261
Y. Zhu et al. (eds.), *Communication, Computation and Perception Technologies for Internet of Vehicles*, https://doi.org/10.1007/978-981-99-5439-1_14

RSU undertakes most of the functions of information interaction between the vehicle and cloud, including receiving and sending data, acting as an information transfer station for Internet of Vehicles, perceiving the state of road, performing signal light operations, etc. The vehicle sends, receives and analyses data through Internet of Vehicles to improve driving safety and travel efficiency. In addition, the vehicle can also perceive surrounding information through various sensors, and exchange information through Internet of Vehicles for overall traffic planning and deployment, especially supporting the deployment of autonomous driving.

Internet of Vehicles plays a vital role in the design of future intelligent transportation. In Internet of Vehicles, the vehicle cooperates with other participants such as vehicles, RSU and other vulnerable road users (e.g., pedestrians and bicycles) to collect and share information. By collecting accident-related data, Internet of Vehicles has the potential to enable global traffic control.

While Internet of Vehicles connects the physical world and the network world, it also faces the risk of information security in cyberspace, and even directly affects the security of the physical world. Throughout the field of computer science, securing systems against malicious attackers has become a fundamental requirement for secure and dependable operation of applications. Nowadays, professional attacks against systems, which are launched by large criminal organizations or even governments, are becoming increasingly common. It can be expected that Internet of Vehicles will be subject to many different types of misbehaviour or attacks, which will be discussed and classified in Sect. 2. If security applications or autonomous driving are considered, attacks on intelligent transportation systems may even cause accidents and possibly endanger the lives of passengers. For instance, given a set of vehicles that are using a platoon with reduced relative safety distance, an attacker could inject a fake message that indicates one of these vehicles is braking, even though this is not the case. In such a situation, vehicles will immediately respond, and in the worst case cause collisions to avoid the non-existent breaking behaviour claimed by the attacker. In response to the threats Internet of Vehicles faces, a lot of research has been conducted to provide appropriate protection for Internet of Vehicles.

In this chapter, we classify and introduce the existing misbehaviour detection mechanisms in Internet of Vehicles, and discuss existing problems. The overall structure is organized as follows. Section 2 discusses cyber threats against Internet of Vehicles, including threats in the vehicle environment and the wireless communication environment. Section 3 introduces and classifies the existing misbehaviour detection technologies. Section 4 discusses unresolved open issues.

2   Threats Against Internet of Vehicles

In this section, we briefly review threats related to Internet of Vehicles. These threats can be classified according to the in-vehicle environment and the wireless communication environment. In the misbehaviour detection mechanisms described later, we mainly focus on the threats to the wireless communication environment.

2.1 Threats to the In-Vehicle Environment

Connected vehicles have a broad attack surface, in which attackers can control the vehicle. Wireless keys, Wi-Fi, Bluetooth, On-Board Diagnostics (OBD) systems, USBs, and automobile applications are examples of entry points for attacks on connected vehicles.

2.1.1 CAN Bus

The CAN bus is equivalent to the neural network of the vehicle, connecting to the various control systems in the vehicle. Its communication adopts the broadcast mechanism. Each connected component in vehicles can send and receive control messages via the CAN bus. Attackers can launch attacks such as forging messages, replaying messages, and denial of service through physical or remote intrusion.

2.1.2 OBD

The OBD is the interface of on-board diagnostic system, and also an important interface for connecting external devices of intelligent connected vehicles to the CAN bus. Attackers can issue diagnostic commands to interact with the bus through OBD. Therefore, external devices are often the source of attacks.

2.1.3 In-Vehicle Operating System and Software

In-vehicle operating systems and software are developed based on traditional computer operating systems and software. Vulnerabilities in traditional computer operating systems and software may also occur in the in-vehicle environment, thereby leading to the risk of similar vulnerabilities in vehicles.

2.2 Threats to the Wireless Communication Environment

Raya et al. [1] defined the threat model in the wireless communication environment from three dimensions:

(1) Based on whether the attacker has a legal identity in the network, the attacker can be divided into an external attacker and an internal attacker. An external attacker launches an attack from outside the network and must bypass system defences such as firewalls or Intrusion Detection System (IDS). An internal attacker is a legitimate member of the network and has the right to access resources in the

network. An internal attacker can carry out almost all types of attacks against the system, which will cause serious damage.

(2) From the attacker's motivation, the attacker can be divided into malicious attacker, where the attacker only tries to damage the system, and purposeful attacker, where the attacker tries to benefit from the attack.

(3) From the attack method, the attacker can be divided into active attacker, which actively sends messages in Internet of Vehicles, and passive attacker, which only sniffs messages in the network.

In the following, we describe the main common attacks against the wireless communication environment.

- **Packet loss attack**: The attacker has enough buffer space and data forwarding ability. However, still discarding data packets, due to its selfish nature or other malicious purposes. The attack will lead to a decrease in data packet transmission rate, and ultimately lead to loss of information.
- **Flooding attack**: The attacker will transmit as many data packets as possible (e.g., forging and replaying data packets) to consume resources of normal vehicles. If the buffer space or bandwidth resources are occupied, the attacked vehicle cannot perform normal communication functions, thereby suffering a denial of service attack.
- **Forged data packet attack**: The attacker can create fake data packets which report, for example, that there are false traffic jams in certain locations to induce vehicles to change their paths. In this case, the attacker can avoid traffic jams for selfish, however resulting in worse overall traffic efficiency. In the autonomous driving scenario, the vehicle will change its behaviour after receiving data packets related to traffic events, that is, fake data packets will cause traffic accidents under specific circumstances, and even threaten the lives of passengers.
- **Sybil attack**: The attacker may steal or forge the identities of legitimate vehicles to jointly report the occurrence of an event, such as traffic jams, to gain a higher degree of trust. The attacker can also launch a denial of service attack through multiple identities. This type of attack can be launched through the cooperation of multiple identities, so as to achieve a higher attack success rate and achieve a variety of malicious purposes.

3   Detection Mechanisms Classification

Although the communication protocols have been relatively perfect, there are still malicious vehicles that do not comply with the protocols in Internet of Vehicles. Malicious vehicles usually use the attack methods described in Sect. 2 to destroy the network or achieve their selfish purposes, leading to a huge waste of network resources. Detecting such vehicles and taking certain measures are very important to maintain network security. However, the high mobility of vehicles increases the challenge of malicious vehicle detection and classification. A possible solution to the problem is to provide vehicles with complex mechanisms that can detect and avoid vehicles with suspicious behaviour. In this section, we divide the misbehaviour detection mechanisms in Internet of Vehicles into three types, introduced in detail in the following subsections. After discussing the various mechanisms of the existing technology, we will provide an analysis summary table.

3.1 Detection Mechanisms Based on Message Content

In Internet of Vehicles, only messages are received by vehicles, therefore it is a very intuitive and effective way to determine their validity based on the content of messages. From the perspective of information security attributes (e.g., authenticity, rationality, and uniformity), we have briefly summarized the relevant mechanisms which can be further subdivided according to their specific approach. An overview of these mechanisms and the example mechanisms we discuss in this chapter is shown in Table 1.

3.1.1 Authenticity

Due to the inadequate management of Internet of Vehicles, when two vehicles communicate or establish a trust relationship, the authenticity of message must be confirmed. This means that the message is indeed sent by a vehicle with a legal certificate, not replayed or sent by a vehicle without a legal certificate. Currently, public key-based encryption schemes are commonly used to realize the identity verification of communicating vehicles [2]. In the initial process of vehicle, the manufacturer registers with the trusted structure to obtain a set of public key certificates and corresponding public/private key pairs. The vehicle then uses its private key to sign the message it creates, and attaches the certificate and signature to the end of message for verification by the recipient. It requires the sender to have a valid public key certificate so that the receiver can authenticate correctly [3].

Azees et al. [4] proposed an effective anonymous authentication scheme for Internet of Vehicles. The proposed scheme was based on the mathematical model of bilinear pairing for certificate issuance and authentication. In addition, the proposed scheme provided effective identity authentication while also preserving the anonymity of vehicle and RSU. Therefore, the proposed scheme could effectively defend against forged identity attacks and external attacks. Meanwhile, the scheme also provided a certain ability of tracking function, since the trusted center could track changes in the identity of suspicious vehicles. In terms of the cost of certificate authentication and identity authentication, the proposed scheme outperforms Boneh et al. [5], Efficient Conditional Privacy Preservation [6], Certificate-less Aggregate Signature [7], Group Signature and Identity-based Signature [8] and Key-insulated Pseudonym Self-Delegation [9]. However, due to the use of bilinear

Table 1 Detection mechanisms based on message content

Mechanism	Category	Advantage	Disadvantage
Azees et al. [4]	Authenticity	Defend against forged identity attacks and external attacks, and provide tracking function	High computational overhead and limited by certification authority
Dua et al. [10]	Authenticity	The two-level authentication reduces the calculation cost and response time	The number of verification steps performed by the CA is difficult to expand
Lo et al. [11]	Rationality	Predefined rules check quick, and are easy to modify and add	The mechanism can only detect attacks that meet rules, and rules are not easy to define
Jaeger et al. [12]	Rationality	The mechanism can predict the movement of the vehicle accurately in the presence of errors	Violate the privacy of the vehicle
Yang et al. [13]	Rationality	The mechanism can effectively determine the location range of the vehicle	High requirements for time delay
Aloqaily et al. [15]	Rationality	Alleviate the high computational performance of machine learning technology through a phased approach	Require real-time communication with TTP
Zaidi et al. [16]	Uniformity	The mechanism is strictly proven by hypothesis testing and has high reliability	The mechanism does not perform well in scenarios with sparse nodes
Zhang et al. [17]	Uniformity	The mechanism considers the geographical relationship between vehicles and events	The mechanism does not perform well in scenarios with sparse nodes
Ge et al. [18]	Uniformity	The mechanism can effectively identify attacks such as black hole attacks, message tampering and identity forgery	Rely on the judgment of trusted centers that cannot be accessed in real time

pairing, the computational overhead of proposed mechanism was still a bit expensive and inefficient. In addition, the mechanism was limited to a Certificate Authority (CA).

Dua et al. [10] proposed a novel secure message communication mechanism based on elliptic curve cryptography and designed a two-level authentication key exchange mechanism. The proposed mechanism divided the vehicles into clusters and performed the first-level authentication between the cluster head and CA. The verified cluster head and the vehicles in the cluster performed the second-level authentication. The cluster head and vehicles exchanged messages after authentication. The proposed mechanism distributed all identity authentication that should be done by the CA to the cluster head, through the two-level authentication and proxy method, so as to reduce the calculation cost and response time. However, in the case of high-density network, the number of verification steps performed by the CA cannot be expanded.

3.1.2   Rationality

In the mechanism based on rationality checking, malicious messages can be quickly and effectively detected through predefined rules. Lo et al. [11] proposed a rule checking model, composed of the check module and rule database. The rule database stored predefined rules, which would analyse and check the data of each field of the message to identify whether the message was valid. These rules could check the reported vehicle location, the vehicle speed, and the time stamp of message, etc. Messages were considered valid only if they conformed to the rules. In addition, these rules were easy to modify and add.

Position prediction can predict the approximate range of next stage of the vehicle position, so as to verify whether the message content is effective. The Kalman filter is the most commonly used method. Jaeger et al. [12] added the position, speed, heading, and size of the vehicle to Cooperative Awareness Messages (CAMs). By using the Kalman filter to track the vehicle, the proposed mechanism could predict the next position range of vehicle, thereby verifying whether the position contained in the CAMs was reasonable. The method realized the function of detecting and correcting the forged data in the CAMs. The method is effective, since the Kalman filter can predict the movement of vehicle accurately in the presence of errors. However, this method violates the privacy of vehicle, since the CAMs that track the vehicle need to be collected, that is, the trajectory of vehicle is known.

Yang et al. [13] proposed a scheme to verify the range of vehicle location based on the distance-bounding protocols. In the distance-bounding protocols, the approximate distance between the verifier and prover could be determined through the round-trip time of cryptographic challenge-response pairs and the signal propagation time. Based on the time division multiple access technology, the verifier and prover first negotiated to select an available channel and time slot to facilitate the next time-critical distance-bounding phase. In the distance-bounding phase, a series of repeated challenge-response operations would be performed, and the time difference of each round would be recorded. If the response error and timeout error of prover were greater than the predefined threshold, the connection was refused. It also meant that the prover was not within the distance range. The distance-bounding protocols can effectively determine the location range of vehicle, however with high requirements for time delay. The mechanism proposed by the author in [14] could estimate the position of node to a limited area according to the upper limit of distance between at least three verifiers and the prover. However, how to trust the upper limit of distance of other verifiers brings duplicate problems.

Detecting malicious messages is essentially dividing messages into normal messages and malicious messages according to certain rules. Multiple classifier frameworks in machine learning technology can effectively solve this problem. Aloqaily et al. [15] proposed a three-stage intrusion detection scheme, based on the Deep Belief Nets and Decision Tree. The proposed scheme divided the vehicles into service-specific clusters. The first stage occurred at the cluster head. The cluster de-duplicated and processed received messages, and then uploaded them to the next stage. The second stage occurred in the Trusted Third-Party (TTP) which acted as a service intermediary. The TTP de-duplicated all cluster head data, extracted features of the data, and classified them according to the Deep Belief Nets and Decision Tree. The third stage occurred at the service provider, repeating the functions of second stage. The machine learning technology requires high computing performance, while it is known that the computing resources of vehicles are limited. Applying machine learning technology in other stages effectively avoids the shortcoming and takes advantage of its advantages. However, the proposed mechanism required real-time communication with TTP, which was limited by the scenario of requesting services.

3.1.3   Uniformity

Messages forged by malicious vehicles cannot always be independently judged whether they are valid, therefore it is necessary to comprehensively check data packets from different vehicles. The uniformity-based mechanism focuses on detecting and resolving conflicting messages. Zaidi et al. [16] compared the surrounding traffic flow reported by a single vehicle with the traffic flow reported by neighbours. The author came from the idea that, the traffic flow reported by vehicles close to each other under the same traffic conditions should be very similar. Therefore, if the message conformed to the described model, it was considered correct, otherwise it was false. The traffic flow used in the paper was measured by vehicle speed and density, and also was an indicator of road congestion. If there was an attacker forging data, there would be a significant difference in traffic flow from normal vehicles and malicious vehicles. The detection of malicious traffic was strictly proven by hypothesis testing and had high reliability.

Zhang et al. [17] calculated the score of accident message created by the vehicle, based on the geographic relationship between the vehicle location and accident location. The calculation of score considered the Manhattan distance between the vehicle location and accident location, and the number of obstacles on the possible path from the vehicle location to accident location. The number of obstacles was defined as the number of intersections on the path. The author defined the value of message that reported the accident as positive, and the opposite was defined as negative. The scores of accident messages created by multiple vehicles were merged, through the weighted voting method to obtain the trust value of accident event. The proposed mechanism used data from multiple vehicles to exclude false messages based on the consistency of most vehicle messages. However, a malicious vehicle would assign a large value when calculating the value of message it created. Therefore, the mechanism depended on the received messages with honest data accounting for the majority.

Similarly, Ge et al. [18] used drones as a research carrier and proposed a provenance-aware distributed trust model, named UAV-pro. The provenance referred to the behaviour history of vehicles that the message passed through during the forwarding process. Based on the certificate signature, the behaviour of message creator and forwarder could be evaluated to generate observational evidence. By collecting observational evidence generated by distributed vehicles, the trusted center could identify malicious nodes in the network and isolate them to maintain the normal performance of network. Simulation experiments showed that the proposed mechanism could effectively identify attacks, e.g., black hole attacks, message tampering, and identity forgery, However, the proposed mechanism relied on judgments made by trusted center, which was not accessible in real time. In addition, the author did not discuss in detail the situation in which nodes with legal identities launch attacks in the network.

3.2 Detection Mechanisms Based on Message Processing Behaviour

In addition to the content of message received by the vehicle, malicious vehicles can also be detected based on the processing of message. Processing behaviours include forwarding, tampering, discarding, and selective discarding, etc. The advantage of such mechanisms is that they are completely independent of the message content. Here, we provide a brief overview of mechanisms in Table 2.

3.2.1 Watchdog

In the watchdog mechanism, each vehicle continuously watches other vehicles it has encountered, to forward data packets and manage the network. Dias et al. [19] assigned an initial reputation score to each vehicle. When the vehicle established contact with others and forwarded messages, it would also receive the neighbour's reputation scores for other vehicles from the neighbour. Then the vehicle updated its

Table 2 Detection mechanisms based on message processing behaviour

Mechanism	Category	Advantage	Disadvantage
Dias et al. [19]	Watchdog	Monitor and evaluate neighbours to calculate trust, and consider the trust of neighbours in other vehicles	Lack of malicious neighbour vehicles or false neighbour trust
Jedari et al. [20]	Watchdog	Identify selfish nodes based on social relationships	Lack of consideration for message content because of possible false messages
Li et al. [23]	Custom rules	Alleviate flooding attacks through rate limiting	Cannot alleviate distributed denial of service attacks
Alajeely et al. [24]	Custom rules	Based on a fast and efficient binary tree to ensure that the message is not tampered with	The mechanism requires all packets of the group to arrive, which is unbearable

score based on the historical score, the score received from neighbours, and the score assigned by the watchdog system. After that, the watchdog system performed appropriate management based on the vehicle reputation score. The proposed mechanism considered the opinions of neighbouring vehicles, however did not discuss in detail the situation when neighbours are malicious vehicles or the opinions of neighbours are false.

Jedari et al. [20] proposed a social-based watchdog system. Watchdog vehicles analysed received messages according to their social relationship to identify the selfish behaviour of nodes in message relaying. At the same time, watchdog vehiclesapplied indirect watchdog information received from other vehicles to improve detection time and accuracy. In addition, the author designed a reputation system, in which watchdog vehicles identified selfish vehicles based on their direct and indirect watchdog information, and distinguished individually and socially selfish vehicles. In addition, the author also added a watchdog evaluation module to avoid being affected by false watchdog information. Watchdog vehicles would check the authenticity of messages before applying indirect watchdog information. The author conducted simulation experiments based on two real data sets, MIT Reality [21] and Social Evolution [22]. Simulation results showed that the detection time and detection rate were better than the contact-based watchdog system.

3.2.2 Custom Rules

Authors in papers added custom rules by modifying the message format to detect misbehaviour of malicious vehicles. Li et al. [23] applied a rate limiting mechanism to prevent flooding attacks. The number of data packets created by the vehicle in each predefined time interval was limited. Meanwhile the number of copies of each data packet generated by forwarding was also limited. In addition, the rate limiting was achieved by adding a rate limiting certificate to the data package, which was issued by a trusted organization, and proving the validity of certificate. The proposed mechanism detected false counts claimed by the attacker through the idea of claim-carry-and-check. This mechanism could indeed effectively alleviate flooding attacks, and the increased performance loss was not much. However, the proposed mechanism cannot alleviate distributed denial of service attacks.

Alajeely et al. [24] applied the hash tree technology to detect message tampering attacks. The proposed mechanism applied the hash value of data packets as leaf nodes. Then two leaf nodes performed XOR operation and hash to obtain the parent node. The above operations were repeated to get the root node hash value. The hash value of root node was placed in a predefined field in the message format header of data packets. After receiving data packets, other vehicles constructed a hash tree according to the same rules, and compared the calculated value with the value in the message header, to determine whether the message had been tampered with. The proposed mechanism builds a hash tree in the form of binary tree, which is fast and efficient under ideal conditions. However, this mechanism requires that all packets arrive. Due

to the characteristics of Internet of Vehicles, it will be very time-consuming to wait
for all packets to arrive, which is difficult to bear.

3.3 Detection Mechanisms Combined with Sensor

When there are few vehicles, it is difficult to distinguish misbehaviour from normal
behaviour only from the message level (e.g., distinguishing normal messages and fake
messages). Sun et al. [25] used on-board sensors to track and measure the acceleration
and direction of other vehicles. After that, they used the extended Kalman filter to
track and predict the dynamic state of target vehicle, including the position and
speed of the vehicle. First, they supposed that the state of target vehicle (including
position and speed) reported by itself was true. If the assumption was true, even if
there was noise interference, the calculated value of extended Kalman filter should
not be much different from the reported value of the target vehicle. Thus, if they
were very different, it was reasonable to believe that the assumption was not valid,
that is, the state of target vehicle reported by itself did not conform to its actual
movement behaviour. After detecting the abnormality of target vehicle, the accurate
position of target vehicle can be obtained based on a trusted neighbour vehicle and
the method of angle of arrival ranging. The proposed mechanism relies on sensor
measurements and location prediction algorithms. However, it may not be so accurate
due to noise interference in practice. Meanwhile, the assumption will not always hold.
The advantage of using a sensor is that the value collected by the sensor comes from
the physical world, and is not affected by the software world.

Yao et al. [26] measured the Received Signal Strength Indication (RSSI) value
in different simulation scenarios in the real environment, summarized and found
four characteristics of RSSI: (a) Channel quality changed with time in Internet of
Vehicles; (b) The propagation model had different physical changes due to different
environments. Since the surrounding environment of vehicle was difficult to perceive,
the propagation model was also difficult to adjust with the surrounding environment;
(c) The RSSI time series of attacking vehicles and the fake Sybil vehicles had very
similar patterns; (d) the RSSI of Sybil vehicles would frequently change suddenly,
while normal vehicles did not. In view of four characteristics, the author adopted similarity measures for time series and change-points detection for time series to detect
Sybil vehicles. The proposed detection method did not rely on any predefined radio
propagation model and the trust relationship of neighbouring vehicles. In addition,
the proposed detection considered the physical characteristics of vehicles that cannot
be hidden. However, if an attacker follows these characteristics to launch attacks,
then the proposed detection mechanism will be useless. Another disadvantage is that
RSSI is greatly affected by the environment.

4 Further Discussions

There are still unresolved issues that need to be concerned.

(1) A robust authentication scheme can prevent unauthorized users from accessing the network. However, it will impose greater computational and communication overhead on participating entities, and violate the requirements of delay-sensitive applications. Therefore, authentication and computing / communication overhead must be weighed. The vehicle must be equipped with a tailor-made authentication scheme that meets the strict requirements of Internet of Vehicles.

(2) The combination of machine learning and intrusion detection is one of the hot research directions. In addition, to fully utilize the potential of machine learning, a sufficient number of representative data sets are required. However, in a large-scale, heterogeneous, and complex system Internet of Vehicles, the generation of representative data sets is a challenge.

(3) 5G-V2X has brought great success in the next generation of Internet of Vehicles, however, it has brought a new attack surface and expanded the scope of security vulnerabilities. Related security research combining 5G features is still of importance.

(4) In the detection mechanism based on cooperation, voting is a commonly used method, relying on the long-term identity of vehicle. However, the privacy protection requires the vehicle to change its pseudonym frequently. Moreover, the frequent use of pseudonyms makes the vehicle have multiple identities at the same time, which is prone to Sybil attacks. In addition, the frequent changes of pseudonyms greatly increase the difficulty of tracking and linking messages of vehicles, which adds additional obstacles to the detection mechanism. Given above, too strict security mechanisms will infringe users' privacy, while too much privacy protection will restrict the security mechanism, leading to lower performance. They need to balance.

(5) Similar to any other information received from neighbouring vehicles, the misbehaviour report received by the vehicle is essentially just another type of information. Therefore, researchers must assume that reports of misbehaviour may also be forged, which raises duplication problems.

5 Conclusion

This chapter summarized the development of misbehaviour detection mechanisms in Internet of Vehicles. First, this chapter introduces common threat models in Internet of Vehicles. Then misbehaviour detection mechanisms designed for Internet of Vehicles are divided into three categories: detection mechanisms based on message content, detection mechanisms based on message processing behaviour, and detection mechanisms combined with sensors. Finally, this chapter discusses problems that still exist in the current stage of misbehaviour detection mechanism. From the analysis of this chapter, it is found that most of existing detection mechanisms are only suitable for determining specific misbehaviours. However, due to the difficulty in distinguishing attack types of malicious nodes, it is urgent to construct a unified misbehaviours detection framework to detect various types of misbehaviours.

References

1. M. Raya, J.P. Hubaux, Securing vehicular ad hoc networks. J. Comput. Secur. **15**(1), 39–68 (2007)
2. J. Petit, F. Schaub, M. Feiri et al., Pseudonym schemes in vehicular networks: a survey. IEEE Commun. Surv. Tutor. **17**(1), 228–255 (2014)
3. Z. Lu, G. Qu, Z. Liu, A survey on recent advances in vehicular network security, trust, and privacy. IEEE Trans. Intell. Transp. Syst. **20**(2), 760–776 (2018)
4. M. Azees, P. Vijayakumar, L.J. Deboarh, EAAP: efficient anonymous authentication with conditional privacy-preserving scheme for vehicular ad hoc networks. IEEE Trans. Intell. Transp. Syst. **18**(9), 2467–2476 (2017)
5. D. Boneh, C. Gentry, B. Lynn, et al., Aggregate and verifiably encrypted signatures from bilinear maps, in *International Conference on the Theory and Applications of Cryptographic Techniques* (Springer, Berlin, Heidelberg, 2003), pp. 416–432
6. R. Lu, X. Lin, H. Zhu, et al., ECPP: efficient conditional privacy preservation protocol for secure vehicular communications, in *IEEE INFOCOM 2008-The 27th Conference on Computer Communications* (IEEE, 2008), pp. 1229–1237
7. Z. Gong, Y. Long, X. Hong, et al., Two certificateless aggregate signatures from bilinear maps, in *Eighth ACIS International Conference on Software Engineering, Artificial Intelligence, Networking, and Parallel/Distributed Computing (SNPD 2007)*, vol. 3 (IEEE, 2007), pp. 188–193
8. X. Lin, X. Sun, P.H. Ho et al., GSIS: a secure and privacy-preserving protocol for vehicular communications. IEEE Trans. Veh. Technol. **56**(6), 3442–3456 (2007)
9. R. Lu, X. Lin, T.H. Luan et al., Pseudonym changing at social spots: an effective strategy for location privacy in Internet of Vehicles. IEEE Trans. Veh. Technol. **61**(1), 86–96 (2011)
10. A. Dua, N. Kumar, A.K. Das et al., Secure message communication protocol among vehicles in smart city. IEEE Trans. Veh. Technol. **67**(5), 4359–4373 (2017)
11. N.W. Lo, H.C. Tsai, Illusion attack on vanet applications-a message plausibility problem, in *2007 IEEE Globecom Workshops* (IEEE, 2007), pp. 1–8
12. A. Jaeger, et al., A novel framework for efficient mobility data verification in vehicular ad-hoc networks. Int. J. Intell. Transp. Syst. Res. **10**(1), 11–21 (2012)
13. A. Yang, J. Weng, N. Cheng et al., DeQoS attack: degrading quality of service in Internet of Vehicles and its mitigation. IEEE Trans. Veh. Technol. **68**(5), 4834–4845 (2019)
14. D. Singelee, B. Preneel, Location verification using secure distance bounding protocols, in *IEEE International Conference on Mobile Adhoc and Sensor Systems Conference, 2005* (IEEE, 2005), pp. 7–840
15. M. Aloqaily, S. Otoum, I. Al Ridhawi et al., An intrusion detection system for connected vehicles in smart cities. Ad Hoc Netw. **90**, 101842 (2019)
16. K. Zaidi, M.B. Milojevic, V. Rakocevic et al., Host-based intrusion detection for Internet of Vehicles: a statistical approach to rogue node detection. IEEE Trans. Veh. Technol. **65**(8), 6703–6714 (2015)
17. X. Zhang, R. Li, W. Hou, et al., Research on Manhattan distance based trust management in vehicular ad hoc network. Secur. Commun. Netw. *2021* (2021)

18. C Ge, L. Zhou, G.P. Hancke, et al., A provenance-aware distributed trust model for resilient unmanned aerial vehicle networks. IEEE Internet Things J. (2020)
19. J.A.F.F. Dias, J.J.P.C. Rodrigues, F. Xia et al., A cooperative watchdog system to detect misbehavior nodes in vehicular delay-tolerant networks. IEEE Trans. Industr. Electron. **62**(12), 7929–7937 (2015)
20. B. Jedari, F. Xia, H. Chen et al., A social-based watchdog system to detect selfish nodes in opportunistic mobile networks. Futur. Gener. Comput. Syst. **92**, 777–788 (2019)
21. N. Eagle, A. Pentland, Reality mining: sensing complex social systems. Pers. Ubiquitous Comput. **10**(4), 255–268 (2006)
22. A. Madan, M. Cebrian, S. Moturu, K. Farrahi, A. Pentland, Sensing the health state of a community. IEEE Pervasive Comput. **11**(4), 36–45 (2012)
23. Q. Li, W. Gao, S. Zhu et al., To lie or to comply: defending against flood attacks in disruption tolerant networks. IEEE Trans. Dependable Secure Comput. **10**(3), 168–182 (2012)
24. M. Alajeely, R. Doss, V. Mak-Hau, Defense against packet collusion attacks in opportunistic networks. Comput. Secur. **65**, 269–282 (2017)
25. M. Sun, M. Li, R. Gerdes, A data trust framework for Internet of Vehicles enabling false data detection and secure vehicle tracking, in *2017 IEEE Conference on Communications and Network Security (CNS)* (IEEE, 2017), pp. 1–9
26. Y. Yao, B. Xiao, G. Wu et al., Multi-channel based Sybil attack detection in vehicular ad hoc networks using RSSI. IEEE Trans. Mob. Comput. **18**(2), 362–375 (2018)

Chapter 15
Security and Privacy-Preserving for Automated Electric Vehicle

Di Wang, Yue Cao, and Naveed Ahmad

Abbreviations

AEVs	Automated electric vehicles
AVP	Autonomous valet parking
LAVP	Long-range autonomous valet parking
SAVP	Short-range autonomous valet parking
AVs	Autonomous vehicles
IoT	The Internet of Things
EVs	Electric vehicles
SG	The smart grid
G2V	Grid to vehicle
V2G	Vehicle to grid
GC	The global control
CP	The car parking
TA	The trusted authority
D/P	The pick-up/drop-off point
SM	The smart meter
ECUs	Electronic control units
OBU	The onboard unit
CS	The charging station
SDN	Software defined network

D. Wang (✉) · Y. Cao
School of Cyber Science and Engineering, Wuhan University, Wuhan, China
e-mail: diwang@whu.edu.cn

Y. Cao
e-mail: yue.cao@whu.edu.cn

N. Ahmad
Department of Computer Science, Prince Sultan University, Riyadh, Saudi Arabia
e-mail: nahmed@psu.edu.sa

© The Author(s), under exclusive license to Springer Nature Singapore Pte Ltd. 2023
Y. Zhu et al. (eds.), *Communication, Computation and Perception Technologies for Internet of Vehicles*, https://doi.org/10.1007/978-981-99-5439-1_15

1 Introduction

Autonomous driving is a milestone and focuses in the global automotive industry. Autonomous vehicles (AVs) sense road conditions through advanced sensors [1]. Access and control modules are used to guarantee secure interaction. The communication module ensures efficient communication between the vehicle and other equipment. Autonomous driving reduces human intervention and improves safety and comfort.

With the prosperity of autonomous driving and significant advancements in the Internet of Things (IoT) technologies, Automated Valet Parking (AVP) has developed steadily. AVP which is a technology with great potential does complete automatic parking according to user requirements. AVP can not only reduce exhaust emissions, but also improve the user's "last mile" driving experience, which has been widely concerned by researchers.

Electric vehicles (EVs) are the best carrier of integration with autonomous driving. More important, automated electric vehicles are not only the integration of technology, but also a novel travel mode in the future. Driven by the global trend of automotive power electrification, intelligent control, and information network, automated electric vehicles have become the frontier hot spot in the field of international automotive engineering and the core competitiveness in the future market.

1.1 Motivation

With the vigorous development of automated vehicles, autonomous valet parking technology is utilized to address issues such as long waiting periods for finding a parking spot and huge costs. Nonetheless, automatic valet parking technology also introduces a good deal of serious security concerns. Specifically, because of the remote control interface, automated electric vehicles are facing a variety of cyber attacks, which threaten the lives of pedestrians and passengers. Furthermore, automated electric vehicles sense the environment and navigation via the perception module and localization module respectively. Once the above modules are damaged or attacked, it will incur serious risks. In addition, AVP is controlled remotely, which helps attackers bypass the security system. Finally, parking service providers can infer the driving style of drivers based on locally stored data, which determines vehicles' location and even identifies passengers' personal information, such as a home address, workplace, and health status. Therefore, a comprehensive review of security solutions and privacy preservation for automatic valet parking is urgently needed.

Due to the limited fossil fuels and the emission pollution caused by combustion, electric vehicles, as the promoters of green transportation, have become indispensable travel tools in daily life. In order to guarantee the safe operation of electric vehicles, drivers send charging requests to the server, resulting in identity privacy leakage. Worse still, the operator must evaluate the power storage capacity by monitoring the status information of electric vehicles, such as location and battery status. Frequent monitoring will cause users to worry about sensitive information being disclosed. In addition, electric vehicles communicate in an open wireless environment, which makes it easy for attackers to eavesdrop or forge data. Last but not least, unauthorized access to data can lead to data privacy leakage. Therefore, a comprehensive review about security solutions and privacy preservation for EV charging is urgently needed.

1.2   Contributions

The main contributions to this chapter are as follows:

(1)   We start with introducing parking modes and charging modes, including short-range autonomous valet parking, long-range autonomous valet parking, grid-to-vehicle, and vehicle-to-grid, to facilitate readers to understand automated electric vehicles.
(2)   We elaborate on key challenges associated with autonomous valet parking and EV charging respectively.
(3)   We present a comprehensive review of the state-of-art research about privacy-preserving for smart parking, autonomous valet parking and EV charging. Then, we highlight the differences among existing schemes.
(4)   We discuss plenty of open problems and challenges as future research directions, to accelerate the future development of automated electric vehicles.

1.3   Structural Organization

The overall structure is organized as follows. Section 2 introduces the background including parking modes and charging modes. The system model, security threats, and security requirements for AVP are expounded in Sect. 3. In addition, Sect. 3 describes the existing privacy strategies in detail. Similarly, the system model, security threats, security requirements, and existing privacy strategies for EV charging are expounded in Sect. 4. Sections 5 and 6 respectively discuss future works and conclusions.

2 Background

2.1 Autonomous Valet Parking Use Cases

Autonomous valet parking has developed with the prosperity of autonomous driving. The classification of automatic driving by the Society of Automotive Engineers (SAE) is shown in Fig. 1. Autonomous driving can be classified into five levels: No automation (L0), driver assistance (L1), partial automation (L2), conditional automation (L3), high automation (L4), and full automation (L5).

   Therefore, autonomous valet parking can be divided into Level 1–level 5 according to the degree of automation. Level 1 is the parking mode in which the driver fully participates. Moreover, level 2 is the parking mode in which the driver remotely controls the vehicle via the smartphone. However, advanced sensor technologies such as ultrasonic sensors and millimeter wave (MMW) radars are applied to avoid obstacles and high-precision maps are utilized to calculate the optimal path to the parking lot in Level 3. Level 4 allows the driver to drop off the vehicle at the entrance of parking lot, and then the vehicle automatically drives to the empty parking space. Level 4 is short-range autonomous valet parking, which requires the driver to arrive at the parking lot, reducing the parking service experience. Hence, researchers have proposed long-range autonomous valet parking, that is, the driver drops off at the Pick-up/Drop-off (D/P) point closing to the destination, and then the vehicle drives from the D/P to the nearby parking lot looking for vacant parking spaces. The book chapter only introduces short-range autonomous valet parking (SAVP) and long-range autonomous valet parking (LAVP) in detail.

2.1.1 Short-Range Autonomous Valet Parking (SAVP)

Due to the imbalance between the supply and demand of parking spaces, Daimler-Benz [2] integrates advanced sensing equipment, communication technology, navigation and positioning systems to provide SAVP services for drivers. The driver parks AV at the entrance of the parking lot, and then millimeter wave radar (MMW Radar), ultrasonic sensor and camera are applied to sense the complex surrounding environment. Dedicated Short Range Communication (DSRC), WiFi and Long Term Evolution (LTE) not only assist the communication between parking lot service providers and AVs, but also help drivers monitor AVs via smartphones. GPS, Beidou

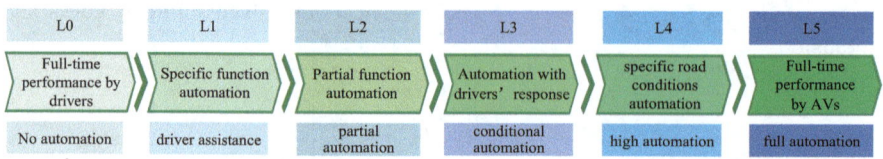

Fig. 1 Autonomous driving classification

navigation and high-precision map utilize AVs' location and parking spaces usage to match parking spaces for AV, and then generate driving tracks so that AVs can be guided to parking spaces.

2.1.2 Long-Range Autonomous Valet Parking (LAVP)

Although SAVP has high automation, it still needs the driver to drive AV to the parking lot, which cannot be fully automated. In order to improve the service quality of AVP and reduce manual intervention, LAVP [3] allows passenger to park AV at D/P. Then passenger walks from D/P to the destination, AV drives from D/P to the parking lot and cruises the available parking spaces. Once the ride request was sent by the passenger through the smartphone, AV will drive from the parking lot towards the D/P adjacent to passengers to pick up the passenger.

2.2 EV Charging Use Cases

In order to promote the rapid and sustainable development of electric vehicles [4], China issued Electric Vehicles Industry Development Plan (2021–2035) [5] in November 2020. The plan clearly stipulates that by 2025, the sales volume of electric vehicles must reach about 20%. By 2035, electric vehicles will become the mainstream development direction of automobile industry transformation. In addition, the United States and the European Union have also taken a series of incentive measures such as increasing subsidies, expanding the scope of subsidies and reducing taxes to promote electric vehicles commercialization. Unsurprisingly, by 2020, the global sales of EVs were 3.24 million along with an increment of 43%. As energy storage units, EVs are connected to the smart grid (SG). Moreover, grid to vehicle (G2V) and vehicle to grid (V2G), as important components of SG, realize the two-way energy trading between EVs and SG.

2.2.1 Grid-To-Vehicle Connection

Due to EVs' mobility and the limited storage capacity of batteries, EVs are inevitable to face challenges about insufficient power and driving anxiety. G2V refers to SG providing charging services to EVs with insufficient power to balance energy demand.

2.2.2 Vehicle-To-Grid Connection

V2G refers to that the EV is connected to the SG as a mobile and distributed standby power source. SG uses incentive mechanism or payment reward promoting EV to sell the excess power to the SG. Next, SG aggregates the energy discharged by EV to realize energy demand response management, energy scheduling and trading, and dynamic pricing during the peak period of power consumption, so as to balance the power load.

3 Secure Autonomous Valet Parking

3.1 System Model

System framework is shown in Fig. 2, which is mainly including passenger, AV, global control (GC), car parking (CP), D/P and trusted authority (TA). The detailed definitions of each entity are as follows.

- Passenger: A passenger who needs to ride, sends a ride request containing identity, destination, parking moment, parking time, and timestamp to GC through the smartphone terminal. After arriving at D/P near the destination, the passenger gets off and walks to the destination.

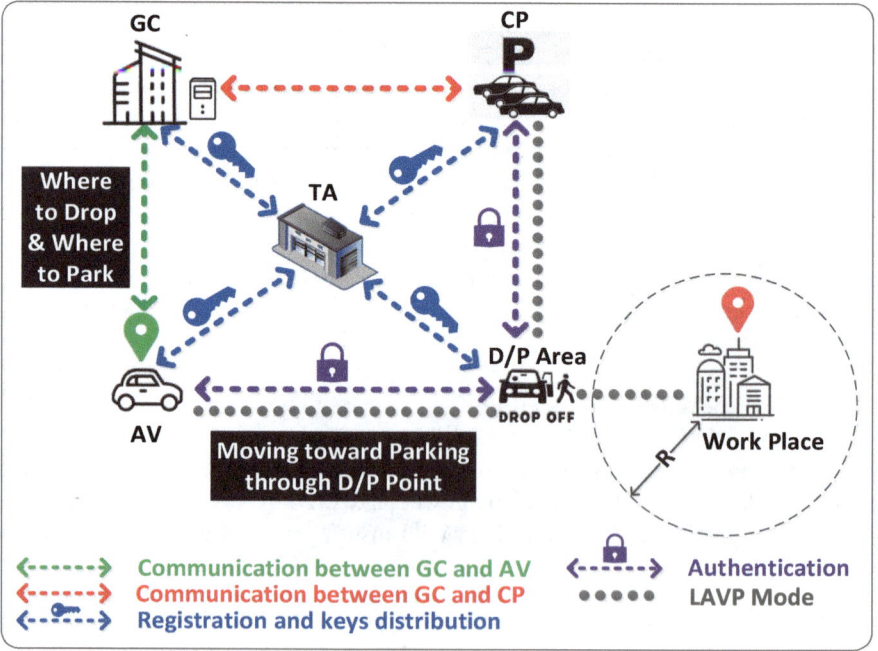

Fig. 2 Autonomous valet parking system model

- AV: Each AV is equipped with sensors and communication facilities for sensing the environment and communicating with other entities. AVs provide AVP services for passengers.
- GC: GC obtains the real-time parking spaces occupation from CP, and then matches the D/P for passengers as well as the best CP for AV based on the destination, parking moment, and parking time.
- CP: It monitors the usage of parking spaces in the CP in real-time and feeds back to GC.
- D/P: It is allowed to stay for a short time to pick up or drop off passengers, but it's not allowed to park for a long time.
- TA: It not only provides registration and authentication for other entities, but also generates system parameters and distributes keys.

3.2   Security Threats

Assuming that TA is credible and not easy to be captured. GC is honest but curious, that is, it abides by the protocol, but is also curious about AV location as well as passenger identity. Security threats in the network are as follows.

- Location Privacy Leakage: Attackers use the eavesdropped or intercepted information to predict passengers' and AVs' locations, threatening the life safety of passengers.
- Double-reservation Attack: Passengers reserve multiple parking spaces at a time to increase the success rate of parking reservations, which greatly reduces the utilization rate of parking spaces.
- Black Hole and Gray Hole Attack: At first, attackers behave the same as honest nodes, but in information transmission, attackers selectively lose some data packets, resulting in incomplete information received by the receiver.
- Broadcast Tampering Attack: Attackers broadcast false information, which causes the true information to be covered up and reduces the success rate of parking reservations.
- Eavesdropping Attack: By eavesdropping on the data forwarded between the passenger and GC, attackers speculate on confidential information such as the passenger's living habits.
- Spamming Attack: Attackers send a large number of spam messages to GC so that bandwidth is occupied, affecting the processing efficiency of parking reservations.
- GPS Spoofing: The attacker generates a stronger signal than the navigation satellite, resulting in inaccurate passenger positioning.

3.3   Security Requirements

Based on the system model and security threats, in order to provide passengers with safe autonomous parking service, the following security requirements shall be met.

- Pseudonymity: GC cannot recognize the real identity of the passenger according to the information sent by the passenger, nor can it recognize the real identity of AV based on the parking reservation forwarded by AV.
- Unlinkability: GC cannot infer whether the two messages are transmitted by the same node.
- Geo-indistinguishability: The location of passengers and AVs should be concealed or obfuscation.
- Double-reservation Attack Resistance: Generate a token for passengers who have successfully reserved a parking space, which ensures each parking space is reserved by only one passenger.
- Availability: AVP reservation service must be robust, that is, even if there are a few malicious nodes in the network, reservation service can be still secure.
- Integrity: Ride requests, parking requests, and parking reservations cannot be tampered with and forged by attackers.
- Mutual Authentication: Each entity realizes mutual authentication with the assistance of TA.

3.4   Privacy Strategies for Parking

3.4.1   Related Works on Smart Parking

Khalid et al. [6] conducted a comprehensive survey on traditional parking and smart parking, and then expounded in detail digital enhanced parking, smart routing, and high-density parking. Besides, AVP is discussed. Finally, Khalid et al. [6] discussed future challenges and research hotspots from the aspects of privacy protection, scheduling, and green parking. Lu et al. [7] proposed a smart parking scheme based on vehicular communications to support privacy protection. The scheme uses pseudonyms distributed by a trusted authority to protect the privacy of vehicles. In order to prevent privacy disclosure when drivers query parking spaces, Ni et al. [8] used zero-knowledge proof to realize conditional privacy protection. Moreover, a data retrieval mechanism based on bloom filters is proposed to improve the probability of vehicles successfully receiving navigation results. Chim et al. [9] used anonymous credentials to prevent attackers from linking the driver's parking query to the destination. Next, the pseudonym is applied to realize mutual authentication and protect the driver's identity privacy. However, malicious vehicles can use the shared master key to obtain the driver's confidential message. Therefore, Cho et al. [10] proposed a privacy-preserving navigation scheme without the shared master key. The scheme provides privacy protection for smart parking to a certain extent, but it does not consider the anonymous payment of parking fees. Zhu et al. [11] used group signature and anonymous identity authentication to provide a new opportunity for private parking space sharing. Lai et al. [12] integrated group signature and differential privacy technology based on mixed zone to protect network nodes' identity privacy and location privacy. Reward mechanism and trust model are proposed to resist free-riding attacks and improve the credibility of nodes, respectively. Based on the PlaaS (Parking Information as a Service) architecture, Safi et al. [13] proposed a parking lock encryption mechanism to prevent vehicle trajectories from leaking, and cloud-based centralized storage assists in digital forensics. An et al. [14] proposed an incentive mechanism for real-time parking space sharing to optimize the service quality such as social welfare and satisfaction of parking service providers under the condition of protecting the destination's location privacy. Garra et al. [15] proposed payment based on electronic currency to protect the attacker from tracing the user's real identity, and in case of dispute, the driver can provide an encrypted payment token as the basis for arbitration.

Blockchain, as a decentralized, open, transparent, and collectively maintained distributed database, was first proposed by Satoshi Nakamoto in 2008 [16]. The prosperity of the blockchain has received extensive attention from industry and scientific researchers [17]. Blockchain technology is also used in the Internet of Vehicles [18]. Zhang et al. [19] used BBS group signature to achieve mutual authentication between drivers and parking service providers, and then smart contracts are utilized to achieve anonymous and fair payment of parking fees. Besides, Vector based encryption is used to protect the privacy of driver parking time. In order to alleviate mismatch issues concerning the supply and demand of public parking spaces, Wang et al. [20] applied blockchain to store and share private parking space messages. Furthermore, pedersen commitment and one-way payment were used to realize anonymity and traceability of transactions, and top trading cycles and chain (TTCC) was used to match parking spaces and drivers. Hu et al. [21] combined BlockChainOpenSource (BCOS) and smart contracts to realize parking management including user registration, search and rent, and payment, meeting the user's requirements for identity authentication and privacy protection. In order to prevent the parking space matching process from leaking the driver's privacy, Li et al. [22] used smart contracts and anonymous credentials to not only prevent the single point of collapse in the traditional parking management system, but also resist attackers from tracking the driver's current location and inferring the user's driving style. In addition, Amiri et al. [23] used consortium blockchain and private information retrieval technology to provide users with available parking space retrieval supporting privacy protection. Parked vehicles can be used as a new paradigm for data calculation and storage. Therefore, Teng et al. [24] designed a parking calculation model based on blockchain for resource scheduling, task allocation, and maximizing system utility. Zhang et al. [25] combined the sub-chain and main-chain to store parking space status information, and applied the linkable group signature (LGS) to prevent attackers from reserving multiple parking spaces at the same time.

3.4.2 Related Works on Autonomous Valet Parking

Huang et al. [26] used zero-knowledge proof and geo-indistinguishable to protect users' location privacy. Proxy re-encryption and reservation tokens are used to achieve data confidentiality and resist double reservation attacks, respectively. In order to prevent AV theft and user privacy disclosure, Ni et al. [27] proposed a two-factor authentication for AV and users based on passwords and mobile devices. In addition, the BBS signature is used to prevent malicious access by attackers. Pokhrel et al. [28] used differential privacy and zero-knowledge proof to ensure the security and privacy of network nodes, integrating machine learning to improve the availability and reservation success rate of the AVP system.

4 Secure EV Charging

4.1 System Model

The system architecture is presented in Fig. 3. Secure EV charging scheme is generally composed of EV, charging station (CS), trusted authority (TA), and global control (GC). Each entity is defined as follows.

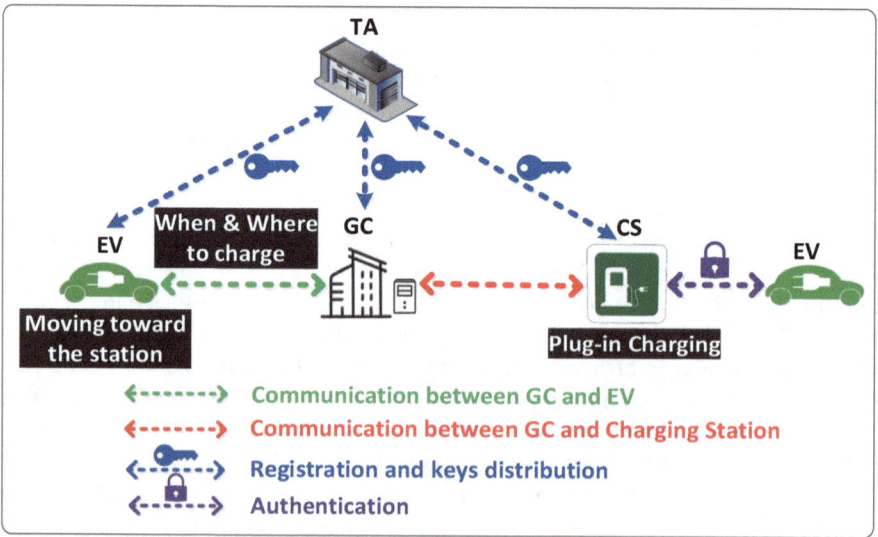

Fig. 3 EV charging system model

- EV: Each EV is equipped with a smart meter (SM) to monitor the battery state of the EV. If the battery status is lower than the threshold, the EV acts as the consumer. Conversely, if the battery state is much higher than the threshold in the peak period, EV can act as the producer.
- CS: Each CS is composed of multiple charging slots. Each charging slot can not only connect to SG to provide charging services for EVs with insufficient power, but also aggregate the energy discharged by EVs to feedback to SG.
- TA: Provide identity registration and authentication for other entities, generate system parameters and distribute keys.
- GC: As a centralized manager, GC can not only monitor the state of the charging slot in real-time, but also process the charging and discharging requests of EVs.

4.2   Security Threats

Assuming that TA is credible and not easy to be captured. GC is vulnerable to hackers, whereas CS is honest but curious, that is, CS abides by the protocol, but CS is also curious about EV's privacy data. Security threats in the network are as follows.

- Denial of Service Attack: Attackers occupy network resources by frequently sending data, resulting in legal EVs being unable to enjoy charge and discharge services.
- Replay Attack: In order to pass identity authentication, attackers repeatedly send the data that is intercepted or eavesdropped on to deceive GC.
- Malware Attack: Electronic control units (ECUs) such as SM and OBU installed on EVs are penetrated, resulting in EVs being remotely controlled by attackers.
- Sybil Attack: Attackers send false information via multiple false identities to GC, which results in GC being misled, network congestion, and energy shortage.
- Unauthorized Access: In order to seek huge economic benefits, attackers unauthorized access GC to remotely control charging slots in CS.
- Privacy Leakage: When an EV conducts an energy transaction, attackers eavesdrop on the power information collected by CS and GC to obtain sensitive information such as EV's identity, location, and user preferences.
- Modification Attack: Attackers tamper with the messages sent by each entity, causing EV and GC to receive false charging station messages and false energy trading messages respectively.

4.3   Security Requirements

Based on the system model and security threats, in order to provide EVs with secure energy trading, the following security requirements should be met.

- Availability: GC must have strong robustness, that is, even if there are some malicious nodes in the network, GC can still ensure secure energy trading.
- Confidentially: Power data and message data are difficult to be accessed by unauthorized persons and have high confidentiality.
- Integrity: Prevent energy trading information from being tampered with and forged.
- Mutual Authentication: GC, EV, and CS realize mutual authentication with the assistance of TA.
- Privacy Preservation: GC and CS cannot infer the location and identity of EVs through energy trading information.
- Access Control: GC musreset hierarchical data access permissions for different entities.
- No-Repudiation: Entities cannot deny that they have conducted energy transactions or sent information.

4.4 Privacy Strategies for Electric Vehicles

4.4.1 Related Works on Cryptography

In order to reduce carbon emissions and promote green and sustainable development, EVs are expected to become the mainstream travel tools in the future. Antoun et al. [29] evaluated the security threats for the EV charging ecosystem from two aspects of household and public charging infrastructure, and put forward corresponding solutions and prospects for the future. For the sake of improving the anti-attack ability of smart meters, Mohammadali et al. [30] proposed the key establishment scheme based on an elliptic curve to support replay attacks resistance. Nicanfar and Leung [31] use the key exchange to improve the security and availability of the system. However, Refs. [30, 31] cannot resist man-in-the-middle attacks. Therefore, Wu and Zhou [32] proposed a key management scheme that can resist man-in-the-middle attacks by combining the symmetric key and public key. Abdallah et al. [33] used pseudonyms to protect the identity privacy of EVs. In addition, signatures are used to realize two-way authentication between the control center and local aggregation.

Due to the heterogeneous devices and limited resources in the Internet of Vehicles, Shen et al. [34] designed a lightweight identity authentication and key agreement protocol for EV charging via hash function and XOR operation A pseudonym update was used to protect the forward security of the session key and the identity anonymity of network nodes. Behalf of solving the problem that the smart grid is not suitable for the network topology with fixed members, Chen et al. [35] used revocable group signature to realize the dynamic management of members. On the other hand, batch verification is utilized to improve the efficiency of information processing, which reduces communication costs as well as computation overhead to a certain extent. In order to match the best charging station for EVs, service providers are inevitable to obtain private information, such as users' location and preferences.

Kumar et al. [36] used lattice-based cryptography technology to protect data privacy. Then, signcryption is used to generate keys for network nodes, reducing the overhead.

To incentivize electric vehicle owners to participate in power transactions, Yang et al. [37] proposed a distributed two-way auction mechanism, and secure multi-party computing is utilized to ensure the privacy of valuation and location. The scheme not only protects privacy, but also meets the incentive compatibility and individual rationality. Charging pads [38], as a common method of dynamic contactless charging, allow EVs to be charged while driving, reducing charging duration and charging waiting time. Li et al. [39] proposed an identity authentication protocol supporting dynamic charging, which uses a pseudonym to cover up the location of EV, and a symmetric key is applied to realize lightweight and efficient authentication. The scheme is low overhead, security, and efficiency. Gope et al. [40] designed lightweight key agreement and identity authentication protocols based on V2G communication, hash function, and XOR operations to ensure communication security and user privacy protection.

4.4.2 Related Works on Blockchain

Bao et al. [41] elaborated that blockchain is widely used in energy applications such as carbon emission, green certification, and charging management because of its advantages of transparency and decentralization. Besides, they discussed the potential value of blockchain in energy security and privacy in the future. For purpose of reducing the response delay, fog computing is used in EV charging systems to improve the efficiency and feasibility of the scheme. Li et al. [42] deployed the blockchain on the fog nodes to achieve efficient processing and response of charging requests. In addition, the key agreement is used to ensure the security and confidentiality of communication. Hassan et al. [43] proposed an energy auction mechanism based on blockchain, applying differential privacy to the V2G auction mechanism to conceal the user's valuation and submission price.

Since the participants in energy transactions are not completely credible and reliable, Wan et al. [44] proposed a distributed energy transaction scheme based on smart contracts and zero-knowledge proof to ensure the fairness and privacy of payment. This scheme has been formally proved that it is secure under the universal composability model. Based on blockchain technology, Danish et al. [45] not only realizes the EV to conduct secure reservation charging without exposing private information to the central controller, but also achieves the EV to select the optimal charging station with the assistance of RSU. Zhou et al. [46] proposed an energy trading and pricing scheme based on a new paradigm of edge computing, to realize secure and efficient energy trading,. Gabay et al. [47] used tokens to realize EV identity authentication. Next, smart contracts and zero-knowledge proof were used to schedule energy, pay fees and protect EV identity privacy. The solution was deployed on Ethereum and proved to have great advantages in gas consumption and overheads. Li and Hu [48] designed a trading framework based on consortium blockchain to protect the security between the smart grid and EV. Moreover, they adopted a heuristic algorithm to minimize the total load and reduce the impact of disorderly charging on the power grid.

5 Future Directions

5.1 Trained Model Privacy

Due to the sharp increase in the number of vehicles, as an important tool for communication and transportation, vehicles are facing challenges such as dynamic changes in network topology, massive data, diversity, and heterogeneity. The above challenges are difficult to meet the requirements of low delay, high reliability, and strong security for intelligent services in the Internet of Vehicles [49]. The emergence of artificial intelligence promotes the vehicle to be more intelligent, to ensure that the vehicle makes accurate judgments and decisions according to the obtained data. The training model requires a lot of real road information and vehicle driving data. On the one hand, it's the research trend to evaluate the data quality and reliability based on redundant data, missing data, and dynamic anomaly detection. On the other hand, most current learning models are based on a centralized structure [1], which has security threats such as single point of collapse, denial of service attacks, and so on. More seriously, the attacker can infer the model parameters or input data according to the model output results. It's still a hassle to train the model on the premise of preventing private data and model parameters disclosure. In addition, because machine learning data has high-dimensional features, data with different features have different sensitivity. It's a research hotspot to design a perfect data sensitivity evaluation system and corresponding data desensitization and privacy protection measures in the future.

5.2 Blockchain Technology

Blockchain, as a trust establishment technology, breaks monopoly and supports value transfer, promotes industrial innovation, and realizes value interconnection [50]. However, the research of blockchain in electric vehicles and autonomous valet parking is still in the infancy and faces great challenges. On the one hand, the existing blockchain and cryptography-based privacy protection schemes require a lot of overhead and storage, which is difficult to apply to large-scale nodes in the Internet of Vehicles. On the other hand, the block capacity, block generation speed, transaction throughput as well as consensus time all affect the performance of the blockchain. It is still an arduous task to improve the performance while taking into account the security of the blockchain network. In addition, the existing research lacks security analysis and attack and defense strategy evaluation about different hierarchical structures of blockchain. Moreover, the research on differentiated privacy protection strategies for different nodes is limited, and single-level security research is not suitable for the overall architecture of blockchain. The key security of transaction payment depends on the trusted hardware of the third-party trusted institution, which also introduces other security challenges. For instance, Without the knowledge of the user, the attacker provides a service interface based on the cracked hardware device. It is a risk of privacy leakage if the user encrypts the data with the public key provided by the attacker. In addition, the vulnerabilities cannot be repaired in a short time or there is no new hardware replacement when there are security risks in key modules, which leads to serious security risks.

5.3 Heterogeneous Communication Environment

The traditional network security policy is fixed and simple [51]. Defense strategies such as firewall and intrusion detection are used to protect the core position in the network. Software defined network (SDN) [52] not only redefines the automobile, dynamically arranges services, and deploys the active defense strategy on the electronic control unit (ECU), but also effectively solves the security risks existing in the traditional network.

However, the dynamic reconfiguration of the underlying network by SDN also introduces new security threats. Firstly, due to the complexity and diversity of the application requirements for the Internet of Vehicles, it's still troublesome to break the barriers among different communication technologies and realize the network controllability. Secondly, it's still a challenge to dynamically and efficiently schedule available network resources on the basis of consumption resource minimization [53]. Then, due to the access of massive devices, heterogeneous network interconnection, huge scale nodes, and limited resources in Internet of Vehicles, it has become the research trend for SDN security to design security architecture including identity authentication, trust assessment, anomaly detection, and risk assessment in the future. Additionally, advanced technologies such as situation awareness, zero trust as well as endogenous safety and security can be combined with SDN security architecture to improve the security of service scenarios.

6 Conclusion

This chapter summarizes the security threats and security requirements concerning AVP and charging management, and then reviews privacy-preserving schemes for smart parking, LAVP, and charging management. Additionally, new security threats towards AEVs are discussed from three aspects containing trained model privacy,

blockchain technology, and heterogeneous communication environment. We hope
that the chapter can provide a theoretical basis for privacy-preserving of AEVs.

References

1. K. Ren, Q. Wang, C. Wang, Z. Qin, X.D. Lin, The security of autonomous driving: threats,
   defenses, and future directions. Proc. IEEE **108**(2), 357–372 (2020)
2. https://www.daimler.com/en/
3. M. Khalid, Y. Cao, N. Aslam, M. Raza, A. Moon, H. Zhou, AVPark: reservation and cost
   optimization-based cyber-physical system for long-range autonomous valet parking (L-AVP).
   IEEE Access **7**, 114141–114153 (2019)
4. Y. Cui, Z.C. Hu, X.Y. Duan, Review on the electric vehicles operation optimization considering
   the spatial flexibility of electric vehicles charging demands. Power Syst. Technol. (2021).
   https://doi.org/10.13335/j.1000-3673.pst.2021.0514
5. http://www.gov.cn/zhengce/content/2020-11/02/content_5556716.htm
6. M. Khalid, K.Z. Wang, N. Aslam, Y. Cao, N. Ahmad, M.K. Khan, From smart parking towards
   autonomous valet parking: a survey, challenges and future works. J. Netw. Comput. Appl.,
   102935 (2020)
7. R.X. Lu, X.D. Lin, H.J. Zhu, X.M. Shen, An intelligent secure and privacy-preserving parking
   scheme through vehicular communications. IEEE Trans. Veh. Technol. **59**(6), 2772–2785
   (2020)
8. J.B. Ni, K. Zhang, Y. Yu, X.D. Lin, X.M. Shen, Privacy-preserving smart parking navigation
   supporting efficient driving guidance retrieval. IEEE Trans. Veh. Technol. **67**(7), 6504–6517
   (2018)
9 T.W. Chim, S.M. Yiu, L.C. Hui, V.O. Li, VSPN: VANET-based secure and privacy preserving
   navigation. IEEE Trans. Comput. **63**(2), 510–524 (2014)
10. W. Cho, Y. Park, C. Sur, K.H. Rhee, An improved privacy-preserving navigation protocol in
    VANETs. J. Wirel. Mob. Netw. Ubiquitous Comput. Dependable Appl. **4**(4), 80–92 (2013)
11. L.H. Zhu, M. Li, Z.J. Zhang, Z. Qin, ASAP: an anonymous smart-parking and payment scheme
    in vehicular networks. IEEE Trans. Dependable Secure Comput. **17**(4), 703–715 (2018)
12. C.Z. Lai, Q. Li, H.B. Zhou, D. Zheng, SRSP: a secure and reliable smart parking scheme with
    dual privacy preservation. IEEE Internet Things J. **8**(13), 10619–10630 (2021)
13. Q.G.K. Safi, S.L. Luo, C. Wei, L.M. Pan, Q.R. Chen, PlaaS: cloud-oriented secure and privacy-
    conscious parking information as a service using VANETs. Comput. Netw. **124**, 33–45 (2017)
14. D. An, Q.Y. Yang, D.H. Li, W. Yu, W. Zhao, C.B. Yan, Where am I parking: incentive online
    parking-space sharing mechanism with privacy protection. IEEE Trans. Autom. Sci. Eng.
    (2020). https://doi.org/10.1109/TASE.2020.3024835
15. R. Garra, S. Martínez, F. Sebé, A privacy-preserving pay-by-phone parking system. IEEE
    Trans. Veh. Technol. **66**(7), 5697–5706 (2017)
16. S. Nakamoto, A. Bitcoin, A peer-to-peer electronic cash system. [EB/OL] (03 Jul 2020). https://
    bitcoin.org/bitcoin.pdf.2008
17. H.B. Zhou, N. Cheng, J. Wang, J.C. Chen, Q. Yu, X.M. Shen, Toward dynamic link utilization
    for efficient vehicular edge content distribution. IEEE Trans. Veh. Technol. **68**(9), 8301–8313
    (2019)
18. S. Biswas, K. Sharif, F. Li, B. Nour, Y. Wang, A scalable blockchain framework for secure
    transactions in IoT. IEEE Internet Things J. **6**(3), 4650–4659 (2019)
19. C. Zhang, L.H. Zhu, C. Xu, C. Zhang, K. Sharif, H.S. Wu, H. Westermann, BSFP: blockchain-
    enabled smart parking with fairness, reliability and privacy protection. IEEE Trans. Veh.
    Technol. **69**(6), 6578–6591 (2020)
20. L.L. Wang, X.D. Lin, E. Zima, C.G. Ma, Towards Airbnb-like privacy-enhanced private parking
    spot sharing based on blockchain. **69**(3), 2411–2423 (2020)

21. J.X. Hu, D.B. He, Q.L. Zhao, K.R. Choo, Parking management: a blockchain-based privacy-preserving system. IEEE Consum. Electron. Mag. **8**(4), 45–49 (2019)
22. Z.P. Li, M. Alazab, S. Garg, M.S. Hossain, PriParkRec: privacy-preserving decentralized parking recommendation service. IEEE Trans. Veh. Technol. **70**(5), 4037–4050 (2021)
23. W.A. Amiri, M. Baza, K. Banawan, M. Mahmound, W. Alasmary, K. Akkaya, Privacy-preserving smart parking system using blockchain and private information retrieval, in *2019 International Conference on Smart Applications, Communications and Networking (SmartNets)* (2019)
24. Y.L. Teng, Y.Y. Cao, M.T. Liu, F.R. Yu, V.C. Leung, Efficient blockchain-enabled large scale parked vehicular computing with green energy supply. IEEE Trans. Veh. Technol. (2021). https://doi.org/10.1109/TVT.2021.3099306
25. Y. Zhang, L. Zhang, B. Kang, Y. Ma, T. Chen, Secure and reliable parking protocol based on blockchain for VANETs, in *2021 IEEE Wireless Communications and Networking Conference (WCNC)* (2021)
26. H. Cheng, R.X. Lu, X.D. Lin, X.M. Shen, Secure automated valet parking: a privacy-preserving reservation scheme for autonomous vehicles. IEEE Trans. Veh. Technol. **67**(11), 11169–11180 (2018)
27. J.B. Ni, X.D. Lin, X.M. Shen, Toward privacy-preserving valet parking in autonomous driving era. IEEE Trans. Veh. Technol. **68**(3), 2893–2905 (2019)
28. S.R. Pokhrel, Y.Y. Qu, S. Nepal, S. Singh, Privacy-aware autonomous valet parking: towards experience driven approach. IEEE Trans. Intell. Transp. Syst. (2021)https://doi.org/10.1109/TITS.2020.3006337
29. J. Antoun, M.E. Kabir, B. Moussa, R. Atallah, C. Assi, A detailed security assessment of the EV charging ecosystem. IEEE Netw. **34**(3), 200–207 (2020)
30. A. MoFhammadali, M.S. Haghighi, M.H. Tadayon, A.M. Nodooshan, A novel identity-based key establishment method for advanced metering infrastructure in smart grid. IEEE Trans. Smart Grid **9**(4), 2834–2842 (2018)
31. H. Nicanfar, V.C.M. Leung, Multilayer consensus ECC-based password authenticated key-exchange (MCEPAK) protocol for smart grid system. IEEE Trans. Smart Grid **4**(1), 253–264 (2013)
32. D.P. Wu, C. Zhou, Fault-tolerant and scalable key management for smart grid. IEEE Trans. Smart Grid **2**(2), 375–378 (2011)
33. A. Abdallah, X.M. Shen, Lightweight authentication and privacy-preserving scheme for V2G connections. IEEE Trans. Veh. Technol. **66**(3), 2615–2629 (2017)
34. J. Shen, T.Q. Zhou, F.S. Wei, X.M. Sun, Y. Xiang, Privacy-preserving and lightweight key agreement protocol for V2G in the social internet of things. IEEE Internet Things J. **5**(4), 2526–2536 (2018)
35. J. Chen, Y.Y. Zhang, W.C. Su, An anonymous authentication scheme for plug-in electric vehicles joining to charging/discharging station in vehicle-to-grid (v2g) networks. China Commun. **12**(3), 9–19 (2015)
36. G. Kumar, R. Saha, M.K. Rai, W.J. Buchanan, R. Thomas, G. Geetha, T. Hoon-Kim, J.J.P.C. Rodrigues, A privacy-preserving secure framework for electric vehicles in IoT using matching market and signcryption. IEEE Trans. Veh. Technol. **69**(7), 7707–7722 (2020)
37. Q.Y. Yang, D.H. Li, D. An, W. Yu, X.W. Fu, X.Y. Yang, W. Zhao, Towards incentive for electrical vehicles demand response with location privacy guaranteeing in microgrids. IEEE Trans. Dependable Secure Comput. (2020)https://doi.org/10.1109/TDSC.2020.2975157
38. H.H. Wu, A. Gilchrist, K. Sealy, P. Israelsen, J. Muhs, A review on inductive charging for electric vehicles, in *Proceedings of the IEMDC* (2011)
39. H.Y. Li, H.Y. Li, G. Dán, K. Nahrstedt, Portunes+: privacy-preserving fast authentication for dynamic electric vehicle charging. IEEE Trans. Smart Grid **8**(5), 2305–2313 (2017)
40. P. Gope, B. Sikdar, An efficient privacy-preserving authentication scheme for energy internet-based vehicle-to-grid communication. IEEE Trans. on Smart Grid **10**(6), 6607–6618 (2019)
41. J.B. Bao, D.B. He, M. Luo, K.R. Choo, A survey of blockchain applications in the energy sector. IEEE Syst. J. (2021). https://doi.org/10.1109/JSYST.2020.2998791

42. H.Z. Li, D.Z. Han, M.D. Tang, A privacy-preserving charging scheme for electric vehicles using blockchain and fog Computing". IEEE Syst. J. (2021). https://doi.org/10.1109/JSYST.2020.3009447

43. M.U. Hassan, M.H. Rehmani, J.J. Chen, DEAL: differentially private auction for blockchain based microgrids energy trading. IEEE Trans. Serv. Comput. 13(2), 263–275 (2020)

44. Z.G. Wan, T. Zhang, W.Z. Liu, M.Q. Wang, L.H. Zhu, Decentralized privacy-preserving fair exchange scheme for V2G based on blockchain. IEEE Trans. Dependable Secure Comput. (2021). https://doi.org/10.1109/TDSC.2021.3059345

45. S.M. Danish, K.W. Zhang, H. Jacobsen, N. Ashraf, H.K. Qureshi, BlockEV: efficient and secure charging station selection for electric vehicles. IEEE Trans. Intell. Transp. Syst. 22(7), 4194–4211 (2021)

46. Z.Y. Zhou, B.C. Wang, M.X. Dong, K. Ota, Secure and efficient vehicle-to grid energy trading in cyber physical systems: integration of blockchain and edge computing. IEEE Trans. Syst. Man Cybern. Syst. 50(1), 43–57 (2020)

47. D. Gabay, K. Akkaya, M. Cebe, Privacy-preserving authentication scheme for connected electric vehicles using blockchain and zero knowledge proofs. IEEE Trans. Veh. Technol. 69(6), 5760–5772 (2020)

48. Y.C. Li, B.J. Hu, A consortium blockchain-enabled secure and privacy-preserving optimized charging and discharging trading scheme for electric vehicles. IEEE Trans. Industr. Inf. 17(3), 1968–1977 (2021)

49. F.X. Tang, Y. Kawamoto, N. Kato, J.J. Liu, Future intelligent and secure vehicular network toward 6G: machine-learning approaches. Proc. IEEE 108(2), 292–307 (2020)

50. X.L. Cao, J.H. Zhang, B. Liu, Review on security, privacy, and performance issues of blockchain. Comput. Integr. Manuf. Syst. 27(7) (Jul 2021)

51. T. Wang, H.C. Chen, G.Z. Cheng, Research on software-defined network and the security defense technology. J. Commun. 38(11), 133–160 (2017)

52. K. Kalkan, S. Zeadally, Securing internet of things with software defined networking. IEEE Commun. Mag. 56(9), 186–192 (2018)

53. L. Chen, F. Li, B.Q. Ren, J.X. Yang, Software-defined internet of things: a survey. Acta Electron. Sin. 49(5), 1019–1032 (2021)